BETTER FOR ALL THE WORLD

"Bruinius deftly combines analysis of how the American quest for moral and social purity prepared people to accept pseudoscience as a basis for national policy with an account of the personal and intellectual development of eugenics' most influential American advocates—Charles Davenport and Harry Laughlin."
—*The New Yorker*

"Highly readable. . . . His story is one worth hearing, and heeding."
—*The New York Times Book Review*

"Brilliant. . . . A masterpiece of American cultural history."
—*The San Diego Union-Tribune*

"Bruinius takes full advantage of his journalistic talents. . . . Well-written and well-researched."
—*The American Prospect*

"Eloquent and troubling. . . . *Better for All the World* deserves to be closely read."
—*Jerusalem Post*

"Stephen King, meet your nonfiction counterpart. Some of the scariest things that one can read these days come straight from the history books, including this comprehensive look at the eugenics/racial purity movement."
—*The Washington Times*

"In his engaging, often moving story, [Bruinius] brings to life an extensive and colorful cast of characters, especially the men (and the occasional woman) who energetically applied English scientist Francis Galton's vision of 'a scientifically organized Utopia' to American society."
—*Chicago Tribune*

"Mr. Bruinius reminds us of a historical irony we are generally happy to forget: that in the early twentieth century, sterilization joined Prohibition and suffragism as one of the favorite causes of the progressive movement."
—*The New York Sun*

HARRY BRUINIUS

BETTER FOR ALL THE WORLD

Harry Bruinius was born in Chicago and attended Yale University, where he studied theology, and Columbia University, where he studied journalism. He is a frequent contributor to *The Christian Science Monitor*, a professor of journalism at Hunter College, and the founder of The Village Quill. He lives in Manhattan.

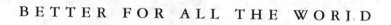

BETTER FOR ALL THE WORLD

A eugenics leaflet prepared by Harry Laughlin, director of the
Eugenics Record Office, ca. 1931.

BETTER
FOR ALL THE WORLD

The Secret History of Forced Sterilization
and America's Quest for Racial Purity

HARRY BRUINIUS

VINTAGE BOOKS
A Division of Random House, Inc.
New York

The Library of Congress has cataloged the Knopf edition as follows:
Bruinius, Harry.
Better for all the world : the secret history of forced sterilization and America's quest for racial purity / Harry Bruinius. — 1st ed.
p. cm.
Includes index.
1. Eugenics—United States—History. 2. Involuntary sterilization—United States—History. 3. Racism—United States—History. I. Title.
HQ755.5.U5B78 2006
363.9'2'0973'0904—dc22   2005044150

ISBN-13: 978-0-375-71305-7

Author photograph © Joyce Ravid
Book design by Iris Weinstein

www.vintagebooks.com

Printed in the United States of America

For My Mother

CONTENTS

CONTENTS

BOOK FOUR: GENERATIONS LOST

ILLUSTRATIONS

BOOK ONE

INTRODUCTION

The buds unfit to mature, fall;
and the weaklings of the flock must perish.

—DR. J. H. BELL

A Simple and Painless Procedure

On a cloudy afternoon on October 19, 1927, as a chilly autumn wind swept down off the Blue Ridge Mountains, rattling the windows of the infirmary at the Virginia Colony for Epileptics and Feeble-minded, Dr. John H. Bell jotted a few notes about an operation he had performed earlier that day. He was the superintendent of this sprawling institution, a campus of regimented brick dormitories and rolling farmland set amid the bluffs overlooking Lynchburg, and one of the country's finest. The morning's procedure was simple, and dozens of such operations had taken place here over the years. But for this patient he wrote with particular care, since it was a case that might draw a bit of attention.

"Patient sterilized this morning under authority of Act of Assembly in 1926, providing for the sterilization of mental defectives, and as ordered by the Board of Directors of this institution," he wrote. "She went to the operating room at 9:30 and returned to her bed at 10:30, recovered promptly from the anaesthesia with no untoward after effects anticipated. One inch was removed from each Fallopian tube, the tubes ligated and the ends cauterized by carbolic acid followed by alcohol, and the edges of the broad ligaments brought together with continuous suture. Abdominal wound was united with layer sutures and the approximation of the closure was good."[1]

The patient lying before him on the operating table that morning was Carrie Buck, a plump, twenty-one-year-old woman who had been under his care at the Colony for over three years. He knew her well. On the day she was admitted, he had been the first to examine her, and he took special note of her dark eyes and slight features, her low, narrow forehead and high cheekbones. He would see her in the Colony's cafeteria, where she was assigned to work, and his words to her were usually cordial and kind. Yet during most of this time, Dr. Bell and Carrie Buck had been, in name at least, legal adversaries.

So, as he finished his surgical report, he decided to add another formal comment: "This is the first case operated on under the sterilization law, and the case was carried through the courts of the State and the United States Supreme Court to test the constitutionality of the Virginia act, and an appeal before the Supreme Court for a rehearing recently having been denied."[2]

It was a momentous day. It had taken over three years to test and litigate Carrie's case, but less than an hour to cut and ligate her Fallopian tubes. But for Dr. Bell, this operation was far more than a legal victory. As a "test case," it had been a carefully orchestrated lawsuit meant not only to sterilize Carrie against her will, but also to protect a bold but controversial social policy he believed would improve the welfare of the nation. Today was the beginning. It was cold, but outside the window, beyond the white, two-tiered veranda on the front façade of the infirmary, Dr. Bell could look out over the Colony and consider the long battle he and other reformers had been fighting for decades.

Beyond the veranda, the parallel rows of austere brick dormitories made this state-run institution look something like a military camp, its geometric precision imposing order on a vast Virginia wilderness, even as magnolias and elms gave the Colony the gentle, pastoral feel of a Southern plantation, peaceful and decorous, like Jefferson's Monticello just hours to the north. Dr. Bell, too, had devoted his life, both as a physician and a scientist, to building a more perfect land. Sterilizing Carrie just may have been one of the most important things he had ever done, and as he considered her surgery he may have even dared to think, as a colleague would later tell him, that "a hundred years from now you will still have a place in this history of which your descendents may well be proud."[3]

Yes, our descendents may well be proud. In the end, real progress in this history, in this quest to battle disease and human suffering, will be found in our descendents, Dr. Bell believed. This was why he had to sterilize Carrie, and this was why he was dedicated to the care of epileptics and feebleminded. To those not familiar with the recent discoveries of "genetics"—a new term in science—Carrie might have seemed a normal girl, if sassy and simple and even a little slow. But Dr. Bell knew she carried within her, like the taint of original sin, the defective "germ-plasm" she would pass on to her children. A defect in her genes made her unusually promiscuous, unable to control herself, and prone to bear a child out of wedlock without shame. It kept her—and her mother Emma before her—from being a productive, law-abiding citizen. And he could say with all confidence, too, that Carrie's illegitimate daughter Vivian, not quite three years old, would follow the same wanton path.

Yes, our descendants may well be proud. Science was revealing the subtleties of nature, the mysterious, microscopic forces that somehow passed on human traits, the genes that kept generations of families mired in poverty, ignorance, and the cesspool of immoral behavior. These families were a breed afflicted by this low-grade mental deficiency called "feeble-mindedness," the primary cause, Dr. Bell and other reformers believed, of the social ills weakening the fabric of the nation. These people weren't insane, exactly. They were not "idiots" or "imbeciles"—terms doctors had long used for people with more noticeable mental defects. No, they were simply a population who behaved like perpetual adolescents—mental dullards without developed moral consciences. To the untrained layperson, someone afflicted with feeble-mindedness might even appear normal. So scientists had just devised a new term for these unmasked defectives: "morons." These people engaged in the unhealthy and antisocial practice of masturbation, which led to sexual impurity and hosts of illegitimate children. These frequented saloons and whorehouses, filled state almshouses and prisons, and worst of all, passed their feebleminded genes to their children.

Biologically speaking, science was revealing perhaps the greatest menace to the future of the human race: fecund, feebleminded females.

This surreptitious menace arose from a strange irony in evolution, Dr. Bell believed. Mankind, the most intelligent and ingenious of creatures, had reached its lofty status through millions of years of toil and struggle. The law of the survival of the fittest had always weeded out the weak and foolish, allowing the best of the race to keep evolving to a higher state. But as this clever species made life less nasty and brutish, as it began to protect the weak of mind and frail of body, the genetic pool began to be polluted. Christian charity and enlightened altruism—and even advances in medical science—had disrupted the natural laws of progress and allowed the weak to live and breed. For Dr. Bell, this irony was made even more ominous as the strong were choosing lives of ease and leisure, shunning large families and the burdens of rearing children. The fitter families of the human race, especially those of the most intelligent Northern European "Nordic" stocks, were reproducing less, while the degenerate stocks were running amok. If nothing were done, the strong might start to lose this battle of genetic survival.

These fears were not simply his own. Many of the world's leaders, Dr. Bell knew, felt the same. Former president Theodore Roosevelt had called this trend "race suicide," and had felt that America's greatness was being threatened not only by rampant poverty but also its cozy affluence. He had once proclaimed, "Some day we will realize that the prime duty, the inescapable duty, of the *good* citizen of the right type is to leave his or her blood behind him in the world; and that we have no business to permit the perpetuation of citizens of the wrong type."[4] The inventor Alexander Graham Bell, the social crusader Margaret Sanger, and the administrators of the Harriman, Carnegie, and Rockefeller philanthropic foundations were each calling for state-sanctioned programs of better breeding. The editorial pages of newspapers such as the *New York Times,* scholars at Harvard, Yale, and Stanford, as well as professional associations of doctors and social workers were each urging the nation's legislatures to quell the tide of "hereditary defectives."

In addition, many British leaders supported compulsory sterilization to purify their nation's genetic pool. When Winston Churchill was home secretary, he had once written to Prime Minister Herbert Asquith urging support for a sterilization bill before Parliament. "The unnatural and increasingly rapid growth of the feeble-minded and insane classes, coupled as it is with a steady restriction among all the thrifty, energetic and superior stocks, constitutes a national and race danger which it is impossible to exaggerate," he explained. "I feel that the source from which all the streams of madness is fed should be cut off and sealed up before the year has passed. . . . [A] simple surgical operation would allow these individuals to live in the world without causing much inconvenience to others."[5]

Now, after nearly twenty years of effort, the case of Carrie Buck provided the most resounding legal affirmation of this theory of genetic engineering. In the Supreme Court case that bore Bell's name, Justice Oliver Wendell Holmes, Jr., feared the United States would be "swamped with incompetence" if women like Carrie continued to have children. "It is better for all the world," he wrote in the majority decision of *Buck v. Bell,* "if instead of waiting to execute degenerate offspring for crime, or to let them starve for their imbecility, society can prevent those who are manifestly unfit from continuing their kind. . . . Three generations of imbeciles are enough."

Yes, our descendants may well be proud. Actually, this idea wasn't new at all, Dr. Bell would muse. "Racial improvement" was a practice as old as the first great civilizations of the world. Didn't remarkably heroic races cast their defective infants in the River Tiber or leave them upon the mountainside to starve? "The idea of elimination, by one way or another, of those who were expected to be disqualified for a certain standard of physical and mental perfection, has come down to us through a great space of time," he would later maintain. "And it persists as strongly in the minds of people today as it did in the minds of the ancient Spartans and Romans. . . . Such efforts to preserve a healthy race, cruel as they may seem, were after all but the pursuit of natural laws: *the buds unfit to mature, fall; and the weaklings of the flock must perish.*"[6]

Sterilization, by contrast, was humane. It was simple and relatively painless. It took nothing from the patient but the ability to pass on the causes of human misery. Far from cruel, forced sterilization represented science and altruism at their most advanced, with goals heroic and noble. Sterilizing Carrie today marked a return to pursuing those natural laws of "elimination"—not in an arbitrary and brutish way, but in a way ordered by science and guided by reason. If science had revealed the congenital, hereditary nature of human imperfection, it was now revealing a path toward restoration.

"It is not foolish to hitch one's wagon to a star, for the unbelievable theory of today becomes the proven laboratory fact of tomorrow," Dr. Bell would explain. "And while perhaps a Utopia may never arise out of our efforts to better our brother's condition in this world in which we live, nevertheless, much that is practical and useful and elevating to all can be developed and carried to a successful conclusion by the simple formula of all who are interested in these things pulling together towards a common goal: a citizenry purged of mental and physical handicaps."[7]

This was his faith, a faith in science and progress, and a faith informed by a long-held vision of American destiny. After putting Car-

Dr. John H. Bell, the surgeon who sterilized Carrie Buck in 1927.

rie's chart in order, Dr. Bell could pick up the day's newspaper and read about the throngs of people in Baltimore cheering Charles Lindbergh, the great American airman who had flown across the ocean alone and was now making a triumphant tour of all forty-eight states. It was indeed a momentous day. But beyond those cheers, Dr. Bell believed America's greatness was much more evident here in the calm, quiet infirmary at the Virginia Colony. America, more than any other nation, held the promise of being a land of innocence, free from the defects of the past. This land could be, as so many others had believed, a city upon a hill, a beacon to all civilizations, so long as its citizens remained vigilant, persevered in virtue, and held to their sense of civic duty. Now the Supreme Court had recognized the wisdom gleaned from science and declared this harmless procedure a constitutionally valid means to combat the country's social ills. Forced sterilization would be effective, Dr. Bell knew, and would help purge from American society those defects found deep within human nature. Today was surely the dawn of a new era, not only for the country, but for all the world.

II.

An Epic Quest in the Modern World

In the early decades of the twentieth century, not long after the technology of surgical sterilization had been devised, state governments throughout the United States began a quest for racial purity that would change the lives of thousands of their citizens. By 1927, before Carrie Buck lay prostrate beneath Dr. Bell's surgical blade, almost 8,500 American citizens had been forcibly sterilized. This "official" figure, taken from informal surveys by proponents of the procedure and representing only what surgeons chose to report, would reach well over 65,000 in the decades to come.[1]

Ill-educated and poor, these people were operated upon and mostly forgotten. But they were first the subjects of methodical research programs in which scientists tried to trace and then eradicate the gene pool that caused what they casually referred to as "the three D's": dependency, delinquency, and mental deficiency. Hundreds of fieldworkers fanned out into the country to visit prisons, mental institutions, and the poor rural hamlets where many of their research subjects dwelled. They collected tens of thousands of pages of data on these subjects' family pedigrees. Armed with this data, which appeared to show a genetic predisposition toward moral deviance and mental deficiency handed down through generations, scientists persuaded state legislatures—and, in the case of Carrie Buck, the U.S. Supreme Court—to enact laws giving states the power to sterilize these genetically "defective" Americans.

Indiana implemented the first sterilization law in human history in 1907; in the next two decades, the United States became the pioneer in state-sanctioned programs to rid society of the "unfit." The procedure, common today as a form of voluntary birth control, was first developed in the late nineteenth century by American doctors who sought to keep male "hereditary criminals" from having children. Later, they expanded this rationale to include any other type of "social undesirables," especially the feebleminded female. California passed its law in 1909, and established by far the world's most aggressive program at the time, sterilizing more than 2,500 people in its first ten years—twenty times more than any other state. Fifteen other states also passed laws before 1927, sparking controversy and outrage and a host of legal challenges. When seven of these statutes were struck down by state courts, the architects of the sterilization programs only conspired to protect their laws from legal scrutiny.

In Virginia, Carrie Buck was chosen to be a "test case," a lawsuit reformers brought against themselves for the specific purpose of bringing their ideas to the Supreme Court. Their plan was a success. After Carrie's sterilization, there was a renewed push to pass new laws, and institutions in states that had already approved the procedure began sterilizing inmates at a dramatically higher rate. In all, at least thirty states, as well as the territory of Puerto Rico, passed laws to quell the tide of hereditary defectives through forced sterilization. After 1927, this American technique of social engineering became the model for laws in Canada, Denmark, Finland, France, and Sweden. In 1933, in one of the first acts of the newly elected government of Reichschancellor Adolf Hitler, the National Socialist Party enacted a comprehensive sterilization law modeled consciously on American legislation.

The number of official, state-sanctioned sterilizations in the United States, however, represents only part of this effort. In states without laws, doctors trained in the latest theories of mental health—or "racial hygiene," as many called it—sterilized untold more of their patients. Often, in the face of public opposition, these operations were recorded simply as "medical necessities" for unspecified "pelvic diseases." The number of appendectomies recorded at state institutions at this time is also suspiciously high. Even after World War II, when the policy of forced sterilization and the science behind it had long been debunked and forgotten, doctors at mental institutions would continue to sterilize patients surreptitiously. Yet, with a sense of moral urgency, the doctors engaged in this hidden work felt they were only doing a great social good. To them, this was their duty and one of the highest callings a person could have: to purify the human race.

The quest to build a better civilization, to battle social ills and human suffering, is in many ways the great epic of the modern world, a story in which men and women, armed with weapons forged by the great god Science, throw off the superstitions of the past and enter a new kind of fray. Revolutions in science, technology, and industry each sparked the hope that human life, now equipped with a new arsenal to overcome the agonies Nature inflicted in the struggle for existence, could progress toward perfection. But instead of the shield of Achilles, Science gave mankind the tool of Method—impersonal, efficient, bureaucratic.

This is a story of desire. This is a story that tells of the passions and unfulfilled longings of individuals, and the conspiracies, betrayals, and ironies that stand behind a scientific program to purify the human race through genetic engineering. It is a story of the lives of two disparate groups of people and the yearnings that bound them together. On the one hand, some of the most enlightened members of the nation brought their zeal and determination to shape a better world. On the other hand, the objects of this zeal were tens of thousands of other Americans, most of them stricken in poverty and living lives of quiet desperation.

The story of Carrie Buck and the bureaucracy of forced sterilization is *in medias res* of a larger story of a science called eugenics. The quest for the Good is at the heart of this science—it is the root of the term, after all, derived from the Greek *eugenes,* meaning "wellborn" or, in effect, "good genes." It was coined in the late nineteenth century by the father of this science, Francis Galton, a wellborn Victorian genius and member of one of the most illustrious scientific families of his day. After his cousin Charles Darwin shook the world with his theory of evolution, proclaiming revolutionary new ideas about the origin and development of living species, Galton began to reflect on the biological differences that made one person a genius and another a dolt. But his own theory soon went beyond studying the origins of intelligence to considering the nature of the Good. Like those who breed good cattle or good peas, he wondered whether human beings could likewise breed themselves and produce a glorious race of biologically superior men and women.

Within any epic story is the specter of the gods, and the Christian God pervades eugenics from the start. *On the Origin of Species* had profound theological implications about human identity and sparked a vociferous response from clergy and scientists both. Darwin's theory, published in 1859, presented in methodic detail the natural, violent forces that shaped the human body—not from the dust of the ground, but from slowly evolving lower forms of life. Man was not "fallen" from a higher state, created perfect through the breath of God, then cast from Paradise because of sin. He was actually rising up from a lowly state, moving toward perfection with the crawl of time. The great tropes of origin and development in the nineteenth century—Darwin's "natural selection," Herbert Spencer's "survival of the fittest," and Galton's "nature versus nurture"—shattered the foundations of human identity, the idea that human nature was immutable and akin to the divine. Salvation could now be a scientific enterprise.

But out of these ruins came an agonizing question. If these new discoveries were true, and people developed no differently than the beasts of the field, then what of human dignity? What happened to dignity when human beings were not created in the *imago dei,* the image of God, but arose instead from *Eoanthropus dawsoni* or other apelike species? What would happen to the Enlightenment's theory of "natural rights" which were said to be endowed by a creating deity?

Science since the Enlightenment had already begun to shake off the unexamined authorities of religion and tradition, replacing them with a process of inquiry and doubt. Yet most thinkers still held to this special place of Man in a divinely created order, a species set apart from the animals. Cartesian doubt reduced all to a disembodied thinking thing and founded knowledge upon the being of the human mind—which was, again, akin to the divine. So when the new "naturalists" studied human beings empirically, just as any other animal, just as an organism of heaving blood and twitching nerves, they seemed to shatter a millennia-long consensus of human uniqueness, even from a humanistic point of view. Only now could Galton take the dramatic step and consider applying the long-established techniques of agricultural breeding to the human species.

Like the evolution of any idea, that of eugenics is complex. As it arose out of the many anxieties caused by the sweeping revolutions of the nineteenth century, it developed into a diverse movement, utilized by people with varying beliefs and motives. But better breeding must assume a better breed, and the "survival of the fittest"—believed to be the brutal and fundamental law of Nature—must assume the presence of the "unfit." Many thinkers began to see the world's interwoven social groups pitted against each other in a great struggle. Whether this was the smart and the dull, the proletariat and the bourgeoisie, or the Aryan and the Jew, by the end of the century the idea began to inform a host of new theories of social life. Natural science allowed many to redefine old, clannish myths of superior ancestry—myths earlier just pride in past accomplishments supported by some god—and change them into the corporeal, biological facts of evolution. Demythologized, the "unfit" were now determined by their genes.

Like a leitmotif in an opera of evolutionary survival, the resounding theme of eugenics became the drone of statistics. Trumpeted in scientific journals and popular periodicals, the growing numbers of paupers and criminals were said to pose a grave danger within society, and the oncoming tide of unfit foreign races, the genetically inferior "semi-barbaric hordes," were said to threaten the gates of civilization itself.

As students of evolution, eugenicists believed the human race must begin to take control of human reproduction and ethnic intermingling. With the rational and objective methods of science, they hoped to breed only the biologically best of the races, and prevent the propagation of the worst. The "mawkish sentiments" and "flimsy shibboleths" of the past, especially those of unenlightened religious authorities, should give way to incontrovertible statistics and scientifically proven theorems, which could sweep away any discomfort methods like sterilization might cause. Goodness and dignity, like the species itself, were neither static nor inherently fixed: they were developing and coming to be. Like cattle and peas, some were better than others, with a greater claim to dignity. Like pigs and flowers, humans could be judged for fitness at the county fair. Like the weaklings of the flock, some should just be made to perish. And so they were.

THIS IS A STORY in many ways uniquely American. Purity and innocence have long been great motifs of American life, and may explain in part why this country, more than any other, first grasped Galton's science and became the pioneer in marriage restriction, forced sterilization, and other methods of eugenic engineering. Since the time of the Puritans, a subtle and pervasive self-understanding has shaped the belief that Americans are a "peculiar people," chosen by God to come to this land of Edenic lushness, where material abundance, good health, and moral purity can reign free. Americans have often defined their civic and spiritual lives through this biblical image, and have remained relentlessly optimistic, ever confident that the burdens of history and the evils of the past can be swallowed up in this New Jerusalem, this paradise regained in America.

But this self-understanding has always carried with it a "sweeping prophecy of doom."[2] The consequences of impurity and disobedience could shatter the foundations of this paradise regained, turning its promise into a curse. On the ship *Arbella*, somewhere in the Atlantic Ocean in 1630, the Puritan minister John Winthrop preached a sermon entitled "A Model of Christian Charity" to a group of shivering families heading out to an unknown world. It was a message of hope and desire, words of promise outlining the settlers' understanding of the epic task before them. This sermon would become a kind of myth, a founding document for the civilization to come, proclaiming America as the new Israel, the radiant, innocent bride entering into a marriage covenant with her God, and bound to keep His precepts. But if she failed, she would be consumed:

> *We shall find that the God of Israel is among us, when ten of us shall be able to resist a thousand of our enemies; when He shall make us a praise and glory that men shall say of succeeding plantations, "the Lord make it like that of New England." For we must consider that we shall be as a city upon a hill. The eyes of all people are upon us, so that if we shall deal falsely with our God in this work we have undertaken, and so cause Him to withdraw His present help from us, we shall be made a story and a by-word through the world. We shall open the mouths of enemies to speak evil of the ways of God, and all professors for God's sake. We shall be consumed out of the good land whither we are agoing.*

Celebrated and critiqued in later American literature, this myth of innocence and purity, this image of a peculiar people called to traverse a spiritual wilderness and conquer a new Promised Land, evolved over the years from Puritan social utopianism to the inner, transcendent solitude of the soul. The purified individual—expressed in the revivalism of the Great Awakening, the individualism of pioneers pressing west, as well as the Transcendentalism of nineteenth-century New England—could transform the American landscape and continue to make this land as a city upon a hill. The image would be cited by politicians and reformers well into the twentieth century, and would contribute to a relentless American drive for technological innovation as well as self-improvement.[3]

This American self-understanding lingered on in the thinking of many of those who promoted eugenics. In America, a number of the movement's leaders were New England Protestants, and, using an evangelical tone which harked back to their Puritan forebears, they proclaimed that the goal of their scientific program was to keep the "American stock" pure by excising the causes of immoral behavior. Their warnings of genetic deterioration echoed Winthrop's prophecy of doom. In effect, they saw the eugenic quest as a way to keep the country from being made a "story and a by-word through the world," and to keep its people from being "consumed out of the good land."

Eugenics first took hold around the time when Carrie Buck was born. The first influential eugenic thinker in America was Charles Davenport, a direct descendant of the Reverend John Davenport, the man who had founded the city of New Haven in 1638 after leading another ragged company of five hundred Puritan settlers to the New World only eight years after Winthrop. Charles's father, Amzi, steeped his son in the traditions of his forefathers, men whose sermons and political discourses helped shape the literary tradition that defined this national self-understanding. Amzi was also a religious enthusiast, a man committed to an intense and solitary relationship with Christ, a relationship he believed could transform the evils of drink and slavery.

But his son Charles became a lover of natural science and a devotee of Francis Galton— a "modern" person, wrestling with the Puritan vision of his deeply pious past. Yet, while rejecting the American faith of his fathers, Charles in many ways simply retranslated it into a secular and scientific form. He was not just one of the first Americans to trumpet the Utopian vision of eugenics, however. He was also a key figure in the development of modern genetic theory. A Harvard and University of Chicago professor, he was the first scholar in the United States to teach the ideas of Gregor Mendel. Later, he convinced the newly formed Carnegie Institution of Washington to provide the funds to establish a Station for Experimental Evolution, a facility that became world-renowned as a center for genetic research.

As a brilliant organizer with a grand vision, Davenport also convinced the philanthropist Mrs. E. H. Harriman, widow of the wealthy railroad magnate, to fund a Eugenics Record Office to study the hereditary problems that plagued the country. To lead this office, Davenport chose Harry Laughlin, a Missouri schoolteacher who had become his friend and protégé. Laughlin was also the son of deeply pious parents, a couple with a faith more progressive than Puritan, yet who both maintained the optimistic hope of American manifest destiny. Laughlin's father was a minister and his mother was a social reformer in the Midwest women's movement; both crusaded for temperance, suffrage, and social change. Laughlin followed in their progressive tradition, devoting himself to eugenic sterilization and becoming the nation's foremost expert in the field.

Davenport and Laughlin, working together on the pastoral campus at Cold Spring Harbor, a small town on the shores of Long Island Sound, did as much as anyone to spread the eugenic gospel in America. While Galton first suggested eugenics become a new "national faith," it was these two men who shaped the movement's particular kinship with religion, making it more than simply a "secular creed." To win public support, many eugenicists began to couch their message in religious terms, producing eugenic "catechisms," sponsoring contests for the best eugenic sermons, and conducting "Fitter Family" and "Better Baby" contests based on good moral and mental hygiene. But this was more than a rhetorical ploy, more than an effort to be wise as serpents and innocent as doves in the face of public scrutiny. As Davenport and Laughlin sent out hundreds of fieldworkers to gather family pedigrees—"pedigrees" essentially tracing the genetic roots of "sin"—they were following a uniquely American theological tradition that demanded citizens of this land be as innocent as a blushing new bride. The eugenic quest could bring a modern Great Awakening.

If Davenport was a researcher and organizer, Laughlin was a zealot and crusader—like his pious mother. The nation's expert in eugenic sterilization, Laughlin also became a leader in the movement to quell the tide of immigrants coming into the country. Appointed the "Expert Eugenics Agent" for the House Committee on Immigration and Naturalization, he played a significant role in the passage of the most restrictive immigration bill in the nation's history. During this same time, too, Laughlin was both a direct and indirect participant in the landmark Supreme Court case *Buck v. Bell,* which made eugenic sterilization a constitutionally sanctioned method to battle poverty and crime. The case, which led to the sterilization of Carrie Buck and countless others, was a vindication of his "Model Sterilization Law," which eugenicists in Virginia had used to shape their legislation. During the trial, they also called upon Laughlin's expertise, and his deposition was even cited by Oliver Wendell Holmes, Jr., in the denouement of Carrie's "test case."

In addition to its effects on American jurisprudence and social policy, Laughlin's law had even greater influence. Though the United States was the pioneer in the legal, administrative, and technical aspects of eugenic sterilization, Nazi Germany borrowed its ideas and applied them in an unprecedented way. One of the first laws passed by the National Socialist government of Adolf Hitler was the "Law for the Prevention of Genetically Diseased Offspring," and its language and structure closely followed Laughlin's Model Law. In less than two years, over 150,000 German citizens were forced to undergo the procedure, preparing the way for the genocide to come. In 1936, when the German sterilization campaign was at its early height, the Nazi regime, through the auspices of Heidelberg University, awarded Laughlin an honorary doctorate for his many contributions to "racial hygiene."

The heaviest burden of the story of eugenics and forced sterilization, perhaps, is this American connection to the master race theories that culminated in the Holocaust. Hovering over history is the almost unbearable fact that the horrors of the twentieth century were not an outbreak of barbarism in Western culture. They were in many ways the consequences of thoroughly modern ideas, especially the notion that society, using the tools of science and technology, could eliminate its supposed imperfections. Some have used the phrase "the banality of evil" to describe how the horrors of the modern world can be a systematic and bureaucratic phenomenon, and how a person need not have the diabolical profundity of an Iago or Macbeth to promulgate evil. The industrial bureaucracy of mass murder in Germany required a measure of calm, rational ingenuity, including careful research, efficient organization, and effective engineering.

Indeed, the specter of the Holocaust has hidden what may have been at first the sincere, well-considered concerns for social betterment by many of the leading scientists of the early twentieth century. The Nazi sterilization law affected mostly ethnic Germans, after all. The genocide that followed has also obscured the fact that American eugenicists— who were recognized by Nazi thinkers as the innovators of the field— were as much interested in creating a "master race" as their German counterparts. American eugenic legislation had begun as a conscious effort to purify the superior "Nordic" race, targeting the genetic impurities within it. At the Nuremberg Trials after the war, Nazi doctors defended their actions by citing American precedents, as well as the majority opinion of Oliver Wendell Holmes, Jr., who had sanctioned the forced sterilization of those he claimed were "manifestly unfit from continuing their own kind."[4] With good intentions and confidence in scientific method, eugenicists had begun their social programs as an effort to improve human life; but this led them to disregard individuals Laughlin deemed a "shiftless, ignorant, and worthless class of people."

Because of this connection to the Holocaust, the history of American eugenics has been in many ways forgotten. Some historians have dismissed it as an extreme "pseudo-science," a momentary "craze" in an era of reform, or a curious "secular faith." The fact that its early incarnation was shaped more by racial and social prejudice than critical observation is true perhaps. But this belies the fact that eugenicists stood firmly within the genesis of modern genetic theory and contributed many basic research methods used to this day. Eugenicists were the first to delve into family histories, seeking the hereditary causes of diseases such as cancer, alcoholism, and Huntington's chorea. Galton himself was the first to study twins and compare the relative influence of "nature versus nurture." His methods of statistical analysis—especially his use of what would later come to be known as the bell curve—revolutionized research in a host of sciences. And it was eugenic thinkers who first developed the idea of the "IQ," or intelligence quotient, and the mental tests used to measure this elusive human trait. Indeed, the original purpose of American IQ testing was to identify and then institutionalize or sterilize the feebleminded poor. The momentary craze has had lasting effects.

Across the United States, from Vermont to Colorado to Oregon, men and women still live with the humiliations of forced sterilization, which continued in some states through the 1970s. "What they did to me was sexual *murder*," says one woman in Denver, who, like Carrie Buck, lost her legal case against the state doctors who sterilized her against her will. "I'm just like a female spayed animal. They made me half a woman. They took my heart and left a stone, you hear me?" In 1955, she filed suit against a state-run hospital in Colorado—a state which had explicitly rejected eugenic sterilization at least five times in its history. This did not, however, prevent the practice. Over the past few decades, many others have also filed lawsuits, and hundreds of victims of eugenics and forced sterilization have sought justice and compensation for their loss of dignity and their ability to bear children.

Every lawsuit filed since World War II has failed, mostly because of the precedent set by *Buck v. Bell,* which to this day has never been reversed by the Supreme Court. In recent years, however, the governors of Virginia, North Carolina, Oregon, and California have each issued official apologies for their states' programs of forced sterilization, humble acknowledgments to the dozens of victims still living today. A few monuments have even been erected. What is rarely discussed, however, is that these programs constituted nothing less than an American quest for racial purity.

After a century in which humankind has probed the mysteries of heredity and discovered some of the secrets of the human genome, the specter of better breeding and eugenics still attracts a host of people who long to remake their imperfect selves and breed a better type of human being. Galton's dream of a glorious race of biologically superior men and women has been revived as scientists begin to map the genome and pore over its marvels. This book is an account of scientists who shared this dream in the past, and the "worthless class" they sought to sterilize. It is neither a polemic nor a moral condemnation of the science of genetics and bioengineering as such. It is not an analytical history of eugenics and involuntary sterilization, detailing the social causes of good intentions run amok. It is a story of what happened.

Behind the banalities of scientific method and modern bureaucracy are age-old passions and human desires. As both the sophisticated and the simple were caught up in this great quest for human perfection, as both wrestled with the ambiguities of love and sex, and as both felt the desire for a healthy family and feared what the future might bring, each found their deepest longings often unfulfilled.

BOOK TWO

THREE GENERATIONS OF IMBECILES

We have seen more than once that the public welfare may call upon the best citizens for their lives. It would be strange if it could not call upon those who already sap the strength of the State for these lesser sacrifices, often not felt to be such by those concerned, in order to prevent our being swamped with incompetence. It is better for all the world, if instead of waiting to execute degenerate offspring for crime, or to let them starve for their imbecility, society can prevent those who are manifestly unfit from continuing their kind. The principle that sustains compulsory vaccination is broad enough to cover cutting the Fallopian tubes. . . . Three generations of imbeciles are enough.

—SUPREME COURT JUSTICE OLIVER WENDELL HOLMES, JR.

The Purity of Our Women

A throng of men, many in cream-colored straw fedoras, were bustling along the old Three Notch'd Road—now called Main Street—in Charlottesville, Virginia. Women on the street moved more slowly, some wearing bonnets, others without, their dresses flared by crinoline petticoats hanging heavy from their waists. "Hurrah for the Fourth! Get ready to enjoy it!" some said to each other amid the clap of horses and buggies on the dusty, macadamized road.[1]

It was July 3, 1906. The town had been buzzing with excitement as people all around reveled in a new sense of anticipation. In years past, after the persecutions of Reconstruction, reverence for the national holiday had fallen into "innocuous deride" all through the South, and most enthusiasm shown for the Fourth was as genuine as a carpetbagger's charm. But even though Confederate veterans and widows were still walking the streets, there'd been a lot of change. Virginia now had a revised constitution, its own people were fully in charge again, and recent government reforms were bringing the state out of the economic devastation wrought by the Civil War. Rail and trolley lines were being built, and even though most of the town's forty-six miles of crisscrossed streets and alleys were still dirt, local politicians were working to get more of them paved with bricks—or, in a new, cheaper, and more efficient method—"macadamized" with crushed gravel. With improved transportation, the town was expanding, too, and on the northwest side new businesses were springing up on Route 29 and Emmett Street. Optimism abounded in the thriving town, and on the eve of the Fourth of July, some were even saying that of all the states in America, Virginia could take special pride in the day, especially here in Albemarle County, where lived "the great Apostle of human freedom, whose intrepid spirit inspired, indeed, dictated, that great Declaration."

Most of the present excitement, however, was over the next day's festivities. For the first time since anyone could remember, there would be a grand horse show and medieval-style tournament in the afternoon, and in the evening, at the Jefferson Park Hotel, a great ball would be held, surpassing anything that had come before, and lookers-on could watch well-dressed ladies and gentlemen dance on the pavilion and hear the orchestra play. Already today, people were heading out to watch the "knights" practice at the county fairgrounds, and a host of visitors, "horseback parties" from neighboring Ivy, Keswick, and Scottsville, were trotting into town to take part in the show.

On Main Street, some of the department stores were selling tickets for the tournament, and there were lines. On the wood-plank porches in front of the stores, where townsfolk milled about, people could gossip or discuss the current news. Politicians in Richmond had just seized control of telegraph rates, regulating the wildly fluctuating prices of the Western Union & Postal Telegraph Company. Some of the men were worried, too, about the resignation tendered the day before by Noah K. Davis, the head of the school of moral philosophy at the University of Virginia, just outside town. He'd just accepted a new post at the Carnegie Institution of Washington, a philanthropic organization set up to support modern scientific research. Andrew Carnegie, the famous steel magnate, had recently endowed the institution with an unprecedented $10 million, explaining his purpose was "to encourage, in the broadest and most liberal manner, investigation, research, and discovery, and the application of knowledge to the improvement of mankind."[2] Northern money, it seemed to most of them, had wooed away Professor Davis, one of their most nationally renowned scholars, and he was to head up to Washington, D.C., to be an ethics adviser on its committee considering proposals for scientific projects.

Stirring the most emotion, however, was news of those Northern "negrophilists," up to their old tricks, trying to fight Kentucky's constitution and make whites go to school with blacks. "This is probably the last of the forty years' effort by Northern people to have social equality between whites and blacks in the South," said an editorial in the local paper. "We cordially concur in the following comment of an exchange on this Kentucky case: 'The South did not yield in her poverty; she will certainly not yield in her renewed prosperity and strength; and the sons and daughters of the old Southerners have inherited the principles of their ancestors. Social equality is not best for either race.' "

In the bustling town, a new mythology of the Old South, along with the "separate but equal" decision of the Supreme Court ten years earlier, was creating a vicious new antagonism between blacks and whites. Even though the regulating role of the state government was becoming a bigger part of their daily lives, federal power, dominated by Yankees, remained a tyrannical enemy for them. But government reforms and growing segregation—which actually belied the "inherited . . . principles" of their genteel, agrarian ancestors, who before Reconstruction were dependent on the peaceful coexistence of slave and master—were starting to rest on a new authority, one stealing in from the industrial North and as subtle and pervasive as the power of the Holy Ghost.

Even the stores on Main Street, which brought in newfangled products from the North, were invoking this power. At James Perley & Sons Furniture, on 103 West Main, a salesman could make an earnest pitch for the new North Star Refrigerator. "A refrigerator is not a luxury but a necessity!" he could say. "Milk, vegetables, meats are kept cold and fresh from day to day, provided you have a refrigerator properly constructed along *scientific* lines. Ten pounds of ice is all you need buy to keep these refrigerators in proper condition. Think of the economy in ice. A cheap refrigerator is the highest priced article you can buy simply because it will melt the ice very rapidly. And remember, we sell these on a small cash payment and small monthly payments, so you have all summer in which to pay for it if you wish."[3]

Down the street, in the stacks of goods shelved in grocery stores, even boxes of crackers, shipped in from Minneapolis, proved their worth with this new authority. On each box was stamped a red and white trademark, an "absolute pledge of quality and purity":

WHAT THE
NATIONAL BISCUIT COMPANY
STANDS FOR

• The scientific, reconstructed baking industry, whereby the goodness and nutrition of biscuit and crackers have been marvelously enhanced.

- The new method of protection by which all dust, dirt and moisture are completely excluded from the package and the freshness and goodness of its contents are carefully preserved.

It was a time of industrialized commerce and growing prosperity, greater mobility and evolving structures of class and race. Charlottesville would celebrate the Fourth, and Southern belles and gentlemen would grace the next night's gala as they had in days long ago. But it was becoming an axiom in this fast-changing world: the methods of science are not only reliable and trustworthy, they're bringing progress, protection from impurities, and, of course, marvelous enhancement.

SOMEWHERE, blocks from the buzz on Main Street, Mrs. Emmett Adaline Buck convalesced quietly after giving birth to a girl the day before. Her area of town was still ramshackle and poor, and few had iceboxes, let alone North Star refrigerators. She was married to Frank Buck, a tinner by trade, and they named their baby Carrie.

Frank and Emma—which was what family and friends called her—had been married ten years, but if they had any children earlier, none had survived. They lived simply and quietly, and few outside their neighborhood and their small circle of acquaintances knew about their daily routines. With their baby Carrie, they were probably not planning to go down to the fairgrounds to see the horse show, or take in the ball at the Jefferson Park Hotel—but they had little money to spend anyway.

It hadn't always been this way. Frank's father Fleming Buck once had property and slaves, but in 1868, after the Civil War, Fleming died, leaving a widow with two small boys and debts. Without free labor, Frank's mother had to sell the property, and as her sons grew to be men in the decades of Reconstruction, receiving little schooling after the sixth grade, the family's former wealth gave way to a life of want.[4] Frank might have made a modest wage when he worked, but it had still been a struggle to earn enough to live over those tumultuous years. In some ways, it was even harder during this time of prosperity and growth, when tin goods from factories up north kept prices low, squeezing out the local makers.

Emma's father Richard Harlow also once owned land. She never knew her mother, who had died giving birth, and she later lost her father, too, when he injured his spine and died. Somewhere, she was convinced, she had $460 in the bank—hers from the sale of her father's property. But though Emma had gone to school up to the fifth grade, and her excellent penmanship might have even been the neatest in her class, she did not deal with banks or understand these matters much. Unlike some of her neighbors, however, she did read and write quite well. Though her married life with Frank was a struggle, with the birth of Carrie and things improving, she might have hoped her life would soon get better, too.

But after Carrie's birth, Frank was gone. Few knew exactly what happened: some said he just up and left, convinced Emma had had an affair and that Carrie wasn't his child. Others explained he was killed in an accident, and doctors would later write on Emma's medical records that she was a widow.[5] In either case, Emma was now alone, a single mother at thirty-three, and with little means of support. She tried to ask around, but she couldn't locate the money she was sure she had, and few friends or family members were able to offer help. So for the next four years, as Charlottesville was becoming more a city than a town, Emma had to rely on the kindness of strangers and whatever organized charity she could find as she struggled to provide for her baby, Carrie.

In 1906, most Virginia communities still maintained a loose-knit system of charity as old as the state itself. From the founding of the original colony, care for the poor had been a part of a churchgoing culture, and it constituted a religious duty of sorts. Most organized relief was still run by local clergy, even when supported by municipal taxes. Town councils funded scattered civic almshouses and gave grants to independent religious charities, and in the last few decades, they'd even helped support some of the new organizations of Christian women devoted to work with destitute homes. Workers called this type of charity "outdoor relief," and it was a system that had remained relatively unchanged for hundreds of years, modeled after aspects of England's Elizabethan Poor Laws of 1601. But Virginia had also been a pioneer in what was called "indoor relief," the state-funded hospitals, asylums, and orphanages that housed those with no other place to go. In fact, Virginia had been the first colony to open a hospital for mental patients in Williamsburg in 1773, which locals came to call "Mad-House" or "Bedlam." It was also the first state to offer free public education for children of the poor and the first to encourage local care for unwanted children born out of wedlock. Virginia social workers were proud of this heritage, and at the annual National Conventions of Charities and Corrections, they were never slow to point out to their Northern colleagues that their agrarian state was the first to build large, state-run institutions.

Their towns, however, still had haphazard schemes for outdoor relief. In Charlottesville, the town council appointed an "Overseer of the Poor"—a position first conceived in Elizabethan times—who worked with the council's Relief Committee. Taking recommendations from the chief of police and volunteer workers, the overseer created charity lists and then coordinated the efforts of the town's churches to distribute aid. Other Virginia towns had different systems. Alexandria appropriated funds for the "Cooperative Charities," a private organization that investigated cases and doled out cash assistance. In Bristol, the mayor had full authority, and he determined who was placed on charity lists and how much the city clerk should give. Fredericksburg provided a small subsidy to the City Mission, a private group of ladies distributing alms. In Richmond, the state capital, the city council appropriated funds for twenty-four separate private charities and one city almshouse, where the superintendent distributed cash to poor households at his own discretion.[6]

This was one part of the world Emma was entering. Charity workers in Charlottesville soon noticed her plight, and placed her and Carrie on their lists for aid. But even as Emma collected baskets of food and small amounts of money, she was finding relief in other ways, too. During a time when a woman with children relied on a man to provide, she took up with different men. Over the years she bore two more children, Doris and Ray, probably from two different fathers. She also contracted syphilis. Some whispered she'd become a prostitute, and her arms showed sores from intravenous injections, drugs for pleasure and escape.

Like so many others, Emma might have remained invisible, just one poor woman with a child on the fringes of a newly bustling society. But Emma and Carrie's lives were becoming more desperate just at the point when leaders throughout the country were radically rethinking relief for the poor. For the first time in centuries, leaders in Virginia were looking for a different approach—a more scientific approach—to provide for the common good with more efficient, effective methods. In Richmond, legislators were already beginning to sketch the blueprints for reform, hoping to revamp this age-old system of outdoor and indoor relief. As she was caught in the transition, however, Emma couldn't know that those calling for the scientific reform of charity were focusing on women, women with stories much like her own.

ON A HUMID DAY in June 1907 Aubrey Strode paced confidently before a jury sitting in a small, stuffy courtroom in Houston, Virginia. He was a thirty-three-year-old state senator, elected just last year, and a marvelous, medal-winning speaker. The young senator was broad-shouldered and handsome, and when he furrowed his eyebrows, his dark, serious eyes looked fierce. Still, Strode's face was soft, like a boy's, and with his wavy brown hair he had the rare trait of being able to look both gentle and grim. The jury trusted him.[7]

The state senator, also a professional attorney, was defending an old friend, Judge William G. Loving, who had shot and killed a boy named Theodore Estes the month before. The facts of the case were not in dispute. Young Estes had taken Loving's daughter Elizabeth for a buggy ride out on a country road, and then in the early evening had dropped her off at her friend's house, where she was staying the night. While there, however, Elizabeth got so drunk she couldn't return home the next day. Knowing her father would be furious, she told him Estes had drugged her and taken advantage of her. Enraged, Loving took his shotgun, found the boy, and shot him to death.

Strode did not even try to prove a sexual assault had taken place. It seemed obvious the daughter had lied. Instead, he appealed to an "unwritten law," an honor code of the Old South in which a daughter's virtue was sacrosanct and a father duty-bound to protect it. It didn't matter if Elizabeth's story were true or not, he told the jury. In the South, the mere suggestion that a man's daughter had been violated would evoke passion and rage. This was a kind of temporary insanity. It was not only understandable, it was natural. So, as Strode gave his closing arguments, he made a special appeal to this unwritten code of honor, which the twelve Virginia men sitting before him would know quite well.

"We have come to you to present the case for the defendant, where we always believed it should be presented—in a court of justice," Strode said to them.

We believe that the people of Halifax and Nelson Counties cherish the same traditions and the same sentiments. The heritage of Virginia is common to all. The test of the fairest and best trial of a man is before a jury, endowed with similar and like passions. . . .

The first principle of law is to put yourself in the place of the man who is before you for trial. That feeling of kindly chivalry and courtesy of the citizens of the south side of Virginia has made it a pleasure to have this case tried in your midst. . . .

The Commonwealth demands the life of the defendant, and for what? I do not undervalue life, but there is something sweeter to all Virginians—the purity of our women. We have written in our laws that if a man attempts to attack one of our daughters he has forfeited his life. You gentlemen of the jury have nothing to do with the truth or falsity of the story Miss Loving told her father. His daughter was his pride. He admired her beauty and her purity.

When he heard that this daughter to whom he was bound by affectionate and devoted ties was brought home drugged and unconscious by a young man to whose courtesy and chivalry she had been intrusted, he labored against his passions and remained away from the sight of young Estes until he heard the story of her ruin from her own lips.

As Strode spoke to the jury, Judge Loving, sitting at the defense table, dabbed his eyes with his handkerchief, while his wife openly wept behind him. The jury was moved. In the gallery, reporters for a number of big national newspapers from up north, including the *New York Times* and *New York World,* scribbled notes as Strode finished his final appeal. They had already dubbed the sensational case "The Fatal Buggy Ride."

The jury didn't take long to deliberate. The next day, the throng of journalists reported Judge Loving's acquittal.

Senator Strode had never been considered a cynical, manipulating lawyer. In fact, he had always cultivated a reputation as a moral idealist. He was known as a crusader who led the fight against local corruption, and even as a lawyer who refused to work to acquit a client he thought guilty.[8] But like any politician, Strode was shrewd; he knew when to be pragmatic, and he recognized that in politics, principles must sometimes bend to necessity. Despite the dubious defense of his old friend and his rhetorical pandering to the jury, the "unwritten laws" of the Old South and the purity of its women really were part of the Virginia heritage he wanted to preserve and defend.

The drama of the Loving case brought Strode a measure of national fame, and though he had already been quite well known in the southern part of his state, his influence was now beginning to grow. A first-term senator, he was part of a new breed of reformers, those who embraced modern science and hoped to use the power of the state to solve social ills. The State Assembly only met every other year, from January to May, but during his first session in Richmond in 1906, Strode had demonstrated that he was a deft law writer and brilliant tactician. His charm, good looks, and oratory skills—which had come to him almost as a birthright—were proving him an effective legislator, and like an evangelist at a tent meeting, Strode moved others with his calm but strongly held beliefs.

Strode had all the qualities of an effective leader. For any person in the public view, an illustrious pedigree was an advantage—especially here in the South, where conversations about family and tradition were pastimes as common as porch swings and lemonade. And Aubrey Ellis Strode's Southern roots ran deep. He was the oldest of eight children, the son of Henry Aubrey Strode, the first president of Clemson University in South Carolina and a Confederate veteran of the Civil War. His mother, Mildred Garland Ellis, had traced her family back to original Virginia settlers, who had come over in 1661.[9] Strode had lived his entire life on the family's Kenmore estate, an old plantation just outside the town of Amherst. Even now with his bride, Rebecca, he still made his home in the family's redbrick manor, a mansion set on a hill overlooking the Virginia countryside.

Strode had inherited this family estate much earlier than expected. His parents both died suddenly in 1898 when he was only twenty-five, leaving him responsible for a younger brother and six sisters. A young man just embarking on a career in law, he became laden with unforeseen responsibilities: running Kenmore, opening a law office in Amherst, and even being a parent to his siblings. He had just begun to become active in politics and was just then preparing for marriage to Rebecca. Under this crush of expectations, he developed a strong sense of duty, a tireless work ethic, and a stoic demeanor in the face of personal tragedy.

But the deaths of his parents also placed on him a certain social stigma, one that could make a shambles of his illustrious pedigree. He discovered this in 1902, four years after his parents died and just before he planned to marry. He had applied for a life insurance policy, as any husband should, but he learned the company was going to reject it. Shaken by the reasons they gave, Strode sat down to write a letter to the longtime family physician, Dr. F. F. Voorheis, requesting his help.

"Dear Doctor: I have an application for life insurance pending which the company is inclined to reject because both of my parents died in hospitals for the insane. The agent here thinks that an explanation of the facts and circumstances might secure the policy. . . ." He then asked Dr. Voorheis to provide detailed answers to a list of questions about his family's medical history, including: "How long you knew my mother's family and its freedom from insanity? The causes of insanity of both my mother and father? How far in your opinion my insurability should be affected by their insanity?"

Dr. Voorheis replied ten days later with the explanation Strode needed desperately. Addressing his letter to the insurance company, Voorheis explained:

> *Mr. H. A. Strode's breakdown began with rupture of a blood vessel in the brain which induced at first partial paresis; this gradually deepened until it ended in almost complete dementia. . . . Mrs. Strode, while Mr. Strode was in above described condition, had a heavy burden thrown on her and was exceedingly anxious and very much worried. At this time she developed a pelvic abscess which was latent as far as symptoms were concerned, not giving the slightest indication of its presence until just before she died. This was the cause of her mental aberration. The abscess burst just before her death (probably a few days) and her mind cleared up completely. . . .*[10]

Although Dr. Voorheis did not explain how a pelvic abscess could cause a mental aberration, the company probably accepted his reasoning. Strode was able to purchase his life insurance policy, since the insanity his parents endured was not, as the company must have worried, a problem passed on to their son. With his pedigree apparently free from hereditary mental taint, he was free to embark on a public career and marry his love, Rebecca. Soon they produced the next generation in the Strode family, a son—William Lewis—born in 1904.

Though steeped in conservative tradition, Strode was enamored of progress, and in this first decade of the twentieth century, he became a champion of reform legislation. This tension between tradition and progress was typical of the larger American Progressive movement: it would never fully embrace European-styled socialism, but, emphasizing a theory of social equality based on individual merit, it would seek to "purify" a corrupted system of patronage and nepotism, and like the prophets who condemned the worship of Ba'al in ancient Israel, urge the country to return to its old, colonial ideals.

As a liberal and a progressive, Strode built a résumé of accomplishments that helped shape the future of his state: he was one of the first defenders of women's suffrage, and during the 1910 session, he wrote the legislation that established a special women's division at the all-male University of Virginia; as a member of the Senate Committee on Moral and Social Welfare, he was the primary force behind the creation of a parole and probation system in law enforcement, advocating state assistance for the poor families of prisoners; he was the leader in the battle to prohibit the sale of alcohol; and years later, as a judge, he was the man to decide that "separate but equal" required Virginia to provide better public facilities for its segregated black citizens.

Finding success in public life, Strode was also a devoted husband and father, happily making a home at Kenmore. But in 1906, the same year Carrie Buck was born, Strode's two-year-old son William succumbed to pneumonia. The tragedy of losing his firstborn must have been devastating for Strode, but he remained as busy as ever. Returning to the legislature after he buried his son—and never once mentioning the incident in his letters—he began to turn his attention back toward the issue that had caused him so much anxiety earlier in his life.

Strode had good reason to worry about the fact that his parents had died in mental institutions. The question of mental defects passed down to children was quickly becoming a key element in the Progressive movement of which he was so much a part. Strode was learning more about the new theories of eugenics, which located the primary causes of poverty and crime in inherited mental defects, including epilepsy and feeblemindedness. The findings of this new science were helping to spark the clamor for reform in the age-old system of charity and corrections. Eugenics and other theories of mental health had been slowly gaining momentum before his parents died, especially in the North, but these ideas were just now beginning to trickle down to Southern states like Virginia. If Strode believed in anything, he believed in family, his Virginia heritage, and the promise of American destiny. The focus of eugenics was on each of these, and with its promises of moral purity and marvelous enhancement, he couldn't help but take a keen interest in it.

THE HEART OF reform is a shift in perspective, an embrace of authority new. By 1906, as Strode was studying the laws of science to help

Senator Aubrey Strode, author of the Virginia sterilization law and the attorney who argued *Buck v. Bell* before the U.S. Supreme Court. Senate photo, 1908.

shape the laws of his state, a half century of rancorous debate over the theological and humanistic implications of Darwin's theory of evolution had begun to change how many viewed the nature of mankind. The mystery of heredity, the biological forces that passed on human traits, was among the many issues Darwin's theory raised, but by this time it had become the great scientific topic of the day. What is this mechanism, this force of heredity? This was the key question in how evolution slowly changed a species and, indeed, the key to unlocking the secret of human destiny. Old myths explaining the origins of human misery were giving way to "the proven laboratory facts of tomorrow."

As many social reformers liked to point out, the locally controlled systems of indoor and outdoor relief had few guidelines other than a vague biblical notion to feed the hungry, clothe the naked, and care for the sick. "The poor shall always be with us," were Jesus' very words, and few, really, ever thought to challenge the enduring fact of poverty. The very word "relief," after all, connoted something less than prevention and treatment.

In the rural past, paupers, town drunks, and even neighborhood brothels could be tolerated as a part of daily life, and local leaders simply tried to stem these problems with fiery sermons against drunkenness and vice—or perhaps a night or two in jail. But at the cusp of a new industrial century, with the influx of immigrants and the growth of congested urban centers, these problems seemed much larger and more ominous. As cities like Charlottesville began to expand, more slums were springing up, more beggars were roaming the streets, and more thieves were lurking in the alleys. These problems were becoming more visible not only in Virginia but throughout the United States. For the past few decades many of the country's leading intellectuals had been becoming alarmed, and Strode could read how social ills were spreading like a literal plague. The problems of Emma Buck and her baby Carrie, before considered just a local nuisance, were now taking on a national and even global significance.

More significantly, however, when it came to care for the poor, the traditional stance of altruism began to shift toward that of efficiency. In an industrial age, efficient structure and organization were known to be the keys to success, and many reformers were now seeking to create centralized, state-run bureaucracies to focus on social problems in a systematic way. State boards and national conferences were being organized around the country, and a new class of scientifically trained experts was meeting to discuss poverty and crime. And as they began to worry that the poor were becoming a dangerous horde, placing greater burdens on society, they also began to wonder whether they could eliminate these problems altogether. Must the poor always be with us?

As their anxiety grew, the same secular forces that drove the expanding, bustling cities also brought a new zeal for science and progress, a vaulting confidence that methodic study and American ingenuity could solve most any problem. And many of the long-held religious assumptions of rural Americans were giving way— or were simply being retranslated—to more modern ideas. Scientific discoveries seemed to show that vice did not just stem from a sinful human heart, needing preaching and patience and alms. Poverty and moral degeneracy, it turned out, were biological rather than spiritual problems, and should be treated as such. As a politician also interested in education reform, Strode may have come across Martin Barr's 1898 address to the National Education Association, which proclaimed: "What, in the beginning was a philanthropic purpose pure and simple, having as its object the most needy and therefore naturally directed toward paupers and idiots, now assumes the proportions of socialistic reform as a matter of self preservation, a necessity to preserve the nations from the encroachments of imbecility, of crime, and of all the fateful heredities of a highly nervous age."[11]

The specter of fateful heredities hovered over almost every discussion of the country's social ills at the end of the nineteenth century. The intellectual milieu was charged with the ideas of Charles Darwin, Herbert Spencer, and Cesare Lombroso, and many scientists at these newly formed conferences thought of life as a struggle for existence, pitting group against group in any given species, including mankind. But like those predestined for either salvation or damnation, some people were naturally endowed with strength, health, and intelligence, while others were born with weakness, disease, and stupidity. In a natural state, the law of the "survival of the fittest" would act as a "purifying process," allowing each species to evolve to higher states of being. But charity and altruism upset this natural purifying process, and ensured the survival of the weak. For the educated elites in America, the words of Spencer had almost come to carry as much weight as the Bible itself, and his critique of poor relief in *Social Statics* (1865) seemed devastating:

> . . . we must call those spurious philanthropists, who, to prevent present misery, would entail greater misery upon future generations. . . . Blind to the fact, that under the natural order of things, society is constantly excreting its unhealthy, imbecile, slow, vacillating, faithless members, these unthinking, though well-meaning, men advocate an interference which not only stops the purifying process, but even increases the vitiation—absolutely encourages the multiplication of the reckless and incompetent by offering them an unfailing provision, and <u>discourages</u> the multiplication of the competent and provident by heightening the prospective difficulty of maintaining a family.[12]

While these ideas, even then being labeled "social Darwinism," did not dominate every intellectual circle in the United States, they were quickly becoming the fundamental assumptions of most scientists and doctors meeting at national conferences. Most were coming to believe that happiness, social repose, and human progress depended on the "purifying process" Spencer described: "The forces which are working out the great scheme of perfect happiness, taking no account of incidental suffering, exterminate such sections of mankind as stand in their way, with the same sternness that they exterminate beasts of prey and herds of useless ruminants. Be he human being, or be he brute, the hindrance must be got rid of."[13]

Although Spencer was an ardent conservative, progressive reformers felt these fateful heredities, the primary causes of human misery, could be studied and controlled. By the turn of the century, the musings of social Darwinism were converging with the applied methods of eugenics, the theory of better breeding formulated by Darwin's aging cousin

Francis Galton. Eugenics explained the heredity of sin, as well as the means to salvation.

Indeed, research seemed to clearly demonstrate that the problems of Emma Buck tended to run in families. One researcher at the Pennsylvania Board of Public Charities explained the standard view in 1893: "The laws of biology, that like begets like, that imperfect seed in parentage cannot produce perfect offspring, that breeding in intensifies and magnifies parental peculiarities, that certain inherited defects or deficiencies induce criminality, and result in pauperism, are well-known, and generally accepted to be as invariable and immutable as the law of gravitation."[14]

As more research was done, scientists began to attribute this "parental peculiarity" to a hereditary mental malady that they loosely labeled "feeble-mindedness." The term had often been used to cover a host of mental illnesses, including everything from severe congenital idiocy to mild insanity. But it was now beginning to take on a specific, technical definition that described the "simple backward boy or girl." The illness of these simpletons was so subtle and so hard to diagnose that it had never before been noticed. Feebleminded people could be literate and fully functional in society, easily blending into the population. But their illness gave them weak wills and poor judgment, making them "easily influenced for evil" and susceptible to antisocial, deviant behavior. The Royal College of Physicians in England soon defined the feebleminded person as "one who is capable of earning his living under favorable circumstances, but is incapable from mental defect existing from birth or from an early age of competing on equal terms with his normal fellows or of managing himself and his affairs with ordinary prudence."[15] While appearing "normal," they were unable to lead moral, productive lives—and this was precisely why the disease was so dangerous. This congenital mental defect, as it was perceived to be, soon became the central focus of many eugenic thinkers, and they began to see it as the primary cause of poverty and crime.

The image of the pauper thus took on a very different shape. Before, recipients of charity were often called the "less fortunate" or "the least of these," people who needed protection from a harsh and brutish world. Now, social scientists started to describe those mired in poverty with such images as "the menace of the feeble-minded" or "the rising tide of incompetence" or "the flood of mental defectives." It was society that needed protection from the feebleminded, they claimed, and not vice versa. Why was society ensuring the survival of the weak, anyway? The logic was inescapable. Since this was primarily a hereditary and biological problem, the looming catastrophic flood of feebleminded paupers could only be prevented at the source: the womb. Across the country, conferences on poverty began to focus their discussions on the feebleminded female. At the 1893 Conference of Charities and Corrections, doctors proclaimed that the symptoms of feeblemindedness were more pronounced in females than in males. Women had weaker wills, they were more easily influenced by their base sexual appetites, and their debauchery posed a particular temptation even to normal, decent men. More ominously, these females seemed to become easily pregnant. After analyzing the 1890 Census, researchers affiliated with the Association of Medical Officers of American Institutions for Idiotic and Feeble-minded Persons reported the problem of "hyper-fecundity" among the feebleminded poor, and began to sound the alarm to the rest of the country.[16]

Many medical officers began to suggest that restrictive marriage laws be used to combat poverty and crime. States should demand a clean bill of mental health before granting marriage licenses, they argued. But this would not prevent illegitimate conception—a significant problem among the underclass. So others began to propose the forced institutionalization of all feebleminded women of childbearing age. Since "like begets like" was an immutable law of Nature, segregating these hyper-fecund women would not only keep paupers off the street, it would also prevent the propagation of more paupers. So by the end of the nineteenth century, some states were building new asylums and "training schools" precisely to keep poor women from becoming pregnant. As one administrator at the Newark Asylum for Women commented, "We owe it not only to the adult imbecile herself, but to humanity and the world at large to guard in every possible way against the abuse and increase of this class."[17]

Emma Buck, a drug user, a mother of illegitimate children, and a person dependent on state aid, was now seen as a member of this menacing class of women. No longer could she remain invisible or just a local nuisance. While states in the economically devastated South had lagged behind the pace of social reform in the North and Midwest, by the time of Carrie's birth in 1906, Emma was swept into a system on the cusp of change.

Like many other states, Virginia first formed organizations to discuss these problems. In 1900, a group of hospital superintendents organized the State Conference of Charities and Corrections, hoping to study Virginia's problems and generate public support for reform.[18] After years of lobbying, members of the conference finally began to convince state leaders. By 1906, Governor Andrew Jackson Montague was urging the State Assembly to create a central authority for social welfare and appropriate funds for new, more modern institutions. At the next legislative session, on March 15, 1908, just before Carrie turned two, Strode helped the assembly create a permanent Board of Charities and Corrections, giving it the authority to sweep through the state, inspect current services, and make recommendations for restructuring. The task presented an enormous bureaucratic and logistical challenge.

The first secretary of the board, the Reverend Joseph Mastin, a Methodist minister and physician from Richmond, began his tenure by visiting hundreds of the local poorhouses, orphanages, and asylums. The conditions he found were deplorable. The criminally insane were housed with epileptics, and low-grade idiots lived in filth with the emotionally disturbed poor. He reported that these institutions were "relics of inhumanity," hovels filled with dirty bedding, vermin, filthy rooms, bad food, lack of water, fallen plaster, and dilapidated furniture all around. He found that local funding for these places was often paltry or nonexistent. At one orphanage in Richmond, he learned, when the woman in charge needed money, she would take a tambourine, station herself at a street corner in the business district, and beg. Even worse, she would teach the children to do the same, and they approached pedestrians, rang doorbells, and methodically collected tambourines full of pennies, nickels, and dimes. All across the state, Dr. Mastin found similar stories. There was, quite literally, no method to this madness.[19]

Aubrey Strode was one of the legislators most interested in this work. He had felt the stigma of madness hovering over his family and he must have seen the bedlam conditions of the hospitals where his parents had died.[20] From the start, Strode directed most of his energy toward reforms in mental health and social welfare, and he wrote a number of the bills that enacted Dr. Mastin's recommendations. By the end of the decade, Strode was widely known as the assembly's foremost expert in mental health reform.

Among his legislative triumphs in 1906, Strode wrote the law establishing a new State Colony for Epileptics. Working closely with two of

Virginia's leading eugenic thinkers, Dr. Albert Priddy and Dr. Joseph DeJarnette, Strode sketched the plans for a new, modern facility. It was meant to provide more appropriate care for epileptics, who, like his parents, did not really belong in the overcrowded hospitals for the insane. The scope of this new institution, as well as its name, would later be expanded to include the functional feebleminded, and it would become a place where this great menace to society could be segregated and treated.

The new Colony was to be built on land in his own district, of course, just outside the town of Lynchburg, and like similar institutions in the nation, it would focus not on those with severe mental defects but on the more dangerous problem: fecund, feebleminded females. The law stipulated that the Colony should take in "those indigent white persons who would be a greatest service to the Colony, who would in the judgment of the superintendent of the colony be most likely to receive benefit from colony care and training, and who are women of child-bearing age from twelve to forty-five years of age. . . ."[21] Virginia would once again be on the cutting edge of mental health; but, as Strode had once said, there was something sweeter to all Virginians: the purity of our women.

ON APRIL 6, 1920, two days after Easter, a day that was cold for early spring, the Virginia "Orange and Blue" defeated Lafayette, 5–1, at the university baseball diamond in Charlottesville. They were cheered on by a small crowd of "Easter queens" snugly wrapped in blankets. A few miles away, at Union Station, Emma Buck was standing on the platform, waiting for the train to Lynchburg.[22]

A man named John T. Dobbs had arrested her during Holy Week. As an officer of the peace, one of his roles was to round up vagrants and prostitutes, take them off the street, and bring them to the courthouse to be booked. After appearing before a judge, these "shiftless" people would normally be committed to state care—either a prison or a mental institution. Emma had been a problem for almost fifteen years, and after a brief hearing and cursory medical examination on April 1, Judge Charles D. Shackleford ordered the forty-seven-year-old woman into the care of the Virginia Colony for Epileptics and Feebleminded. She remained in a jail cell over Good Friday and Easter, but on this Tuesday she was boarding the train for the three-hour ride to the Colony.

Some ten years earlier, when Emma and her daughter Carrie were on the city's old charity lists, it was John T. Dobbs who helped take Emma's children and place them into foster care. Reformers had passed new child protection laws, allowing the state to intervene when single mothers like Emma seemed unable to care for their children. Foster homes were hard to find, however, and Officer Dobbs and his wife, Alice, agreed to take four-year-old Carrie into their small, stone house on Grove Street—an "act of kindness," they called it. The Municipal Court, which had replaced many of the responsibilities of the Overseer of the Poor, agreed to the arrangement, and Carrie was sent to live with the Dobbses and their daughter, a girl just a few years older than Carrie. If Emma tried to keep her children, no one made a note of it.

On the train to Lynchburg ten years later, though escorted by a social worker, Emma was alone. When she arrived, an orderly picked her up from the train station and brought her to Ward V, in one of the regimented brick dormitories at the Colony. Alice Jones, the ward's charge attendant, saw her first, and began the standard commitment procedure. She took off Emma's tattered clothes, bathed her head and armpits, and anointed her pubic hair with a mercurial balm. She jotted on the chart that Emma's physical condition was "very good," with no disease findings other than poor teeth and an "abdomen very prominent." Yet Miss Jones also noted that Emma had been treated for syphilis and that her arms still showed sores from intravenous injections. On the chart, she listed everything Emma brought with her: "waist skirt, over shoes, 1pr shoes, 1pr hose, 1pr coat, hat, underskirt, 2, shirt 1." Next to this she scribbled, "Clothes in very bad condition." In her coat pocket, Emma also had $4.50, and she was allowed to keep it. After dressing her in a clean set of clothes, Miss Jones led Emma to her new bed.[23]

The modern facilities at the Colony provided its charges with some of the best care and treatment in the nation, and Emma must have felt relieved to have a clean bed, clean clothes, and meals three times a day. She was put to work in the sewing room; during her free time, she just sat quietly and crocheted.

One day not long after she arrived, Emma was taken to a room to see R. L. Brown, a social worker at the Colony. There was a table with two chairs, and he asked her to sit down. In front of her, on the table, was a printed sheet of paper with a list of questions. At the top it read: "Record Sheet for the Stanford Revision of the Binet-Simon Tests."

A lot had happened in the decade since the Colony first opened its doors in 1910, and mental tests had become an innovative new tool for those who worked with the feebleminded. Measurement and classification were essential steps in scientific method, as any researcher knew, and social scientists had been working to develop an accurate, objective measure to diagnose and classify types of mental deficiency. Henry H. Goddard, the head of a "Training School" for backward and feebleminded children in Vineland, New Jersey, first experimented with his own tests, but after hearing about the work of Théodore Simon and Alfred Binet in France, he traveled to Paris to study their methods. In 1908, Goddard brought their intelligence test to the United States, translated it into English, and published it for researchers in his field. A scholar at Stanford University in California, Lewis Terman, also interested in these tests, spent years researching and revising the concept. In 1916, Terman published a new version of the Binet-Simon test, calling it the Stanford Revision. Within four years, it became the standard tool in institutions around the country, and scientists believed it was the most accurate way to diagnose feeblemindedness. In fact, the primary purpose of Terman's intelligence test was "efficiency" in the task to weed out and eliminate the great menace to society. As he wrote in his book *The Measurement of Intelligence* in 1916:

> *In the near future intelligence tests will bring tens of thousands of these high-grade defectives under surveillance and protection of society. This will ultimately result in curtailing the reproduction of feeble-mindedness and in the elimination of an enormous amount of crime, pauperism, and industrial inefficiency. It is hardly necessary to emphasize that the high grade cases, of the type so frequently overlooked, are precisely the ones whose guardianship is most important for the State to assume.* [24]

All inmates at the Virginia Colony had to take Terman's test.

R. L. Brown was specially trained to administer the Stanford Revision. On the sheet before him was a series of question groups, six in each, categorized by chronological age. For preverbal years, the tester must simply observe the physical behavior of the subject. Beginning with "Age: Three Months," the tester was to look for normal signs of intelligence, such as "carries hands or object to mouth," or "able to grasp small blocks in right and left hands," or "reaction to sudden start, such as telegraph key or hand clap." By "Age: Two Years," the tester looked for the ability to point out objects in a picture, or imitate simple movements. The question groups moved on year by year, becoming more difficult. The verbal testing process was also fairly simple: the tester would simply ask questions until the subject could no longer give correct answers consistently. If the questions became too difficult after "Age: Twelve Years," for example, the subject was classified as having a "mental age" of twelve years. If a six-year-old could reach this level, she was precocious and bright. If a forty-year-old could answer no further, she was dull, and classified as feebleminded.

Brown began by asking Emma the questions for Age 7, marking results on the form.

Point with your left finger. Now your right. Correct. Repeat this series of numbers after me: 3, 1, 7, 5, 8. Correct. Can you tie a bow knot, like tying a shoe? Correct. What is the difference between a fly and a butterfly? A stone and an egg? Wood and glass? Correct. Can you make a picture of this diamond? Emma picked up the pencil on the table and traced out a crude parallelogram, her hand trembling Correct.

Emma was feeling self-conscious, and she giggled nervously as he asked her these questions. Brown made a note of this at the bottom of the page, and continued with the questions for Age 8.

Can you count backwards from 20 to zero? Incorrect. What should you do when you break something? What should you do when you are in danger of being tardy? What should you do if a person hits you? These were tricky questions for Brown to judge, but Emma answered them satisfactorily, and he marked for each, Correct. How are these similar: wood and coal; apple and peach; iron and silver; ship and automobile? Incorrect.

Now, Brown asked Emma, "Can you write your name?" Next to the parallelogram she had just traced out, Emma wrote, "Mrs. Emmett Adaline Buck." The first "M" on "Mrs." had a sweeping curl at the bottom left side, and though her hand trembled as she spelled out her full, formal name, each letter was perfectly correct, in standard cursive form. Not only was she literate, her polished penmanship was better than most of the doctors' scrawl that covered her records. Brown then gave Emma a list of vocabulary words. She knew most of them, so Brown marked, Correct. Now, to the "Age: Nine Years" section.

What are the days of the week? What are the months of the year? Correct. Can you count by 3's? Emma giggled, and said, "Three, six, nine, twelve, fifteen—" Good, he interrupted. Correct. Can you make change? If a stamp costs 4 cents, and you pay with a 10-cent dime, how much change do you get? Incorrect. Repeat these numbers to me, backwards. 6-5-2-8. Incorrect. What rhymes with day? What rhymes with mill? What rhymes with spring? Correct. Emma was giggling nervously.

This was the last question she got right, however. She could not give the definitions for most of the vocabulary words for "Age: Ten Years." She got three of six questions in Age Nine correct, and four of six questions in Age Eight correct. But at the top of the record sheet, Brown jotted, "Mental Age: 7. Conclusion: Low Grade Moron."

For many social scientists, the discovery of the "moron" was the first great breakthrough of intelligence tests. The lack of precision had always plagued eugenicists; but now, since doctors believed they could easily diagnose their patients' precise mental age, they were able to establish this new technical category. The word "moron" was invented by Henry Goddard to give a specific label for the most dangerous class of feebleminded people—those like Emma. Greek for "foolish" or "stupid," it distinguished them from "idiots," who were by definition severely retarded and unable to speak, and "imbeciles," who could speak and function at the level of a four-year-old. Feebleminded morons were adults who had "mental ages" of about seven to twelve years, and, as Goddard had explained, they appeared "normal" to most people. In an enormously influential study, *The Kallikak Family* (1912), he wrote:

> A *large proportion of those who are considered feeble-minded . . . are persons who would not be recognized as such by the untrained observer. They are not the imbeciles nor idiots who plainly show in their countenances the extent of their mental defect. They are people whom the community has tolerated and helped to support, at the same time that it has deplored their vices and their inefficiency. They are people who have won the pity rather than the blame of their neighbors, but no one has seemed to suspect the real cause of their delinquencies, which careful psychological tests have now determined to be feeble-mindedness.* [25]

Whether Emma's neighbors pitied or blamed her, her vices and inefficiencies would no longer be a problem for them. Not as neighbors, that is. Segregated in a state-run institution, Emma was now a burden to them as taxpayers. Intelligence tests could only diagnose the problem; they could not eradicate it.

Solutions, though, were being sought. In the decade before she was admitted, the science of eugenics was evolving in leaps, and once again, Emma was entering a system on the cusp of change.

When the Virginia Colony opened in 1910, the first superintendent was Albert S. Priddy, a colleague and close friend of Aubrey Strode. As a doctor and social scientist, Priddy had been swept into the eugenics movement at the turn of the century. He had been a successful doctor in Keysville, a small, rural community in the southern part of Virginia, and later became the superintendent of Southwestern State Hospital, a traditional medical facility. Priddy had also been active in politics, and he had served two terms in the Virginia House of Delegates. He had been on the House Committee on Prisons and Asylums, and had helped draft legislation for the improvement of the state's penitentiaries. During his time in the legislature and as a superintendent, Priddy also became close friends with Dr. J. S. DeJarnette, the longtime superintendent of Western State Hospital, and an early supporter of the new theories of eugenics.

By the time Emma arrived, Priddy had made the Colony into a social weapon. Instead of care, its primary purpose was to be a place to attack the hereditary causes of mental deficiency, and protect the society at large. It was his original vision, presented in the first Annual Report for the Virginia State Colony for Epileptics in 1910:

> *The epileptic remains with us always, alike the poor, as one of the most pitiful, helpless and troublesome of human beings, with their various and numerous afflictions, and worst to contemplate is the fact that of the known causes which contribute to the development and growth of epilepsy, that of bad heredity is the most potent, and with the unrestricted marriage and intermarriage of the insane, mentally defective, and epileptic, its increase is but natural and is thus to be reasonably accounted for.*[26]

But must the epileptic, alike the poor, always remain with us? While alluding to these words of Jesus, Priddy began to urge his old friends in the state legislature to continue to enact eugenic reforms, and sweep away this biblical fatalism. The epileptics and poor were not only with us, they were becoming legion, and something must be done. In the next year's annual report he wrote again that "it is reasonable to anticipate a rapid increase in epileptics and mental defectives. Therefore, it seems not inopportune to call the attention of our lawmakers to the consideration of legalized eugenics."

By 1914, Priddy asked the State Assembly to expand the Colony's mission to treat feeblemindedness, that great menace to the human race. When the law was revised, and the name of the institution officially changed to the Virginia Colony for Epileptics and Feebleminded, Priddy began to focus his efforts on prevention rather than simple treatment.

While he was trying to develop a state-of-the-art facility, utilizing the newest therapeutic drugs and treatment methods, his dedication to the care of epileptics and feebleminded was simply a means toward an end, and not an end in itself. In his annual report of 1915, he wrote that "this blight on mankind is increasing at a rapid rate. . . . Unless some radical measures are adopted to curb the influences which tend to promote its growth, it will be only a matter of time before the resulting pauperism and criminality will be a burden too heavy to bear." The goal was not "relief" per se, but purity and perfection for all the world. If society could rid itself of its genetic undesirables, its ills would diminish greatly. And for Priddy, the "radical measure" and by far the most efficient method to combat this blight on mankind was obvious: surgical sterilization.

Like other aspects of mental health, Virginia had been an early pioneer in forced sterilization, as Priddy knew quite well. In 1910, the same year the Colony had opened its doors, reformers had introduced a eugenics sterilization bill. The surgeon at the Virginia Penitentiary in Richmond, Charles Carrington, was one of the first in the nation to experiment with the new surgical procedure, and he had been writing papers for a national audience, arguing that sterilization would help solve the problem of crime.[27] In an article entitled "Hereditary Criminals—The One Sure Cure," he quoted the Bible, pointing out that "the sins of the fathers shall be visited upon the children." But he subtly subverted the biblical message, arguing that sterilization, "the one sure cure," could take away this divine curse on humanity.[28] Even the Reverend Joseph Mastin, the Methodist minister who was leading the reforms of Virginia's system of charity and corrections, had been advocating the sterilization of the feebleminded. But though the Virginia Medical Society supported the 1910 bill, and though the House Committee on Prisons and Asylums recommended its passage, it was easily defeated. Aubrey Strode, who had followed these issues closely in the legislature, later remembered how "public sentiment was very much against the idea."

While Virginia rejected its bill, however, other states had already legalized forced sterilization. California, coordinating the efforts of its seven institutions, was sterilizing hundreds of feebleminded inmates. Other states with laws were beset by legal problems, and courts were striking down eugenic legislation in some parts of the country.

But this didn't prevent Dr. Priddy from sterilizing inmates anyway. Strode had revised the law governing the State Colony in 1916, giving the superintendent the discretion to conduct whatever medical and surgical treatments he deemed necessary for the inmates' well-being. Dr. Priddy took advantage of this vaguely worded authority. In the 1916 Annual Report, he wrote that he had sterilized at least twenty young moron women and "four males showing vicious and dangerous tendencies."[29] He was also sterilizing women suffering from undefined "pelvic disorders." But in one of these cases, Dr. Priddy was beset by legal problems of his own. One of these women accused him of sterilizing her without her consent, and her angry husband decided to sue.

These were the facts at Priddy's 1918 civil trial: In September 1916, a Richmond man named George Mallory left home to work at a sawmill in another county. Work was hard to find, and he had a large family to support. His wife Willie and their eight children stayed behind in their house in Richmond, with little means to live. One night, when two apparent family friends were visiting, the police conducted a raid on the house and accused Mallory's wife of operating a brothel. Together with her two male guests and each of her eight children, Willie was taken into custody. The younger children were placed in care of the Children's Home Society, but Willie and her two oldest daughters, Nannie and Jessie, were sent to the City Detention Home. Here, a "Commission of Feeble-mindedness" diagnosed each as having the condition and committed them to the Virginia Colony. George Mallory, living miles away, was never informed.

Dr. Priddy sterilized Willie after she had been detained for six months, then released her. He did the same to her daughter Jessie. Nannie was kept in the institution, but had not yet been sterilized. George Mallory filed a lawsuit against Priddy in October 1917, seeking damages for his wife's lost wages and for their pain and suffering. But he also sought the release of his daughter Nannie. Angry and frustrated at what had happened to his family, he wrote a letter to Priddy in November, after he filed suit:

Dear sir one more time I am go write to you to ask you about my child I cannot here from her bye no means. I have wrote three or four times cant yet hereing from her at all we have sent her a box and I don't know wheather she received them or not. I want to know when can I get my child home again my family have been broked up on false pertents same as white slavery, Dr what business did you have operating on my wife and daughtr with out my consent. I am a hard working man can take care of my family and can prove it and before I am finish you will find out that I am. . . . What cause did you have to operateing her please let me no for there is no law for such treatment I have found that out I am a poor man but was smart anuf to find that out—I had a good home as eny man wanted nine sweet little children—now to think it is all broke up for nothing I want to no what you are go to do I earn 75$ a mounth I don't want my child on the state I did not put her on them. if you dont let me have her bye easy terms I will get her by bad she is not feeble minded over there working for the state for nothing now let me no at once I am a humanbeen as well as you are I am tired of bein treated this way for nothing I want my child. . . .

> Very Truly
> Mr. George Mallory

Though Priddy was incensed by the letter, and wrote back threatening to have him arrested, the court agreed with Mallory and ordered Nannie released. But at the civil trial, Dr. Priddy testified that the procedure was a "medical necessity," even though Willie claimed there was nothing wrong with her at the time. The jury did not believe the testimony of the poverty-stricken woman, however, and acquitted Dr. Priddy of any wrongdoing. He carried the authority of medical science, and this far outweighed the semi-literate ramblings of the Mallorys. Yet, after the trial, the judge in the case took Priddy aside and suggested he stop sterilizing patients until there was a law that gave him explicit authorization to do so.[30]

Though Priddy did continue to sterilize some patients for vague "pelvic diseases" over the next few years, he was determined to get a eugenic sterilization law for his state. Along with Dr. Joseph DeJarnette and Aubrey Strode, he was waiting for the right time to once again introduce a sterilization law to the Virginia legislature.

EMMA BUCK was to spend the rest of her life at the Virginia Colony, segregated as a "low-grade moron" from the rest of society for over twenty-four years. She was well cared for, and those who kept her chart noted that she was quiet and well behaved, eating and sleeping well, and spending most of her time alone "doing crude knitting." Of course, every night they were giving her 1.5 grams of the depressant Phenobarbital, a breakthrough anticonvulsant for epilepsy—a disease Emma had never had. When the transverse arches on her feet gave her trouble, the Colony provided her with special shoes, which gave her "a great deal of relief."[31] When she was sick, she was cared for in the infirmary, and she could walk out onto the porch of the veranda, and watch the magnolias and elms sway in the breeze.

Dr. Albert Priddy, the first superintendent of the Virginia Colony for Epileptics and Feeble-minded, 1912.

Emma died of pneumonia on April 15, 1944, at the age of seventy-one. Colony administrators sent a wire to her daughter Carrie, but they never received a response. Four days later, orderlies took Emma's body to the Colony Cemetery and buried her in grave #575. Just two weeks later, Emma's other children, Doris and Ray, came to inquire about their mother, whom they hadn't seen for a long time. Though Doris had also been an inmate at the Colony, administrators had made no effort to reach her or her brother. When the two children heard the news, they were devastated. They weren't even informed their mother had been sick. But it was a tumultuous time. Headlines in the papers were describing the great Allied push into Europe, the impending liberation of Rome, and the sure fall of the evil Fascist empire. So in the end, they simply deferred to the authority of the Colony, and left in tears. One office worker, describing Doris and Ray's distraught reaction to their mother's unexpected death, made a final entry in the file of Emma Buck:

"However, they were most considerate and accepted the explanation."

A Forgotten Gravestone

In the fall of 1912, over fifteen hundred students, both elementary and high school, poured through the doors into the Midway School, a large three-story brick building at the intersection of Main Street and Old Lynchburg Road. Sitting atop "Vinegar Hill," it was one of the highest points in Charlottesville, and students who walked had a bit of a climb. During the Civil War the building served as a hospital, but at this time it was a free public school for white students only, even though the Vinegar Hill neighborhood was quickly becoming an important center of commerce for the city's segregated black community.[1] Here, amid the din, six-year-old Carrie Buck had her first day of school.

Aubrey Strode had long called for better funding for schools like Midway. In the past few years, the state treasury had been carrying a surplus, and as Strode campaigned for reelection, he was promising to devote this extra cash to improving public education. Strode had always held to the principle that wealth should be redistributed for the common good. "To tax the rich community to educate the children of the poor community is as right and just as it is to tax the rich man to educate his poor neighbor's children, and upon this bed rock [*sic*] principle the whole free school structure rests," he would proclaim on the stump, arguing that even the poor deserved to go to excellent schools and receive an opportunity to succeed. This was the essence of American liberalism, the beginning of a new idea of meritocracy, where class (but not race) would no longer determine success. And as a reformer, he believed a strong, central authority should ensure this would happen throughout the state.[2]

Partly because of Strode's relentless efforts to improve Virginia's welfare, Carrie had a more stable life away from her mother, Emma. Her foster parents John and Alice Dobbs provided for her, bought her clothes, and were now making sure she got an education. In fact, years later, the Dobbs even took her out of Midway and sent her to McGuffey, a much smaller elementary school that emphasized a special curriculum of reading, grammar, and the simple moral lessons of *McGuffey's Eclectic Reader,* the most famous textbook in the nation, outsold only by the Bible.

Carrie liked going to school, but sometimes she'd cut class and walk around town with friends. She also liked to write notes to boys in her class when she had a crush. But her teachers were satisfied with her work, and in her progress report in the 1918 school year, when she was in the sixth grade, her teacher wrote that her performance was "very good—deportment and lessons," and again recommended her promotion. By the time Carrie was twelve, she was a gangling and awkward tomboy, sneaking to the river to fish with the boys or climbing Charlottesville's many hills.

She never went to the seventh grade, however. The Dobbses may simply have maintained the common belief that a sixth-grade education was all a girl like Carrie needed, or they may have worried her rambunctious demeanor needed closer supervision. So, despite her normal progress, the Dobbses stopped sending her to school. They kept tight control of her life, and expected her to keep the house in order. Carrie's teenage years became a time of "endless work" and "servant's chores," and more and more she felt she was not a member of the family.[3]

Carrie was growing to be a quiet, docile teenager, lonely and isolated, though not without characteristic bursts of temper. She was taller than most girls her age, and with her dark hair she looked a little like her mother. Carrie's eyebrows, low on her forehead, made her look dark and brooding, and it didn't help that she rarely went out into the busy city anymore. She did go to church with her foster parents on Sundays, and they allowed her to sing in the choir.

In the summer of 1923, when she just turned seventeen, Carrie's life again changed. She was pregnant. She told her foster parents that, while they were away, their nephew Clarence forced himself on her when she was alone in the Grove Street house. For John and Alice, this was a humiliating scandal. But whether they believed her story or not, it would have been shaming for them to have an unmarried pregnant girl under their care, especially as the father was a close relative. Complicating matters even more, their own daughter, who was married now, had also just become pregnant. So the Dobbses felt the best thing to do with their foster child, for whom they had cared for over thirteen years, was just to send her away.

This would not be difficult. After over two decades of reform based on the science of eugenics, it would be easy to show that Carrie was afflicted by the congenital defects passed on to her from her mother, now an inmate at the Colony for Epileptics and Feeble-minded. Carrie's baby, too, they worried, would most likely carry the same mental taint. John Dobbs understood this well, since part of his job was to take vagrant feebleminded men and women off the streets—just as he had with Emma Buck.

John and Alice brought Carrie before Charles D. Shackleford, the same municipal judge who had committed Emma three years earlier. The Dobbses knew exactly what to say. They testified that Carrie had started to show symptoms of epilepsy and feeblemindedness when she was ten or eleven, and that she was a strange, ungovernable girl, given to "hallucinations and outbreaks of temper," as well as various "peculiar actions." Though they had agreed years before to care for Carrie as an "act of kindness," they testified she was now putting a financial strain on the family. Judge Shackleford appointed two Charlottesville physicians to examine the girl—one of them was the Dobbses' own family doctor—and both agreed she was "feebleminded within the meaning of the law." On January 23, 1924, Shackleford ordered Carrie to be delivered to the superintendent of the Colony for Epileptics and Feeble-minded, and assigned the case to a Red Cross social worker, Caroline Wilhelm.[4]

There were delays, however. After a problem with the paperwork transferring Carrie's legal custody to the Virginia Colony, Miss Wilhelm wrote to Superintendent Albert Priddy, urging him to expedite Carrie's transfer as quickly as possible. Since the Dobbses' other daughter was also expecting a child, Wilhelm explained that Alice did not want to look out for Carrie while the lawyers were handling the problems. "As Carrie Buck is expecting her baby about the middle of April, it is very important that she be admitted to the Colony before that time if it can be arranged."[5]

Dr. Priddy, however, wrote back to say he was not convinced due process of law had been observed in Carrie's case. He complained about a minor problem in the paperwork, but the question was moot anyway. "We make it a rule to positively refuse admission of any expectant mothers to the Colony," he told her. "You will have to make some provision to keep her until the child is born and disposed of and then on notification we will take her when the law has been complied with in committing her."

Carrie gave birth to a baby girl on March 28, 1924, and either she or her foster mother Alice named her Vivian Elaine. In May, Wilhelm again wrote to Dr. Priddy, telling him the baby was born and Carrie was ready to be transferred. But she also had a concern: how should they "dispose of" Vivian now, since she was probably defective like her mother? "It has been difficult for us to decide what disposition to make in the case of Carrie Buck as we feel that a baby whose mother and grandmother are both feeble-minded ought not be placed out in a home for adoption," she wrote. "However, the people who have had Carrie in their home ever since she was a little girl, are willing to keep the baby with the understanding that it will be committed later on if it is found to be feeble-minded also. We are therefore anxious to send Carrie to you as soon as possible and should be glad to know when you can receive her."

Dr. Priddy reminded Wilhelm that the law required any feeble-minded person to be at least eight years old before commitment. But if Vivian were found to have the condition, he would certainly receive her at that time. As for Carrie, "The early morning local train comes into Union Station at 9 a.m. and is the most convenient one for us to meet and whoever comes with her could return on the 12 o'clock train." So, on June 4, 1924, Carrie and her social worker Caroline Wilhelm took the early train to Lynchburg, and two-month-old baby Vivian remained behind, under the care of John and Alice Dobbs.

IN THE SUMMER OF 1922, Colonel Aubrey Strode was beset with grief. Rebecca, his wife of nineteen years, had been killed in an automobile accident, and he was now alone at the Kenmore Estate with his four young children. After he buried Rebecca, he immersed himself in his work, as he usually did during difficult times. In a few weeks he placed a notice in the local paper, seeking a housekeeper and governess for his little ones.[6]

It had already been a very hard year for Colonel Strode. He had assumed his military experience and new title would be a boon to his political career. During the Great War, his childhood friend Irving Whitehead had helped him secure a prestigious post in the Army, and he served in France as a lawyer under the great General John J. Pershing, Commander in Chief of the American Expeditionary Force. But this didn't help his political aspirations as he had expected. A year before, he lost his bid for the U.S. Senate, and this year he failed again to gain the Democratic nomination for the local U.S. congressional seat. Both Whitehead and Dr. Albert Priddy, two of Strode's closest friends, tried to support his campaigns as much as possible, and Dr. Priddy even sent carloads of Virginia Colony inmates to swell the crowds at his campaign rallies. Though Strode was still an influential figure among Virginia's political elites—who, as was common in the genteel world of the South, always referred to him as "Colonel"—he realized he no longer had his earlier popular appeal. Yet at the time his wife was killed, he was still helping Priddy in the long battle to reform Virginia's mental health system, and was still following the science of eugenics as it spread its message of better breeding and the need for compulsory sterilization.

After one of the most difficult years of his life, however, Colonel Strode was falling in love again. For the last few years, he had been a friend and mentor to a young social worker, a twenty-five-year-old woman named Louisa Hubbard. When she was working for the Red Cross in Lynchburg four years earlier, she had written him a letter, asking if she could use a room in his law office. She had been working near a rubber factory, and the smell was unbearable, so a friend suggested she contact the Colonel and ask whether she could use some of his space. "Does this request seem very 'nervy'?" she wrote to Strode. "I'll have to have a room that can be made a little more livable than this one, but I shouldn't imagine we would harm anything, yet there will be the use of the stove and wear of things in general, and unless the plan is perfectly satisfactory to you please don't feel that it's your 'duty.' "[7] Strode didn't, and after he agreed to the arrangement, they grew to be friends.

Even when Miss Hubbard accepted a Red Cross post in Greenville, North Carolina, they continued to correspond. As a social worker, Hubbard was very interested in the eugenic reforms in North Carolina, and she became friends with Dr. Arthur Estabrook, a researcher from the Eugenics Record Office in Cold Spring Harbor, New York. Estabrook was investigating the hereditary problems of the "free issues," or "Ishies"—a term many Southerners used for folks with one black and one white parent. Miss Hubbard was fascinated with the ideas of eugenics, and she always asked questions about the "degeneracy" of those she often had to see. Young, idealistic, and bubbling with enthusiasm, she corresponded often with both her newfound older mentors.

Soon after Rebecca was killed, however, the friendship between Strode and Hubbard became romantic. In less than a year, they were speaking of marriage. But Louisa had a terrifying concern: she had had stomach ulcers in the past, and now she was afraid her problems might have made her less than a woman. Could she have children? Could any man really love her if she were sterile? This worry so consumed her that Aubrey finally told her she should see his friend. Having children wasn't very important for him at this stage in his life— he was almost fifty, after all—but he arranged an appointment for his fiancée to see an expert in the field, Dr. Priddy, who would examine her womb and let her know if she could bear a child.

Louisa was giddy about the idea. "I feel everything will work out all right if Dr. Priddy will just find out truly if I can and that there is not too great a risk," she wrote to Aubrey. "Do please let me know just as soon as you hear anything from him." She went to her appointment, and after the exam she wrote Aubrey a quick note of anticipation. "Oh—I do hope Dr. Priddy will let us hear from him definitely very soon. Of course we think we know it's all right and are acting accordingly, but dearest man, we must *know* in so far as we can."

When Louisa got the letter from Dr. Priddy, the news was good. She could barely contain herself, feeling an enormous sense of relief. She wrote to Aubrey,

You simply can never know the way I wanted you to take me in your arms and just hold me the minute that I finished Dr. Priddy's letter! I'm yours—darling I'm yours!! I love Dr. Priddy more today than I ever have—and he has certainly been splendid and thoughtful in every way, and I just know that he will be one of the best friends that "Col. Strode" and the future Mrs. Aubrey E. Strode have.

I sat down with a dictionary and have reread Dr. Priddy's letter a number of times, and am sending it to you to read. "Assuming that you and Col. Strode love

each other"! I like that—guess he's questioning you as long as you were not so frank as I. Maybe he thinks I did the proposing too![8]

During this year, the same issue was consuming Colonel Strode in another way. As he prepared for marriage and helped his bride-to-be get through her fears, he was also working very closely with Dr. Priddy on a bill Priddy had asked him to write. Priddy had long been an ardent proponent of eugenic sterilization, and for over a decade he had been lobbying his friends in the assembly to pass a law that would prevent unfit "moron girls" from having children. He had been waiting for the right time, ever since he was sued by George Mallory, to introduce a law. So earlier that year, when Governor E. Lee Trinkle told him the time was ripe to push a bill through the legislature, Priddy knew the man to write the law should be his friend, the state's expert in mental health legislation.

As Strode prepared the bill, Dr. Priddy began to lobby in earnest again. In the Colony's Annual Report for 1923, he explained to the legislature that if feebleminded women were not sterilized, the state would be obligated to keep them in institutions all through their fertile years—a fate far more cruel, he said, than the simple operation.

Each day I work as the custodial caretaker of delinquent high-grade moron girls. . . . {This} impresses me with the gravity of the responsibility that managers of institutions for the feebleminded assume in enforcing morality in their actions, or rather restraining them from overt acts of immorality. If they are to be kept from indulging in sexual immorality it means they are to be kept a lifetime in institutions under the strictest custody. This, to any fair-minded thinker must appear to be a cruel and unjust degree of punishment for their weaknesses. . . .

If they are to be kept in institutions and supported at the expense of the State for the child-bearing period covering at least 30 years, to prevent them from bearing children to increase the population of mental and physical defectives and dependents, it certainly seems more humane and just to them to give them the benefit of a milder and less severe method of attaining the desired end. . . .

Therefore, every reasonable and fair-minded person must concede that the withdrawal of the right to propagate their kind could and should be given to society in such cases of females as have demonstrated their constitutional mental and moral inability to use the right of childbearing as a blessing to humanity rather than a curse.

Across the country, the responsibility to enforce morality, or restrain immorality, had become a primary mission for institutions like the Virginia Colony. It was not only society's right but its duty, Priddy believed, to determine which females would be a blessing and not a curse to humanity. Since sterilization was the "less severe" way to protect the country from these women, he was anxious to see Virginia pass its own law.

As Strode studied the sterilization laws of other states, however, he found many had been declared unconstitutional on the grounds that they were "class legislation," affecting only poor inmates of state institutions. Some courts had also ruled they did not provide due process of law, while others declared sterilization meant "depriving a person of the natural right of procreation beyond the power of the state legally to take away."[9] For the past two decades, the procedure had also evoked vehement public resistance, especially from traditional Catholics and fundamentalist Protestants; the latter were also fighting the teaching of Charles Darwin's theory of evolution in the public schools.

Yet after the Great War, the mood in Virginia and the rest of the nation slowly began to change. The patina of time gave eugenics a greater aura of scientific authority, and eugenic ideas gained momentum among some of the more sophisticated circles of American society. Popular syndicated writers such as H. L. Mencken and Albert E. Wiggam actively encouraged the science of human betterment, and in state fairs all across the nation, eugenic societies spread their gospel through popular "Better Baby" and "Fitter Family" contests. At the same time, Congress was working to restrict immigration on eugenic grounds, and reformers were rallying public support to curtail the foreign "menace," also causing social ills. With this cultural shift, the public and the Virginia legislature might just be more amenable to sterilization.

Strode's job was to write a law that could pass constitutional scrutiny, and not simply public opinion. As he explained to a friend, his job was to cure "such defects as I could in the form of the Acts declared invalid by the courts, trusting that the growth of knowledge of the laws of heredity and eugenics and changing public sentiment might bring a more favorable attitude for the Legislature and the courts."[10] While he was courting Louisa Hubbard, he began to study the writings of Harry Laughlin, the superintendent of the Eugenics Record Office in Cold Spring Harbor, and the man considered the nation's foremost expert on legalized sterilization. When Strode drew up his legislation, he considered the "Model Law" Laughlin had written for state legislators, a law designed to avoid the constitutional pitfalls that had plagued sterilization over the decade.

In the session of 1924, when Strode's bill was introduced, it passed overwhelmingly. Governor Trinkle's instincts were correct. In the Virginia Senate, Strode's former colleagues approved the bill unanimously, 30–0. In the House of Delegates, it passed 75–2. But despite this overwhelming support, Dr. Priddy and other supporters were still concerned. Even as they celebrated their great victory, they wanted it to be foolproof, free from the threat of lawsuits from yokels like George Mallory. So the board of directors again turned to Strode and asked him to take to court a "test case."

For the administrators of the Virginia Colony, a new arrival provided an ideal subject for this test case. Carrie Buck and her mother, Emma, were inmates of the Colony, and since Carrie had just given birth to a girl, these three generations of feebleminded females could provide concrete evidence illustrating eugenic theory. As Strode later recalled to a friend, "With the very active and helpful cooperation of Doctors A. S. Priddy and J. S. DeJarnette [the test case] was done, having as the subject of the litigation Carrie Buck, a typical 19-year-old, feeble-minded patient of the Colony having an illegitimate infant already giving evidence of feeble-mindedness, and Carrie's mother also being a feeble-minded patient at the Colony."[11]

WHEN DR. J. H. BELL, the Colony's resident doctor, examined Carrie the day after she arrived, he noted that she was "dark and slight, with a low narrow forehead and high cheekbones." After the standard admittance procedure, he recorded that her health was good and her body free from "eruptions." She was sent to Ward FB9, one of the Colony's two-hundred-bed dormitories, and assigned to kitchen duty.

Carrie knew that her mother was in a ward somewhere in the Virginia Colony. She had never had a relationship with her, but still, she wanted to find her. In her free time between shifts in the Colony's kitchen, Carrie finally located Emma working in the sewing room in Building V. Carrie brought her mother small treats from the kitchen, and they talked of the weather and gossiped about the other girls in the Colony. On certain weekends, Carrie was even allowed to take furloughs to the Dobbses' home, where she could see Vivian, not yet six months old.[12] When she told her mother about her little infant girl, she simply referred to her as "Baby."

Like Emma, Carrie took the Stanford Revision of the Binet-Simon Test. John and Alice Dobbs's testimony and the cursory medical exam had been the primary evidence in her commitment hearing, but the social worker examining Carrie had some misgivings about some of their claims. "Papers state she has some outbreaks of temper and hallucinations, but this is doubted. At the time of examination there is no evidence of psychosis—she reads and writes and keeps herself in tidy condition." He also reported that her "practical knowledge and general information" was normal for those of her class, "which is below the normal average." But on the test, Carrie only showed a "mental age" of nine years. "Conclusion: Middle Grade Moron."[13]

A few months after her arrival, Carrie's court-appointed guardian, R. G. Shelton, a local Lynchburg attorney, told her she must go before the board of the Virginia Colony. She had to have an operation. This was to be the first step in her constitutionally required due process, and the board was planning to hear testimony to determine whether or not she should be sterilized, as Strode's new law stipulated. In December 1924, Carrie went before the board, accompanied by her guardian, Shelton. At the table sat Aubrey Strode, representing the Colony, and Dr. Priddy. This time, it would be the board of directors, not the super-

Carrie and Emma Buck at the Virginia Colony for Epileptics and Feeble-minded, 1924. Photograph by Arthur Estabrook, a social worker from the Eugenics Record Office.

intendent, making the final decision. The only person giving testimony, however, would be Dr. Priddy.

From the start, the testimony for Carrie's case was careless. As Strode led the questioning, Dr. Priddy testified that Carrie was a feebleminded woman of the "lowest grade Moron class." Though the Colony's social worker himself had reported her literacy, her education, and her general mental health—and tested her as "Middle Grade," not the "lowest grade"—Sheldon never once brought any of this into the hearing, or pointed out Dr. Priddy's exaggerations. Indeed, oral testimony in Carrie's case would constantly contradict the facts, and no one would ever in the three years of litigation come to point out the slipshod factual claims.

But the point of the test case was not to protect Carrie's rights. It was the broad, sweeping vision of eugenics that was really on trial. With his signature argument for sterilization, Dr. Priddy testified about Carrie's "moral" problems and how sterilization was the one sure cure:

> *She is a moral delinquent but physically capable of earning her own living if protected against childbearing by sterilization. Otherwise she would have to remain in an institution for mentally {sic} defectives during the period of her child-bearing potentiality covering thirty years. The history of all such cases in which mental defectiveness, insanity and epilepsy develop in the generations of feeble-minded persons is that the baneful effects of heredity will be shown in descendents of all future generations. Should she be corrected against child-bearing by the simple and comparatively harmless operation of salpingectomy, she could leave the institution, enjoy her liberty and life and become self-sustaining.*[14]

Future generations. Life. Liberty. These were the themes of eugenic sterilization, and Dr. Priddy was laying out a simple version of the science's social aims. Shelton's cross-examination was cursory. He simply asked for clarifications, and never referred to any evidence in Carrie's written records. Finally, Colonel Strode turned to Carrie and asked, "Do you care anything about having this operation performed on you?"

Carrie hesitated for a moment, and looked around the room. Trying to be as formal as she could, perhaps even coached a bit on what to say, she awkwardly responded, "I have not, it is up to my people."

The board's decision came quickly, and Carrie was adjudged a "probable potential parent of socially inadequate offspring" and ordered sterilized. "[The] welfare of the said Carrie Buck and of society will be promoted by such sterilization," it declared.

The first stage of the test case was a success. Of course, there was really never any doubt at the outcome—it was the board who had asked Strode to take to court a test case, after all—but it prepared the way for the next stage in the process. Now, they had to orchestrate a lawsuit on behalf of Carrie—against themselves.

Officially, R. G. Shelton asked Irving Whitehead to represent Carrie in her suit against Dr. Priddy. It was a curious choice, however, since Whitehead was one of the original board members of the Colony, and had himself, years earlier, approved some of Priddy's requests to sterilize Colony residents. After he left the board, he had also acted as the Colony's attorney before the State Board of Hospitals in various matters. But even more startling, a building on the Colony's campus had been named after him. The Colony would pay Whitehead's costs, and his checks would be signed by Priddy, the very person he was now helping to sue. But for this test case, these staggering conflicts of interest posed no problem. Who would actually object?

Colonel Strode and Dr. Priddy began to prepare for their important day in court. Both were convinced that if they could win this case, the decision would be a resounding victory, not only for the future of Virginia but for all the states seeking social betterment. So they worked to gather the scientific evidence that would be the cornerstone of their "defense." Using the methods of the Eugenics Record Office, the nation's state-of-the-art scientific research facility, they sent out a memo to all the social workers who had had contact with Carrie, explaining the process of how to gather necessary data.

Dr. Priddy also wrote to Harry Laughlin, superintendent of the Eugenics Record Office, and asked for his direct involvement in the case. The long letter explained how Strode had used his Model Law for the successful passage of the sterilization bill, and how, though there were still grave doubts over whether the courts would sustain the law, he hoped "with a progressive enlightenment as to the needs of eugenics, they might get away from their constitutional moorings within a few years, and give decisions, which would meet this pressing need." Then, in one of the most astonishing claims of his letter, he explained what he saw as the real mission of the Virginia Colony for Epileptics and Feeble-minded:

Our purpose is to use, as I have been trying for some ten years, this Institution as a kind of clearing house to give these young women educational, industrial, and moral training, sterilize them and send them out to earn their own living, and permit them to enjoy "Liberty and the peaceful pursuit of happiness" guaranteed in our Declaration of Independence, and to relieve the State of this enormous burden.[15]

Priddy was not being disingenuous. His earnest, patriotic tone was quite sincere. As a "clearing house," the regimented rows of dormitories would hold inmates until they could be sent to the infirmary, sterilized, and then released. Yet the liberty guaranteed by the great Declaration did not, for Dr. Priddy, include the pursuit of happiness derived from having children, the happiness so longed for by people like the recently married Mrs. Aubrey E. Strode.

But Dr. Priddy also hoped Laughlin would provide expert testimony at the trial, and he quickly filled him in on the facts of the case:

> *Now, as to our test case, I am very sorry I can not make you out a genealogical tree such as you would like to have, but this girl comes from a shiftless, ignorant and moving class of people, and it is impossible to get intelligent and satisfactory data. . . . She is well-grown, has a rather badly formed face, of a sensual emotional nature with a mental age of nine years; is incapable of self support and restraint except under strict supervision.*

Laughlin could only agree to give a deposition, since he could not find the time to come to Virginia to testify. He did, however, recommend Arthur Estabrook, one of his social workers who had been doing research in North Carolina at this time, and who happened to be a mentor to Strode's new wife.

Though Harry Laughlin never examined or even saw Carrie, parts of his deposition became the most important evidence at the trial. Testifying under oath about Carrie and her kin, he stated, "These people belong to the shiftless, ignorant, and worthless class of anti-social whites of the South. They are an ignorant and moving class of people, and it is impossible to get intelligent and satisfactory data." Carrie, he also testified, "is well grown, has rather badly formed face, of sensual emotional reaction. . . . She is incapable of self-support and restraint, except under strict supervision."

The fact that Laughlin really knew nothing about Carrie might explain why he just lifted the facts from Priddy's letter. As for the borrowed language? Well, Laughlin was a very busy man.

IT WAS A WINDY November day in 1924, and the stark, leafless branches of two tall elms and an old, gnarled oak swayed furiously above the Circuit Court of Amherst County. There was a certain grandeur about this courthouse, standing in a tiny town less than twenty miles from Lynchburg like a federal shrine deep within Virginia's countryside. It was a simple brown brick building designed in Colonial Revival style, a type of architecture recalling New England more than Dixie. With a cruciform foundation and a grand pedimented gable on the front façade, it stood on the highest point in Amherst, where the previous wood-frame courthouse had burned to the ground in 1872. In the center of the roof, above the window lintels and soffits painted white, an octagonal belltower stood amid the swaying branches, like the steeple of a church.[16]

As Carrie Buck walked into the courthouse on this morning, she passed a twenty-foot marble obelisk only steps from the door. It had been placed there two years earlier by the United Daughters of the Confederacy. The monument's inscription read: "To the memory of the sons of Amherst County who from 1861 to 1865 upheld in arms the cause of Virginia and the South, who fell in battle or died from wounds, and survivors of the war who as long as they lived were ever proud that they had done their part in the noble cause." With Carrie were two of Amherst's sons: her attorney Irving Whitehead, who grew up just miles from this town; and Colonel Aubrey Strode, whose Kenmore Estate was also just a few minutes' drive from here. The monument might have a special meaning for Strode, too. His late father was one of these proud veterans.

Presiding over this court was Judge Bennett Gordon, an old Amherst attorney who had known both Strode and Whitehead since they were boys. He had watched them as they grew from young, small-town attorneys to influential Virginia citizens who had argued cases before the Supreme Court of the United States. Now they were back in Amherst, opposing counsel in the case *Buck v. Priddy.*

Judge Gordon's courtroom was traditionally austere. His bench, on a platform looming over the gallery and the two tables in front of it, was set off by a cherrywood railing with thick, round-curved spindles. On the railing, near the court clerk, a row of two-inch notches had been carved on top—one for every person hanged in the town's commons. As the clerk announced that the court was in session, Carrie Buck, the attorneys, and a collection of witnesses rose as Judge Gordon entered and took his seat.[17]

Normally, the trial would take place before a jury, but Whitehead waived this right. Normally, the plaintiff would present her case first, but Whitehead had no witnesses to call, no evidence to present, no case to make. So it was Colonel Strode who began with the defense. He had a simple plan, and he had eleven witnesses to examine as he presented his case. The first six would describe firsthand the peculiar and degenerate character of the Buck family tree. The last five would be the experts, the social workers and scientists who would outline the case for eugenic sterilization. The plan barely focused at all on the facts of Carrie's own medical case.

Strode called his first witness, Anne Harris, a Charlottesville District Nurse who remembered working with Carrie's mother, Emma, over a decade ago. After a few preliminary questions, Colonel Strode asked her what she knew about the family.

"Well, I know that Emma Buck, Carrie Buck's mother, was on the charity lists for a number of years, off and on—mostly on—that she was living in the worst neighborhoods, and that she was not able to, or would not, work and support her children, and that they were on the streets more or less. . . . Emma was absolutely irresponsible. She did not have any idea of providing for herself and her children. She was literally on the streets with her children, and the numerous charity organizations worked for her at different times, but all that was done for her was to give her relief."

After such hazy recollections, Mrs. Harris went on to testify about Carrie's sister Doris, whom she simply described as a "stormy individual" with an "ungovernable temper."

When Strode concluded his questions, Irving Whitehead stood up to begin his cross-examination. However, instead of challenging the witness's testimony, Whitehead actually asked for further information. First, he got Harris to testify that Carrie was illegitimate—which Strode had failed to do—and though she had no concrete evidence of the fact, Whitehead did nothing to contradict Harris's vague surmise. A distinguished trial attorney, Whitehead even allowed the witness to bring in new, damaging testimony against his client, when he asked an open-ended question:

"Now what are the—what about this girl, Carrie, herself? Is there anything about her? Is she incorrigible?"

"I really know very little about Carrie after she left her mother. Before that time she was most too small . . ."

"So far you know, you know nothing about her after the Dobbs took her?"

"Except one time when she was in school, in the grammar grade, the Superintendent called me and said she was having trouble with Carrie. She told me that Carrie was writing notes, and that sort of thing, and asked what she should do about it."

"Writing notes to boys, I suppose?"

"Yes, sir."

"Is writing notes to boys in school, nine or ten years old, considered anti-social?"

"It depends on the character of the note."

"Did you see the notes?"

"Yes, sir."

"Well, if the note is not altogether proper, is it evidence of anti-social —"

Harris interrupted him. "For a child ten years old to write the notes she was writing, I should say so. . . . If a girl of sixteen had written that kind of note, she ought to have been sent to Parnell—Isle of Hope," she exclaimed, referring to the school for wayward girls. Whitehead had no more questions.

Colonel Strode next called four schoolteachers to testify about Doris, Ray, and some of Carrie's cousins. He was trying to build the case that the Buck family line was replete with dull, stupid, and backward students. For over an hour, Judge Gordon listened to tales of misbehavior, poor performance, and "peculiar" personalities. Sometimes, the teachers equivocated, and Strode became frustrated and impatient. He had paid for these four teachers to come all the way from Charlottesville to testify. Of course, the most logical thing at this point would have been to call Carrie's teachers, too, or cite *her* school records, but neither Strode nor Whitehead offered a single piece of evidence about Carrie's education. Though his client had attended Charlottesville schools for seven years—never held back once—Whitehead never rebutted the testimony being given about her family by pointing out these facts. Both attorneys surely knew about them, however, given all the work they had put into locating the teachers and records of her siblings Doris and Ray.

During his examinations of the teachers, Strode seemed ill-prepared, often becoming exasperated with their unclear testimony. He also seemed not to know beforehand what these witnesses would testify. He must have been relieved to start the "expert" testimony by people trained in the science of eugenics—and people he knew quite well. He first called Caroline Wilhelm, the Red Cross social worker who had brought Carrie to the Colony and had handled Vivian's case with the Dobbses. (The Dobbses, to Strode's chagrin, refused to take part in the trial.) After the usual preliminary questions, he asked:

"From your experience as a social worker, if Carrie were discharged from the Colony still capable of child-bearing, is she likely to become the parent of deficient off-spring?"

"I should judge so. I think a girl of her mentality is more or less at the mercy of other people, and this girl particularly, from her past record. Her mother had three illegitimate children, and I should say that Carrie would be very likely to have illegitimate children."

"So that the only way she could likely be kept from increasing her own kind would be by either segregation or something that would stop her power to propagate," Strode said, leading her along. "Is she an asset or a liability to society?"

"A distinct liability, I should say."

"Did you have any personal dealings with Carrie?"

"Just in the few weeks between the time when the mission was held and when I brought her to Lynchburg."

"Was she obviously feeble-minded?"

"I should say so, as a social worker."

"Did you know her mother?"

"No, I never saw her mother."

"Where is the child?"

"The child is with Mr. and Mrs. Dobbs. They kept the child."

"How old is the child?"

"It is not quite eight months old."

"Have you any impression about the child?"

"It is difficult to judge probabilities of a child as young as that, but it seems to me not quite a normal baby."

"You don't regard her child as a normal baby?"

"In its appearance—I should say that perhaps my knowledge of the mother may prejudice me in that regard—but I saw the child at the same time as Mrs. Dobbs' daughter's baby, which is three days older than this one, and there is a very decided difference in the development of the babies. That was about two weeks ago."

"You would not judge the child as a normal baby?"

"There is a look about it that is not quite normal, but just what it is, I can't tell."

When Irving Whitehead began his cross-examination, he did not challenge any of these vague, subjective observations. The crux of the case was whether Carrie would produce feebleminded offspring, and Wilhelm's breezy testimony was the only evidence so far showing that the eight-month-old Vivian was a "defective" child. But Whitehead did ask for some clarifications.

"This baby you are talking about is Carrie Buck's baby?"

"Yes, sir."

"What other baby was the comparison made by?"

"Mr. and Mrs. Dobbs', who have had Carrie since she was three years old. They have a daughter who has a baby three days older than Carrie's."

"You say the baby of Carrie does not measure up to the Dobbs'?"

"Not nearly."

"Neither one of them can talk?"

"No."

"Can they walk?"

"No."

"In what way do they differ?"

"Mrs. Dobbs' daughter's baby is a very responsive baby. When you play with it, or try to attract its attention. . . . It is a baby that you can play with. The other baby is not. It seems very apathetic and not responsive."

Whitehead had no further questions.

After calling another social worker who had worked with the Buck case, Strode summoned Dr. Joseph S. DeJarnette, superintendent of the Western State Hospital at Staunton for thirty-six years, and Dr. Arthur Estabrook, the fieldworker from the Eugenics Record Office. Both explained, in a very simplistic way, the exciting implications of genetic research—especially Mendel's laws of crossbreeding.

Strode's final witness was the defendant himself, Dr. Albert Priddy. By this time, Priddy was a very sick man, suffering from Hodgkin's disease. Strode began his questions with matters of Carrie's family history, but then asked him about the benefits of sterilization.

"Would you think her welfare would be promoted by such sterilization?"

"I certainly do."

"Why? And how?"

"Well, every human being craves liberty; she would get that, under supervision. She would not have a feeling of dependence; she would be earning her own livelihood, and would get some pleasure out of life, which would be denied her in having to spend her life in custodial care in an institution."

"Would you think the public welfare would be promoted by her sterilization?"

"Unquestionably. You mean society in its full scope?"

"Yes, sir."

"Well, in the first place, she would cease to be a charge on society if sterilized. It would remove one potential source of the incalculable number of descendants who would be feebleminded. She would contribute to the raising of the general mental average and standard."

"Well, taking into consideration the years of experience you have had in dealing with the socially inadequate, and more particularly with the feebleminded, what, in your judgment, would be the general effect, both upon patients and upon society at large, by the operation of this law?"

Carrie Buck's "pedigree."

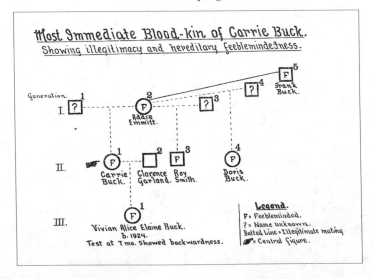

"It would be a blessing."

"Of course, these people, being of limited intelligence, lack full judgment of what is best for them, but, generally, so far as patients are concerned, do they object to this operation or not?" Strode asked, simply ignoring the very premise of the lawsuit.

"They clamor for it."

"Why?"

"Because they know that it means the enjoyment of life and the peaceful pursuance of happiness, as they view it, on the outside of institution walls. Also they have the opportunity of marrying men of their mental levels and making good wives in many cases."

Under cross-examination, Dr. Priddy testified that sterilization did not interfere with the "sexual passions" of the patient. He had always argued for the patient's "right" to enjoy the pleasures of sex. He went on to explain how he had sterilized certain women because of a "medical necessity," and how afterwards they were able to be discharged and live happily married lives. But while saying this, he casually mentioned to Whitehead that he probably knew the women he was talking about, alluding to Whitehead's participation in those sterilization hearings. This made Whitehead nervous, however, so he quickly told the court reporter, "Put in there that I knew them through being a member of the Special Board of Directors."

Whitehead's final question—the final question of the trial—was an unabashed boost for the defense rather than for his client. "Doctor, I understand you to say that if this girl could be sterilized, the Dobbs' home would be open to her?"

"I understand they want her back."

"And the only thing to prevent her having an independent home is her child-bearing capacity?"

"Yes, the Dobbs' home would be open to her to return."

It was a blatant lie, however. But how could Whitehead know that? After he sat down, Strode rested his case. Since opposing counsel, his friend, had presented no case for the plaintiff, no final summation would be necessary.

TWO MONTHS AFTER Dr. Albert Priddy testified at his own trial, he died of Hodgkin's disease. He did not live to hear Judge Gordon's ruling in his favor in February 1925, a decision that upheld the validity of the order to sterilize Carrie Buck. After the decision came down, Colonel Strode wrote to Dr. Bell, who had assumed Dr. Priddy's position as superintendent of the Virginia Colony, and asked him to take on the role of the defendant in the case. Dr. Bell quickly agreed, and in the Virginia Court of Appeals, and later in the U.S. Supreme Court, the case was known as *Buck v. Bell.*

Irving Whitehead had put little effort into the case on the trial level, since he was part of the friendly nature of this test case. But on appeal, he presented a more vigorous legal argument against the notion of state-sanctioned involuntary sterilization. The point of the test case, in the end, was to make the procedure constitutionally protected, so Whitehead brought up all the serious objections. His case was based on three points. First, he argued that the sterilization law deprived a citizen of the right to procreate without due process of law. Second, since the law applied only to institutionalized epileptics and feebleminded persons, it violated the Fourteenth Amendment, which demanded equal protection of the law for all citizens. Finally, Whitehead argued sterilization violated the Eighth Amendment by inflicting cruel and unusual punishment. Each of these had been successful arguments on the state level, and had caused seven sterilization laws to be ruled unconstitutional in the previous decade, so these were the issues the Supreme Court needed to address.

He had never attacked the facts of the case, however, such as the key presumption that there were three generations of feebleminded Buck women. And, as in the lower courts, his effort hardly matched the thoroughness of Colonel Strode's. Whitehead presented a five-page brief to the court, while Strode presented a forty-page brief. The Appeals Court quickly ruled in favor of the defendant this time as well, and all involved must have been relieved when the Supreme Court agreed to hear Carrie Buck's final appeal. Their plan was falling into place perfectly.

In front of the U.S. Supreme Court on April 22, 1927, Whitehead emphasized that the law violated a person's constitutional right to full bodily integrity, based on the Fourteenth Amendment. And he even gave a chilling prediction as he pleaded Carrie's case:

> *If this Act be a valid enactment, then the limits of the power of the State (which in the end is nothing more than the faction in control of the government) to rid itself of those citizens deemed undesirable according to its standards, by means of surgi-*

cal sterilization, have not been set. . . . A reign of doctors will be inaugurated and in the name of science new classes will be added, even races may be brought within the scope of such regulation, and the worst forms of tyranny practiced.

Whitehead's words, a vigorous and bold indictment of the dangers of eugenics, and a warning of the tyranny science could wield if it were given this power to decide who is "desirable" and who is not, might have been somewhat disingenuous. Yet even at that very moment, foreign governments in Europe were watching the example of the United States, and, indeed, beginning to inaugurate a reign of doctors as they legalized the forced sterilization of their "undesirable" citizens. Whatever Whitehead really felt, his words would prove to be prophetic.

The Supreme Court was not moved by the argument. The task of writing the opinion was given to Oliver Wendell Holmes, Jr., already considered by many the greatest jurist the country had ever produced. Now in his mid-eighties, a veteran of the Civil War like Strode's late father, he was known for his principles of judicial restraint and individual liberty. He had once proclaimed that freedom should never be curtailed without a "clear and present danger" to the welfare of the state. As he sat down to write the opinion in *Buck v. Bell,* however, what struck him most, as a legal matter, was that the state could already compel its citizens to undergo vaccination, a procedure necessary to preserve public health. This argument was first suggested by Harry Laughlin. And besides, the country often compelled its citizens to fight in wars and give their lives. Why should it be unconstitutional to compel these people, who sap the strength of the country, to undergo sterilization? Giving up one's ability to bear children for the social good was far less a sacrifice than giving one's life.

The Supreme Court ruled 8–1 that the Constitution did not prohibit the compulsory sterilization of a U.S. citizen. Hence, the Commonwealth of Virginia could proceed to sterilize Carrie without her consent. In the closing lines of his majority opinion, Holmes wrote:

We have seen more than once that the public welfare may call upon the best citizens for their lives. It would be strange if it could not call upon those who already sap the strength of the State for these lesser sacrifices, often not felt to be such by those concerned, in order to prevent our being swamped with incompetence. It is better for all the world, if instead of waiting to execute degenerate offspring for crime, or to let them starve for their imbecility, society can prevent those who are manifestly unfit from continuing their kind. The principle that sustains compulsory vaccination is broad enough to cover cutting the Fallopian tubes. Three generations of imbeciles are enough.

The test case was a resounding success. The final, dramatic words of Carrie's case, of course, contained one obvious factual error. Carrie and her mother had been diagnosed "morons," not imbeciles. To the very end, the facts of this case were hardly the point.

Holmes was particularly proud of this decision, since he had been interested in eugenics for years. A few days later, he wrote to his friend Harold Laski, the famous British intellectual and socialist: "I wrote and delivered a decision upholding the constitutionality of a state law for sterilizing imbeciles the other day—and felt I was getting near the first principle of reform. . . . I have no respect for the passion of equality, which seems to me merely idealizing envy." A few months after that, he again told Laski about his experience writing this majority decision. "Cranks as usual do not fail. One letter yesterday told me that I was a monster and might expect the judgment of God for a decision that a law allowing the sterilization of imbeciles was constitutional." Laski replied with a similar joke and a sneer: "Sterilize all the unfit, among whom I include all fundamentalists!"[18]

Buck v. Bell began a new era in eugenics. State legislatures, urged on by the relentless lobbying of people like Harry Laughlin and newly formed eugenic societies, began adopting more sterilization laws. In 1927, Indiana passed a new law, since its original, pioneering statute had been struck down by state courts. North Dakota passed legislation that year, and in 1928, Mississippi did so as well. During 1929, nine state legislatures revised or enacted new sterilization laws. In 1931, bills were introduced in ten states, becoming the law in five. By the end of that year, twenty-eight states would have sterilization laws that could survive constitutional attack, based on *Buck v. Bell*. And in just four years after the decision, the total number of sterilizations in the United States would double from the previous two decades.

The decision was also a victory for the father of eugenics, Sir Francis Galton, who had died almost fifteen years earlier. It was his ideas that had shaped the case heard before the U.S. Supreme Court, and it was his vision, first formed some sixty years ago, that inaugurated the progressive movement to purify the human race through better breeding—including the forced sterilization of "undesirables" like Carrie Buck.

Now his ideas were evolving into a worldwide crusade, especially in the liberal democracies of the Western world, where the startling successes of science, technology, and industry were bringing unprecedented cultural change. As Galton had once predicted, "I cannot doubt that our democracy will ultimately refuse consent to that liberty of propagating children which is now allowed to the undesirable classes, but the populace has yet to be taught the true state of these things. A democracy cannot endure unless it be composed of able citizens; therefore it must in self-defense withstand the free introduction of degenerate stock."[19]

With the decision in *Buck v. Bell,* the democracy of the United States heeded Galton's words, refusing the liberty of bearing children to the undesirable classes. Six years after the decision, the newly elected government of Chancellor Hitler, too, was trying to teach its populace "the true state of these things," instituting a sweeping eugenics sterilization program not very different from Virginia's, and based on Harry Laughlin's Model Law.

Yes, our descendants may well be proud. America would endure. The country could now ensure it would be composed of able citizens. Three generations of imbeciles were enough.

ON DECEMBER 8, 1928, a bitterly cold day in Bland, Virginia, Carrie Buck sat down to jot a letter to Dr. Bell in a chilly upstairs room in the house of A. T. Newberry and his wife. She cooked and cleaned for the Newberrys, making five dollars a month, and though she argued a lot with the grandmother, she was doing quite well since leaving the Colony. But on this day, she was walled in her cold room, mired with doubt, fretting for her future.

> Dearest Dr. Bell,
>
> I will write you a few lines to let you know how I am getting along as I haven't wrote you since I have been away. I am getting along very well. I guess a lot of the girls have gone away since I have been up here.
>
> Dr. Bell I am expecting Mrs. Newberry to write you about some trouble I have had but I hope you will not put it against me and have me come back there as I am trying now to make a good record and get my discharge.
>
> You promised it to me in a years time but I guess the trouble I had will throw me back in getting it, but I hope not. Give Mrs. Berry my love and best regards.

I have a real nice home and you don't know how much I appreciate it and they are just as good to me as can be. . . . We have had lots of cold weather up here. Guess I will close for this time.

I am your sincerely,
Carrie Buck[20]

As Dr. Bell read Carrie's letter, he knew her neat handwriting, lucid words, and warmth for her friends were not uncommon for a middle-grade moron girl. In fact, this was why she was so dangerous—she would seem normal to the untrained layperson. Now Carrie was on parole, but she had been hoping for a full discharge from the Colony ever since she left its dorms. Many of the girls she used to know were leaving. They, too, were being sterilized and given parole. It was the practice at the Colony to release these girls and send them to wealthy Virginians needing housekeepers, and Dr. Bell was constantly receiving inquiries from families seeking live-in servants. The Colony, which kept legal custody of the young women, only asked the families to provide room and board and pay five dollars a month. It wasn't slavery, after all. If they misbehaved, however, they would be sent back to the Colony. Carrie would do anything to keep this from happening, and she was smart enough to try to preempt the news she knew Dr. Bell was just about to get.

When she was first paroled in October 1927, just one week after she was sterilized, Carrie was sent to the wealthy owners of a Belspring lumber mill, the Colemans. But they sent her back to the Colony after only two months. Dr. Bell then asked the Dobbses if they would take her in again, as Dr. Priddy had testified they would. They refused. They had wanted nothing to do with her trial, and now they wanted nothing to do with her life. Then, in February 1928, Dr. Bell received a letter from A. T. Newberry, a Bland cattle rancher. "I'm really anxious to get a good girl from your Institution," he wrote. Dr. Bell sent him Carrie, and she had been living in his house for almost a year now.

But just as Carrie worried, Dr. Bell received a letter from Mrs. Newberry the day after he received hers.

Dear Sir:
    In regard to Carrie Buck I have never seen a better girl to work, and is as obedient with Mr. Newberry and myself as can be, and tries so hard to please us with her work and we like her very much. But she is begin-

ning her adultery again, (I say again for I believe she is an old hand at the business.) I feel that this has been her downfall before. I advise her as best I know how, and try to impress upon her the importance of living a clean pure life. She promises me she will conduct herself right, or try as best she can, as she hates the idea of going back to Lynchburg again.

So if she lives right we will give her a home as long as she wants it. But we cannot have this conduct in our home.

I feel so sorry for her to think that she cannot live a purer life. We have been nice to her. Treat her like one of the family. Have gotten her nice clothes, and has as good a room as we have in the house all her own. . . . [We] look after her welfare as we would a daughter as we realize the temptations of today.

But if she conducts herself right from now on she may stay on and on. She deeply grieved [*sic*] for fear of going back to the Colony.

She is also worried over getting her discharge. I don't know just how she would do if discharged. I fear she would go to the bad. . . .

Still concerned about the purity of women like Carrie, Dr. Bell agreed with Mrs. Newberry, and did not give Carrie the discharge she longed for. But to Carrie's relief, the Newberrys didn't decide to send her back. Despite her loneliness, which was somehow made worse by their kindness, the kindness shown only to a perpetual guest, she tried to be as pure as she could. In just a month, however, Mrs. Newberry had another worry. Carrie was thinking of getting married, perhaps to the same man with whom she was involved. Mrs. Newberry didn't mind, really, but that would leave them without a housekeeper. So she wrote to Dr. Bell: "Carrie is doing nicely now. Don't know for sure yet whether she will marry or not. But in case she does, please keep a good girl in reserve for me as I will want one."[21]

CARRIE DIDN'T MARRY that year. Neither did she get the discharge she desired. For the next five years, she continued to keep house for the Newberrys. She still wrote to Dr. Bell, and the correspondence between them was cordial and kind. She was also writing her mother, Emma, telling her about her life, sending her things when she could. But on May 14, 1932, Carrie finally did marry a man named William Eagle, a carpenter and deputy sheriff in the town of Bland. He was a sixty-three-year-old widower, thirty-seven years her senior.

After she married, Carrie wrote to Dr. Bell, telling him how she joined the Methodist Church with her husband, and how she now had a pig and a nice garden with onions, lettuce, cabbage, and peppers. She was singing in the choir again, as she had when she was a girl in Charlottesville. She also wanted to bring her mother to live with them in Bland. Dr. Bell had no objections—it would save the Colony money—but he said Carrie and her husband would have to pay Emma's train fare and expenses. In the middle of the Great Depression, Carrie and William had little money to spare, so Emma remained the rest of her life in her Colony dorm.

Carrie lived with William Eagle for the next twenty-four years in Bland. After he died, she moved to another part of Virginia and married a man named Charles Detamore. Stricken by poverty, she and Charles worked as laborers on farms and orchards, and later Carrie took a job caring for an elderly couple. In 1970, they moved back to Charlottesville, living in a single-room cinder-block shed with no running water or electricity. They paid no rent on this hovel, but ten years later, after Carrie was hospitalized for exposure and malnutrition, social workers placed her and her husband in a state-operated nursing home—a modern institution of "indoor relief."[22]

Carrie died on January 28, 1983, in this state home in Waynesboro, Virginia. By the time of her death, more than 65,000 American citizens had been officially sterilized and forgotten. In her last years, historians and journalists, discovering she was still alive, met with her, asked about her life, and told parts of her long-forgotten story. Carrie was buried back in Charlottesville in a standard grave plot; her small gravestone, which she shares with her second husband, reads simply: CARRIE E.

A few people attended her funeral on a cold, drizzly day. Not too far from them, in the rolling hills of crosses and shrines in this Virginia cemetery, a small, one-foot arched stone stands next to a five-foot obelisk. The obelisk marks the gravesite of John and Alice Dobbs, Carrie's foster parents from years ago. The small arched stone next to it reads:

VAED

MARCH 28, 1924–JULY 3, 1932.

Vivian Alice Elaine Dobbs, Carrie's only child, had died of an intestinal infection after a bout with the measles when she was eight years old. During her brief life, she had attended the Venable School, one of the better public schools in Charlottesville. She completed four semesters before she died. Written on her mostly forgotten grade reports is that Vivian, the third-generation imbecile noted by Justice Holmes, had made the honor roll.[23]

BOOK THREE

THE SINS OF THE FATHERS

*The LORD, the LORD, a God merciful and gracious, slow to anger, and
abounding in steadfast love and faithfulness, keeping steadfast love for
thousands, forgiving iniquity and transgression and sin, but who will by no
means clear the guilty, visiting the iniquity of the fathers upon the children
and the children's children, to the third and the fourth generation.*

— EXODUS 34:6–7

Hottentots in Kantsaywhere

Sometimes, when he walked down the street or traveled through the English countryside, Francis Galton liked to play a game. He kept in his pocket a crude paper cross, and with a needle mounted on a thimble on his thumb, he pinpricked holes to chart the beauty of women he saw. When he walked by a woman he found beautiful, he punched a hole at the top of the cross. If the woman was just average, he punctured the outstretched arms. If unattractive, the long bottom strip.

He was only half-serious, of course, but in his own desultory way, he was trying to put together a "beauty map" of the British Isles by taking a statistical sampling from various regions. He gathered quite a few of these icons of data, crosses riddled with holes and showing at a glance the distribution of women "attractive, indifferent, or repellant." He found London ranked highest in beauty, and Aberdeen lowest.[1]

Galton was handsome himself, and having inherited a small fortune from his father, he was independently wealthy and able to indulge in most of the things that pleased him—and that propriety allowed. He was a famous explorer of Africa—a contemporary of David Livingstone—and his books, vivid accounts of adventure in tropical lands, had made him something of a celebrity. Galton's looks and charisma befitted this fame: he was tall and powerfully built, an impeccable dresser, clean-shaven except for the wide bushy sideburns that extended to his jaw. With thinning blond hair, bright blue eyes, and, like his father, a physical trait that allowed him to arch one eyebrow, his face often displayed an arrogant mirth. This was only enhanced by his thin upper lip, which naturally pressed down, V-shaped, and seemed to give him a perpetual smirk. With his wit and vaulting confidence, he was also a remarkable storyteller, and at dinner parties he liked to regale his guests with accounts of his exotic escapades, titillating them with an allusive, ribald humor and a touch of the profane. And as he told his tales, Galton often turned to the topics that consumed him most: women and statistics.

He had had an eye for female figures most all his life, and it was often fun for him, living in the time of Queen Victoria, to startle people with stories that bordered on the bawdy. Remembering the days when he was a sixteen-year-old medical student, traveling through Europe with his mentor to observe surgical techniques, Galton was fond of telling the "long-lost Fritz" story. "I accompanied Bowman to a lunatic asylum in Vienna," he could recall. "In those days I was particularly shy and sensitive, and a consciousness of even the least unconventionality made me blush to an absurd degree. In one of the female wards a young, buxom, and uncommonly good-looking female lunatic dashed forward with a joyful scream, she clasped me tightly to her bosom with both her arms, calling me her long-lost Fritz! *Tableau*—Amusement of the others, myself pink to the ears!"[2]

Galton often tended to blush around buxom women, although, ever the Victorian scientist, he didn't hesitate to stand back and observe. As a young man, now forgoing medical school to seek adventure in Africa, he wrote to his brother, Darwin Galton, and described the voluptuous women he found. After explaining how he tracked and shot a hyena through the backbone, he quipped:

> *Talking of back bones, as I have just left the land of the Hottentots, I am sure you will be curious to learn whether the Hottentot Ladies are really endowed with that shape which European milliners so vainly attempt to imitate. They are so, it is a fact, Darwin. I have seen figures that would drive the females of our native land desperate—figures that could afford to scoff at Crinoline, nay more, as a scientific man and as a lover of the beautiful, I have dexterously even without the knowledge of the parties concerned, resorted to actual measurement.*

He could have approached the Hottentot women in person and measured them himself, Galton boasted, if only he had been able to speak their language. But the point of the story was to show Darwin his surreptitious and scientific cunning. "Here I should have blushed, bowed, and smiled again, handed the tape and requested them to make themselves the necessary measurement as I stood by and registered the inches—or rather yards. This however I could not do—there were none but Missionaries near to interpret for me, they would never have entered into my feelings and therefore to them I did not apply." Instead, with his characteristic smirk, Galton stood at a distance, held up his sextant, and began to gather his data from afar. "As the ladies turned themselves about, as women always do, to be admired, I surveyed them in every way and subsequently measured the distance of the spot where they stood—worked out and tabulated the results at my leisure."[3]

From lunatic asylums to the African bush, the fixations of this smirking adventurer led Galton to an idea that would make him even more famous, world renowned as a brilliant scientist as well as an explorer. He was living in a revolutionary time, an age of *Pax Britannica,* industrial innovation, and scientific discovery, and his sophomoric obsessions—which would actually last well into his old age—slowly evolved into a novel inspiration. Sex and reproduction were important themes in the sciences of his day, especially in biology and the "natural" sciences, and the imagination of this scientific man and lover of the beautiful would grab hold of them, turning bawdy stories into a sweeping scientific vision to purify the peoples of the world. During a time when thinkers were using the calm, methodical techniques of empirical observation to discover the fundamental truths of human existence, Galton's great contribution would be his incessant urge to measure, count, and correlate.

Even before he presented his novel idea, this urge contributed much to the methods of science. Insatiably curious, Galton pursued an array of interests throughout his life, and he constantly made maps and jotted down statistics. For his maps of uncharted territory in southern Africa, the Royal Geographical Society awarded him a gold medal, their highest honor. As an explorer, he also dabbled in meteorology, drawing weather maps and charting temperature and pressure. Analyzing this data, he became the first person to discover the phenomenon of the "anticyclone"—a term he invented. And his fascination with personal identification and statistical differentiation made him a pioneer in forensic fingerprinting. His methods, first refined by Scotland Yard, were later used by police investigators in countries across the globe.

But it was the ideas of his cousin, Charles Darwin, that struck Galton like a light on the road to Damascus—a city where he once lived an "oriental life," actually, with his three pets, two monkeys and a mongoose. While abroad, he had long been fascinated by the physical and mental "peculiarities" of peoples like the Hottentots, and had long felt, too, the unassailable truth of Shakespeare's words, that the "island of England breeds very valiant creatures; their mastiffs are of unmatchable courage." After the publication of his cousin's *Origin of Species,* Galton's casual observations, and his thoughts that there might be more than a poetic analogy between the breeding of dogs and men, began to take a systematic shape. Armed with his own penchant for the new methods of statistical analysis, he pored over the theories of his cousin's stunning book; and after a life of scattered interests and obsessions, he finally proclaimed, for the first time, a startling idea: *Genius is inherited.* Man's natural abilities, especially intelligence, are biologically based traits passed on in generations of families.

But even more shocking, Galton proclaimed the logical application of this apparent discovery. Through careful, scientific selection in sex and marriage, mankind could purify the process of human reproduction and create an even greater and more valiant race. Desirable traits could be multiplied, and undesirable traits could be extinguished.

"Eugenics," he called this science. He invented the word from the Greek *eugenes,* which meant, he said, "good in stock," or "hereditarily endowed with noble qualities."[4] But he intended his program to be more than scientific research. Eugenics was an "applied" science, a practical research program with profound social and political implications. It was, as its Greek prefix implied, a quest for the Good, a vision of men and women who could move themselves toward a higher state of perfection. Galton thus stood at the genesis of a new endeavor in human history. His ideas helped prompt a study that would evolve into "genetics"—a term not yet invented—as well as a social program of controlled marriage, immigration restriction, and forced sterilization. In effect, his was the first notion of human "bioengineering," another term for a later age.

But marriage, sex, and reproduction had never been within the realm of science. The state, of course, might sanction the social contract between a husband and a wife, but most people still believed there was something sacred, something profound and perhaps even divine about the mystery of sexual intimacy and the charge to "be fruitful and multiply." This was the realm of the Church. Yet for Galton, a member of a new breed of British intellectuals calling themselves "agnostics," the science of eugenics was literally a new theology that could become nothing less than a religious creed. "In brief, eugenics is a virile creed, full of hopefulness, and appealing to many of the noblest feelings of our nature," Galton said near the end of his life.[5] It was a virile doctrine that Galton hoped would riddle with holes the religious strictures that prevented a rational treatment of sexual reproduction and human destiny. But, like any statement of faith, it was also a source of hope, a vision of paradise for a man wracked by a legion of inner demons.

IT BEGAN WITH A SIMPLE INTUITION. Farmers and horticulturists had known for centuries that through careful crossbreeding they could produce pedigrees of plants and animals that maintained the strongest, most desirable traits. Indeed, the axiom "like produces like" was as old as human understanding. Even in the Bible, Jacob, the crafty son of Isaac who stole his brother's birthright, used selective breeding techniques to produce a stronger strain of sheep and cattle— and thus made himself a wealthy man. Yet when Galton considered these things as a young, frustrated student at Cambridge University, beset by fits of insecurity and an actual breakdown, he made a mental leap that few before had thought to take.

Francis Galton, 1860.

"It is a first step with farmers and gardeners to endeavour to obtain good breeds of domestic animals and sedulously to cultivate plants, for it pays them well to do so," he recalled thinking during these years. "The question was then forced upon me—Could not the race of men be similarly improved? Could not the undesirables be got rid of and the desirables multiplied?"[6] It was a simple, obvious question, really, yet no one had ever taken it seriously. No one would dare compare a human being to a lowly plant or animal.

But why not? It was just a passing thought back then, and it would be years before Galton could pursue this question earnestly, confronting the deep cultural resistance to the idea that Man could be bred like any farmyard beast or fancy flower. Still, already during these Cambridge years, the ideas of natural science, supported by an epistemology of strict empiricism, were slowly beginning to sever the links that protected mankind's lofty place on a Great Chain of Being.

At first, Galton noticed what seemed to be a curious fact: In his class, the high achievers, the brightest and the best, often had fathers and grandfathers who were much the same. It was a fact that could not be mere coincidence, he thought, nor explained by chance or excellent education. These high achievers must have been born with this mental capacity, just like blond hair or blue eyes or light or dark skin. Conversely, the exceptionally ignorant also seemed to breed in families. Could they be got rid of? Why couldn't there be more high achievers and fewer dullards? "Evidently the methods used in animal breeding were quite inappropriate to human society, but were there no gentler ways of obtaining the same end?"[7]

The gentler ways of modern life. All around Galton, new technologies of the Industrial Revolution seemed to promise a world of greater ease and leisure, lessening the toil and labor of a nasty, brutish world. Science was progress. Its possibilities were practically unlimited, and firm, rational thought, applied to human breeding—couldn't it produce a race of men and women with superior, desirable traits?

But the time was not quite ripe for Galton's simple intuition, for even "gentler ways" would disrupt what many believed to be the unique identity of human beings. While he studied mathematics and medicine at Cambridge, his older cousin Charles was still formulating ideas from his voyages on HMS *Beagle,* and even though it was a time of intellectual turbulence, in which old beliefs were being reinterpreted or cast aside, the Bible was still a primary source for ideas of origin and destiny. The Christian notion of mankind's special place in the world, a species created in the image of God, the lord of nature, and just a little lower than the angels, still dominated the thinking of most European intellectuals, even if they did strip it down to its barest, nonreligious concept. In addition to this, while Galton was in school, Romantic notions of the human soul were propelling thinkers to "intimations of immortality." While rejecting traditional religious ideas, these thinkers also resisted the cold, myopic reductions of reason and science. "Speaking generally, most authors agreed that all bodily and some mental qualities were inherited by brutes, but they refused to believe the same of man," Galton recalled of the time. "Moreover, theologians made a sharp distinction between the body and mind of man, on purely dogmatic grounds."[8] To be sure, the dualism between mind and body had been a part of Western philosophical tradition for millennia, and both Galton and his older cousin Darwin would face complex, deep-seated resistance to their ideas—and not simply from the Church.

Even so, Galton did not bring forth his intuition *ex nihilo*. Tradition was under assault. Dogmatic religious beliefs had already been challenged by the Enlightenment's rigorous doubt of the ancient authorities, especially the "revealed" truths of Scripture. Two centuries before, the Catholic Inquisition had forced Galileo to recant his observation that the Earth moved about the Sun—a fact that confirmed Copernicus's theory and seemed to contradict the Bible and diminish the central place of mankind in the cosmos. Galileo's trial and humiliation later became a rallying cry for freedom of thought and the independence of reason. A few decades after this trial, the Dutch philosopher Baruch Spinoza wrote his *Tractatus Theologico-Politicus,* an anonymous work that was one of the first to critically analyze the biblical texts, ushering in a new era of systematic scrutiny of the ancient authorities. Enlightenment thinkers began to demand the Bible be read like any other book and analyzed with independent rational judgment, rather than predetermined dogmatic guides.

By Galton's time, an impressive critical tradition had arisen in European universities, and scholars in England, France, and Germany were engaged in a meticulous historical-critical analysis of Scripture. Yet, insisting on a stance of inquiry and rigorous doubt, Enlightenment thinkers also seemed to attack the very essence of faith. For almost fifteen hundred years, since the time of St. Augustine, most philosophers of the West understood their calling as "faith seeking understanding." After the time of Galileo and Spinoza, however, this seamless intellectual tradition in Europe had begun to come apart. Thinkers drove a wedge between matters of faith and reason, creating a gulf that widened with every new discovery. The German philosopher Gotthold Lessing called this "the ugly, broad ditch which I cannot get across, however often and however earnestly I have tried to make the leap."[9]

Although most biblical critics were working in Germany, where a tradition of close reading of Scripture had reigned since Luther, German Romantic poets and novelists brought these ideas to Galton's England. While British divines were wrangling in disputes over Catholic and Anglican theology, the novelist George Eliot translated into English David Friedrich Strauss's *The Life of Jesus Critically Examined,* a relentless and rhetorically brilliant attack on the historical reliability of the Gospels. Samuel Taylor Coleridge was also one of the first to introduce historical-critical ideas. In a series of letters, *Confessions of an Inquiring Spirit,* first published after his death in 1840, when Galton was just entering Cambridge, Coleridge attacked the delusions of traditional faith, and replaced them with an individual, inner intuition of the divine. "In short, whatever *finds* me," he wrote, "bears witness for itself that it has proceeded from a Holy Spirit."[10]

The wedge between faith and reason could cut both ways, and, as Spinoza's pious contemporary Blaise Pascal had said, "the heart has its reasons which reason cannot know." The Romantics reveled on the other side of the ugly, broad ditch; they celebrated the individual self and the intuitions of the heart, which transcended the limitations of science. Reason, they felt, could be as oppressive as any religious dogma. For them, there was something inhuman about the disinterested nature of scientific investigation and technological innovation. The barons of industry, who improved their factories to make production more efficient, brought not "marvelous improvement" but a landscape of "dark Satanic mills." William Wordsworth mocked "misled" scientists as "joyless as the blind," and accused them of waging "impious warfare with the very life of our souls," reducing life to a corporeal process devoid of its higher essence. In *The Excursion* (Book Fourth, Despondency Corrected, verse 27), he sang:

> *And they who rather dive than soar, whose pains*
> *Have solved the elements, or analyzed*
> *The thinking principle—shall they in fact*
> *Prove a degraded Race?*

As the movement waned in the second half of the nineteenth century, Galton would set out to upgrade the human race with science. Still, in the 1840s, the time was still not ripe for his simple intuition. Natural science, a relatively new discipline rising up from the foundations of English empiricism, was only just beginning its disinterested ascent, preparing the way for his idea of eugenics. Despite Romantic and religious protests, science was starting to take on an irresistible cultural authority, and the notions of "natural philosophy" and "natural history," before practiced by secluded clerics and aristocrats, were now evolving into a recognizable guild of working natural scientists. The great Royal Society and its journal *Philosophical Transactions* had always been at the center of English science, but at the cusp of the nineteenth century, specialists in a host of new fields were forming their own organizations. In 1807, scientists established the Geological Society; in 1820, the Astronomical Society; in 1830, the Royal Geographical Society; and in 1831, the British Association for the Advancement of Science.

As the Industrial Revolution improved publishing technology and rendered newspapers, journals, and books the first inexpensive mass media in modern society, these associations also began to publish their own "Proceedings." Now, not only could scientists communicate with each other more efficiently, the general public could read and discuss their new ideas. Books on science caught the imagination, and throngs of people purchased such works as George Combe's *The Constitution of Man in Relation to External Objects*, first published in 1828. Selling an astonishing and unprecedented 350,000 copies, Combe's book proclaimed that mankind was subject only to natural laws, and not the will of any deity. The key to the "good life" was to study and submit to these natural laws, rather than the Bible. In 1858, the *Illustrated London News* noted that "no book published within the memory of man, in the English or any other language, has effected so great a revolution in the previously received opinions of society. . . ."[11]

The traditions of science and industry—as well as religion—defined Galton's heritage. His father, Tertius, grew up in a clan of wealthy Quaker merchants who, despite being members of a fiercely pacifist faith, made their fortune manufacturing guns. His mother, Violetta, was a Darwin, and her father, Erasmus, was not only a well-respected naturalist but also the most famous physician in all of England. Galton's parents called their precocious child "little Frank"—he was the youngest of their six surviving children—and they were grooming him to be a doctor, just like his famous grandfather.

Erasmus Darwin was a fat, stuttering, pockfaced man with an enormous nose, who limped badly. But he was a genius. Even King

George III sought the services of the brilliant physician. Erasmus fiddled with inventions, too, building such things as a speaking machine, which used silk ribbon and a bellows to utter words. But he was also one of the most influential natural scientists of the early nineteenth century, and he delighted readers by expressing his ideas in popular verse.

It was Galton's grandfather who translated into English the work of the Swedish botanist Carl Linnaeus, who had devised a system of taxonomy organizing all of nature into a hierarchy of categories. This system, along with its binomial Latin names, would become the standard method for scientists to classify plants and animals, organizing all known living things into a chart that included kingdom, class, genus, and species. (It also was a primary reason thinkers began to classify human beings into a new concept called "race.") Yet, as Erasmus considered the nature of sex and reproduction over time, he was also one of the first to challenge the assumption that this hierarchy was static and immutable. He wrote encyclopedic poems about natural science, such as *The Botanic Garden,* a charming work that described in vivid anthropomorphic detail the "sex lives" of plants. But in his magnum opus, *Zoonomia, or the Laws of Organic Life,* he set out a theory of evolution in which he claimed all living creatures sprang from a single "living filament," a great first cause "endued with animality" and possessing "the faculty of continuing to improve by its own inherent activity, and of delivering down those improvements by generation to its posterity, world without end."[12] As the massive Darwin forebear wrote in another poem, *The Temple of Nature,*

> *Organic life beneath the shoreless waves*
> *Was born and nurs'd in ocean's pearly caves;*
> *First forms minute, unseen by spheric glass,*
> *Move on the mud, or pierce the watery mass;*
> *These, as successive generations bloom,*
> *New powers acquire and larger limbs assume;*
> *Whence countless groups of vegetation spring,*
> *And breathing realms of fin and feet and wing.*[13]

When Charles Darwin expanded his grandfather's nascent ideas and presented a systematic account of the birth of organic life, he would help shatter the slowly cracking dogmatic wall that divided man from brute. Despite religious doctrines, Enlightenment views of the mind, or Romantic intimations of a divine self, Darwin's observations revealed that, far from being created "a little lower than the angels," mankind was actually closer to the domestic creatures it knew how to breed.

Galton's college intuition lay dormant for decades. Failing to make honors and despising the practice of medicine, he went abroad, traveling throughout the British Empire, making a name in other ways. During his travels, however, he began to think again of inherited mental traits, especially the ostensible differences between Englishmen and their darker-skinned subjects. But it was Darwin's *Origin of Species* that really brought him back to the possibility of better breeding.

In 1869, Francis Galton wrote his older cousin to thank him for his scientific masterpiece, which followed in their grandfather's path. He described an emotion a number of thinkers were experiencing after reading Darwin's work: "The *Origin of Species* formed a real crisis in my life; your book drove away the constraint of my old superstition, as if it had been a nightmare, and was the first to give me freedom of thought."[14] Galton later recalled how he "devoured its contents and assimilated them as fast as they were devoured, a fact which may be ascribed to an hereditary bent of mind that both its illustrious author and myself have inherited from our common grandfather, Dr. Erasmus Darwin."[15] By breaking down the barriers that had always distinguished *Homo sapiens* from the beasts of the field, *Origin of Species* ushered in the possibility of a new way of thinking about these "hereditary bents of mind."

The time was ripe. Indeed, even as Galton was writing this letter to his famous cousin, he was preparing to publish a book of his own.

"I PROPOSE TO SHOW in this book that a man's natural abilities are derived by inheritance, under exactly the same limitations as are the form and physical features of the whole organic world," Galton wrote at the opening of *Hereditary Genius, An Inquiry into Its Laws and Consequences* (1869). "Consequently, as it is easy, notwithstanding those limitations, to obtain by careful selection a permanent breed of dogs or horses gifted with particular powers of running, or of doing anything else, so it would be quite practicable to produce a highly gifted race of men by judicious marriages during several consecutive generations."

The debt to his cousin was obvious. The organic world was governed by a natural process, and mankind, too, was subject to these laws. While certain "social agencies" were influencing the evolution of human nature—some working to degrade it and others to improve it—Galton announced that it was his idea to investigate these influences, consider the natural causes of genius and stupidity, and to apply the experience with animal breeding to the human race. "I conclude that each generation has enormous power over the natural gifts of those that follow, and maintain that it is a duty we owe to humanity to investigate the range of that power, and to exercise it in a way that, without being unwise toward ourselves, shall be most advantageous to future inhabitants of the earth."[16]

It was a duty we owed to humanity. He presented this bold and sweeping moral conclusion at the outset, but the investigations that followed were also revolutionary. Galton had already presented a short version of his ideas four years earlier in a two-part piece in *Macmillan's Magazine* called "Hereditary Talent and Character." Here, for the first time, he introduced the method of "pedigree analysis," tracing human traits through family trees and looking for signs of inheritance. In the *Macmillan's* piece, he admitted he could not prove there was a biological mechanism for heredity. In fact, he wasn't even looking for one. But by making educated inferences based on statistical probability rather than empirical observation, Galton believed he could prove his hypothesis.

It was an entirely new and innovative way of approaching a biological question. Statistics was a fledgling field, having more to do with concrete census counts and industry figures than abstract analysis. Naturalists never had much use for such numbers. And questions about human heredity, implied in Darwin's theory, had barely begun to be asked, since there were few ideas of where to look for its physical causes. So Galton's approach was innovative in two ways. Not trained as an empirical biologist, Galton pursued his curiosity indirectly, first studying the pedigrees of men of genius, trying to show the simple fact that "talent and peculiarities of character are found in the children, when they have existed in either of the parents, to an extent beyond all question greater than in the children of ordinary persons."[17]

In *Hereditary Genius,* Galton combined his method of pedigree analysis with a newly discovered mathematical law. The German mathematician Carl Friedrich Gauss had recently derived a "law of error," which showed how variations in physical properties—say, the size, shape, or distance of any given phenomenon—would be distributed normally within a certain range. According to the law, most measurements would cluster around the "mean," like iron flakes to a magnet, but there would also be a regular, descending number of deviations on either side. Though Galton explained this law by showing the distribution of adult male height on a straight bar graph—a strip of paper punched with holes—when it was later displayed on a graph with two axes, this "law of error" formed a neat and useful "bell curve."

Most mathematicians using Gaussian distribution had been physicists and astronomers interested in discovering the convenient mean of their measurements—throwing out the "errors" and anomalies—in order to have a single number for their equations. Galton, however, was interested in the anomalies. "This is what I am driving at—that analogy [of height] clearly shows there must be a fairly constant average mental capacity in the inhabitants of the British Isles, and that the deviations from that average—upwards towards genius, and downwards toward stupidity—must follow the law that governs deviations from all true averages."[18]

But he had a problem. Mental capacity—or intelligence—was not as discrete as weight or height, so how could it be measured? Galton used two clever estimates to chart this elusive human trait. First, he categorized men according to "reputation." Gathering data from a biographical handbook called *Men of the Time,* he determined that there were approximately 250 "eminent men" for every million in the British Isles. He then read the obituary lists published in the January 1, 1869, issue of *The Times.* Here he found that of the approximately 210,000 deaths of men that year, 50 of them could be considered "eminent"— the exact same ratio as 250 to a million! Going back to read obituary lists of years gone by, he found again, incredibly, that the proportion of eminent men to the total population, reduced to a common denominator, was consistent throughout: one for every 4,000.

With numbers and statistics, Galton conveniently reified the abstract concept of "eminence" and "reputation" in order to measure them. But his second method was a little more concrete. He analyzed the test scores of men who passed the honors mathematics test at Cambridge (a feat he himself had failed to achieve when he was there). First, he determined the Gaussian distribution of these scores, and then he compared them to figures he gleaned from two studies, one on the chest width of Scottish soldiers, and the other on the height of French military recruits. The test scores of the Cambridge honors students seemed to fall into a similar pattern as the physical traits. Though mental capacity could not be measured per se, Galton believed the test scores indicated a remarkable analogy, and allowed him to make an inescapable inference: "Now, if this be the case with stature, then it will be true as regards every other physical feature—as circumference of head, size of brain, weight of grey matter, number of brain fibers, etc.; and thence, by a step on which no physiologist will hesitate, as regards mental capacity."[19]

Now, having determined a regular pattern from these statistics, Galton charted, in effect, the "bell curve" of intelligence for all 15 million adult males in the United Kingdom. Using the ratio of one "eminent" man for every 4,000—as well as the converse, one "imbecile" for every 4,000—he created six ascending and six descending "grades" of mental capacity on either side of the mean. Separating his population into six age groups, beginning with ages 20–30 and ending with ages 70–80, he determined the total number of men within each mental grade. There were almost 2 million men aged 30 to 40, for example, so there would be approximately 500 geniuses and 500 dullards in that group, according to his ratio of one to 4,000. About 1 million of these men would have average mental capacity; about 600,000 would be just above and just below the mean; and so on, the numbers diminishing in a regular, predictable pattern.

But was this pattern hereditary? Having established his scale of intelligence, Galton turned again to his study of pedigrees. In a series of chapters, he analyzed the family trees of judges, statesmen, writers, scientists, divines, as well as great oarsmen and wrestlers and others. His working hypothesis was simple: the closer a person was related to an eminent man, the more likely he himself would also be eminent. Gathering biographical data from Foss's *Lives of the Judges,* and constructing elaborate family trees and statistical charts, he thought his hypothesis was confirmed. A son was more likely to be eminent than a grandson or nephew, for example, and vice versa. "Speaking roughly, the percentages are quartered at each successive remove, whether by descent or collaterally," Galton wrote. "The statistics show that there is a regular average increase of ability in the generations that precede its culmination, and as regular a decrease in those that succeed it. . . . After three successive dilutions of the blood, the descendants of the Judges appear incapable of rising to eminence."[20] Galton explained this in part by the fact that intelligence was actually based on a "triple foot-ing" of three distinct, inherited traits: capacity, zeal, and vigor. Since each of these was discrete and independent—and necessary for genius—the probability of inheriting the highest level of each was quite remote.

Galton simply repeated this method for the other categories of his eminent men. In the final section of his book, however, he also added a summary on "the comparative worth of different races," again relying on the Gaussian law of distribution. He casually compared the Negro race with the Anglo-Saxon, and with no methodical analysis, asserted that the darker-skinned races had a much lower mental mean on the curve of "natural ability." Of course, given his definitions of eminence, how could it be otherwise? Yet scientists would point to this ostensible statistical phenomenon, gleaned from similar definitions of ability and intelligence, well into the next century.

Galton also discussed the ancient Greeks, one of the "ablest races" in human history. He wondered why this great civilization declined, and gave two main causes. First, their "social morality" grew lax, making marriage "unfashionable" and leading to fewer and smaller families. Second, he claimed that immigration in the small Greek city-states began to pollute the purity of this noble race. If the Athenians had maintained their racial purity, Galton explained, if they had multiplied and spread across the globe, displacing "inferior populations," they would have made human civilization unimaginably greater. The demise of the ancient Greeks should now serve as a warning for En-glishmen: if they did not work to raise the mean of their own natural ability, high as it already was, and if they did not work to preserve their own racial purity by curtailing immigration, the British Empire, too, would start to decline.[21]

In the *Macmillan's* piece, Galton imagined what extraordinary effects it would have on the human race if those who possessed the "finest and most suitable natures, mental, moral, and physical," united in mar-riage, and thus raised the mean of natural capacity. In a flourish of char-acteristic wit, he described a scene where the best and brightest girls and boys, determined by a state-run system of competitive examina-tions, would be joined in matrimony. Every year, the winners of the competition, ten "deeply-blushing young men, all of twenty-five years old," would marry ten women, twenty-one years old, at a solemn festi-val in Westminster Abbey. The superior couple would be given £5,000 and charged to "yield children who would grow into eminent servants of the State." This was written with a smirk, of course, perhaps even a blush; but Galton really believed "if a twentieth part of the cost and pains were spent in measures for the improvement of the human race that is spent on the improvement of the breed of horses and cattle, what a galaxy of genius might we not create!"[22]

At the end of *Hereditary Genius,* Galton was less jocular, but his hope for a galaxy of genius remained much the same. Marriage, he said, "held in as high honour as in ancient Jewish times, where the pride of race was encouraged," was still most essential for racial health, and the fittest of the population should be urged to join together and have large families. But his science also had political implications. He envisioned a modern society where the hereditary aristocracy was swept away, and the most influential citizens would earn their livings from "professional sources" rather than inherited wealth. While his own family had not been aristocrats by blood, he, of course, had inherited his own small fortune. Yet this irony was tempered by the fact that Galton saw himself as a pioneer proposing a new future, based on a new meritocracy of intelligence and skill. He believed nature was the key factor in natural ability, but he also recognized the importance of nurture, too, and hoped society would be shrewd enough to bring the two together, making an efficient production, as it were, of breeding talented citizens, who were essential for the well-being of the country.[23]

These were momentous ideas. Galton's article and book not only helped set in motion a scientific enterprise that would evolve into intelligence testing, the statistical analysis of complex human behavior, and even modern genetic theory; they also established a new vision of the social Good, a vision thousands would use to develop scientific reforms they believed would be better for all the world. British socialists were drawn to his criticism of the irrational hereditary aristocracy, and began to use his ideas to try to break down the social barriers of the class system. Progressive reformers in the United States—where the tradition of social equality, as described by Tocqueville, naturally gravitated toward the idea of meritocracy—also took notice of Galton's vision and began to champion his message of racial purity. But this message had a "negative" side.

Genius had its converse in stupidity, and so, marriage had its converse in celibacy. Though Galton had mocked the Catholic tradition, and lamented the long history of European celibacy, which he felt pre-

vented the propagation of intelligent men, in the case of the "undesirable classes," he now saw it as having an important social function. In
the Good Society, the weak and dull "could find a welcome and a refuge
in celibate monasteries or sisterhoods," and live out their days in tran-
quility, not reproducing their kind.[24] The state would house these
undesirables, tainted by their inner natures, and eliminate the problems they caused by keeping them from breeding.

The technology of forced sterilization was years away, but the idea of
keeping the "unfit" from having children would begin to resonate with
a host of thinkers. In the welter of social upheaval at the end of the
nineteenth century, as Wagner's operas were trumpeting triumphant
myths of national pride and Herbert Spencer's philosophy was proclaiming ideas of natural social conflict, a new zeitgeist of Difference,
girded by science, was sweeping through the Western world. Galton's
work would have a significant and sophisticated audience.

Even his cousin, infamous now throughout the Western world, was
impressed by this idea. After Charles Darwin read *Hereditary Genius,* he
quickly jotted the author a note:

> My dear Galton,—I have only read about 50 pages of your book (to
> Judges), but I must exhale myself, else something will go wrong with
> my inside. I do not think I have ever in all my life read anything more
> interesting and original—George [Darwin's son], who has finished the
> book, and who expressed himself in just the same terms, tells me that
> the earlier chapters are nothing in interest to the later ones! It will take
> me some time to get to these later chapters, as it is read aloud to me by
> my wife, who is also much interested.
>
> You have made a convert of an opponent in one sense, for I have
> always maintained that, excepting fools, men did not differ much in
> intellect, only in zeal and hard work. I look forward with intense inter-
> est to each reading, but it sets me thinking so much that I find it very
> hard work; but that is wholly the fault of my brain and not of your beau-
> tifully clear style—Yours most sincerely, Ch. Darwin.[25]

It was indeed very hard work to conceive intelligence as a biologically inherited trait—just as it was hard work to believe mankind
descended from apelike brutes. Dozens of popular periodicals reviewed
Galton's book, including the *Daily News,* the *Saturday Review,* and the
Morning Post. Though they sometimes praised its originality, reviewers
criticized how it ignored the role of environment and education, and

they generally found the ideas too sweeping to be true. The religious press offered scathing critiques, of course, and anonymous reviewers in the *Catholic World* and the *British Quarterly Review,* a Congregationalist journal, wrote biting condemnations of the author's assault on the human soul. Yet Victorian scientists responded much like Darwin. The great naturalist Alfred Russel Wallace, who helped develop the theory of natural selection, gave a positive assessment in *Nature,* writing that many "who read it without the care and attention it requires and deserves, will admit that it is ingenious, but declare that the question is incapable of proof. Such a verdict will, however, by no means do justice to Mr. Galton's argument."[26]

It is often the questions as much as the theories that drive scientific innovation. While many realized the weakness of Galton's argument, his questions presented naturalists with an unavoidable problem. What sort of biological mechanism drove evolution? What, exactly, passed on the strong traits that helped ensure a species's survival? Scientists like Darwin, Wallace, and others recognized that the mystery of heredity was an important frontier in the ongoing discovery of human development. The mathematician Galton seemed to show them that even intelligence was a product of biological evolution; but what passed on this or any inherited trait from generation to generation? It was a question on which their theories could hinge.

Yet amid the intellectual and nationalistic din of the time, Francis Galton was writing in torment. Behind the smirk, behind the disinterested analysis of punched holes on a graph, his mind was haunted by the very ideas he was proclaiming. Women, statistics, his illustrious family pedigree, the meaning of sex and marriage, the duty to bear offspring with superior mental traits—What will *I* leave to this world? he must have wondered. What will *my* legacy be? It was becoming an agony he almost could not bear.

THE BREAKDOWN OCCURRED just before his forty-fourth birthday. He had just published "Hereditary Talent and Character," but in the winter of 1866, Galton fell into a depressed, incapacitated state that numbed both his body and his mind. He could not work. He could not read. He could not entertain the guests who often came to his home, a spacious town house at 42 Rutland Gate in Knightsbridge. He could only lie in bed and stare blankly at the walls surrounding him.

His house was starting to feel like an asylum. The front drawing room, where his elegant Swiss butler Gifi welcomed visitors, was painted in white enamel, giving it a light and airy feel. Galton had collected expensive pieces of furniture, each from different periods and styles, but as they sat in the empty white room, they seemed to haunt him with a sense of meaninglessness. The dining room was even worse. It was long and grand, and he had made it into a shrine of sorts, a place to display his devotion to science. He placed a bookcase at one end of the narrow room, and on the side, near the front window, his desk, so all who came to dine at the Galton residence could see where he worked and wrote. It was here in the dining room, too, that he hung portraits of his heroes and illustrious colleagues: his cousin Charles, of course; his friend Herbert Spencer, with whom he sometimes played billiards; the famous botanist Joseph Hooker; the mathematician William Spottiswoode; as well as many others. But without the laughter, the storytelling, the wine and the food, the faces in this empty room seemed to look down and mock him.[27] Walled in at home for weeks on end, Galton could only feel this strange, mysterious anguish, the sense of purposelessness and despair that breakdowns often bring.

But more than anything else, perhaps, his mind may have been tormented by his marriage. After his emotional collapse, he could barely speak to his wife of thirteen years. Louisa—or Loui, as he liked to call her—was so different from him. She seemed to be everything he never cared to be. While he had tolerated her deep piety, it always annoyed him. He considered himself an "agnostic," a term recently coined by his good friend Thomas Henry Huxley, one of Darwin's most ardent defenders, and meant to describe a secular stance that was something less than systematic atheism. While Loui often prayed, Galton was a man who snipped that religion "crushed the inquiring spirit, the love of observation, the pursuit of inductive studies, the habit of independent thought."[28] Loui was also the complete opposite of the voluptuous, sinuous women over whom he had always obsessed. She was thin and gaunt, her hollow, milky eyes were plain, devoid of mirth, and her round nose and hanging jowls made her even less attractive. She was a proper Victorian, too, wearing layers of crinoline, keeping her neck and ankles always covered. And, as a family member later recalled, "any comfort which might have given pleasure to [Galton's] leisure hours was often denied him by his wife."[29]

In his debilitating despair, however, Galton might have blamed himself, and not his wife, for their profoundly unhappy marriage. How easy it was to feel like a failure in this condition he was in. How easy it was to lie in bed and recall his youth, the mistakes he may have made, the improper carousing, the lack of moral strength that just might be the cause of his present misery. His breakdown also might have brought him back to an awful day in Cambridge, when an earlier nervous breakdown made him abandon the life his parents had always planned for him. Why did he feel this way? Was he a natural-born hysteric, with an inherited tendency to suffer "from giddiness and other mental maladies prejudicial to mental effort"?[30]

He had been a precocious child, the youngest in his family, and though he now considered the Christian faith as relevant as a flat-earth society, his parents had brought him up in a household pervaded by family tradition and religious devotion. He was unusually bright. He could recognize the alphabet by eighteen months, could read a book called *Cobwebs to Catch Flies* at age two and a half, and could write and do arithmetic by age four. His older sisters doted on him endlessly, and everyone in the family was constantly giving the prodigy lessons in some subject. When Violetta once saw her four-year-old son carefully saving pennies, she asked him why. "Why to buy honours at the University," he responded. Even as a boy, little Frank, as he was called, had felt his family's heavy expectations.[31]

He still could see the family manor on the old Galton estate near Sparkbrook, a small town outside Birmingham. It had been "a paradise to my childhood," Galton noted. The estate was called "The Larches" for the two towering larch trees at the entrance. Three acres of gardens, designed by his great-grandfather and improved by his grandfather, were bordered by rolling farmland and hills. The central section of their magnificent house was three stories high, with a wing extending on either side. The left wing, with a huge bay window, had a terrace that looked out over the gardens. The right wing was near the stable and brewhouse, and in the back were sheds for chickens, pigs, and cows. Behind these sheds were fields, where the children rode their ponies.[32]

It was a paradise lost, however, by the time Galton was ten. His parents moved out of The Larches, since they no longer needed such a vast space—and the country was giving way to urban life. In his later years, Galton remembered the mansion built by his forefathers and wistfully mused, "The house, which was once a centre of refined entertainment, gradually lost its charm of isolation; later on, it wholly ceased to be attractive as a residence. It was then leased by my father to the proprietor of a lunatic asylum, because, as he remarked, no one in his senses would live in it."[33]

There was something haunting about the past. Though Galton could read Latin texts by the time he was eight, he hated the rigors of school. His father had packed him off to elite boarding schools, but the boy detested the corporal punishment and cruel discipline of the masters. By the time he was sixteen, his parents sent him to the General Hospital in Birmingham, where he would apprentice as a doctor. In a few years, he went on to King's College Medical School in London. Though he also disliked the study of medicine, he compiled a near-perfect record at the prestigious college. "Hurrah! Hurrah!! I am 2nd Prizeman in Anatomy and Chemistry. I had only expected a certificate of honour. Hurrah! Go to it ye cripples!" he wrote to Delly, his adoring sister, who was confined to a couch because of a curvature of the spine.[34]

Then came the summer of 1840. Galton was eighteen and weary, really, of school and study, but he was becoming obsessed by two themes in his life: travel and the beauty of foreign women. With the consent and encouragement of his father, Galton planned to go to Germany during the summer to study organic chemistry. But once there, he quickly wrote his father to tell him he was "determined to make a bolt down the Danube and to see Constantinople and Athens." He was on his first adventure, alone, and as he traveled south through Europe, he dutifully wrote describing all he saw—especially the women. He told his father of the pretty Hungarian girl, Marie, whom he met on the train to Linz, a town "universally famous for the beauty of its fair sex." Sailing down the Danube, he wrote about the packed boat, with "only one pretty girl and she would hold down her eyes."

When Galton finally arrived in Istanbul, he was filled with awe. The dome of the Hagia Sophia, the long arcades of ancient aqueducts, and the minarets of a thousand mosques captured his imagination, and he was thrilled by the throngs of dark-eyed Jews, Armenians, Greeks, Albanians, Franks, and Circassians. He described it all to Tertius, writing: "The seraglios are splendid, ditto palaces, such a great deal of trellis work about them, and then there are cypresses, and the veiled ladies just looking out between folds of gauze and very pretty eyes they have too; then there are the Greeks, I never saw such black eyes in all my life. I should like to put one of them into a rage; they must look splendid then. I saw the women's slave market today—if I had 50 pounds at my disposal I could have invested in an excessively beautiful one, a Georgian."[35]

When Galton returned to England, his older cousin Charles, having just published his *Beagle Journal* and been appointed a fellow of the Royal Society, suggested that he supplement his medical studies with mathematics. Galton jumped at the chance, and since by law he could not receive a medical degree until he was twenty-one anyway, he convinced his father to send him to Trinity College, Cambridge, to study the riddles of numbers.

Galton studied hard, but the competition at Cambridge was fierce. There were so many brilliant men here, Galton thought, and he began to feel a sense of insecurity. Though he enjoyed the social life, he often spent nights on end poring over his lessons. He skipped teatime to study, and even bought a machine called the "Gumption-Reviver" to stay awake late into the night. The contraption was a funnel on a six-foot stand, filled with water, dripping regulated drops onto a cloth band around his head. When the cloth became too sodden, water dripped down onto his shirt—which was a good thing, Galton thought, because "damp shirts do not invite repose."[36] But when it came time to take the first major exam for his bachelor's degree, a test students called the "Little Go" (distinguishing it from the final "Great Go"), Galton's endless study could only rank him in the second class, while many of his friends made first-class honors.

The headaches began soon after. Without warning, heart palpitations and dizziness would come upon Galton like a thief in the night. The nervousness and stress became unbearable, and in the fall of 1843, Galton had his first full-fledged nervous breakdown. He had to drop out that term, and eventually settle for a "poll degree," a recognition of his passing grades, but with no honors.

Twenty-five years later, as he lay incapacitated by a more serious breakdown, Galton could recall how his life changed drastically after this first bout of mental anguish. The following year, his father Tertius succumbed to severe asthma and gout. He had been so devoted to his son's studies, so involved in shaping his career; but after Tertius died, Galton took his inheritance, and like the prodigal son, pursued the one thing that made him happy. "I was therefore free, and I eagerly desired a complete change," he later recalled. "Besides, I had many 'wild oats' yet to sow."[37] He was going to travel and seek adventure.

Despite the breakdown at Cambridge, Galton was growing into a tall, powerful man, with exceptional physical prowess. He left England in the fall of 1845 and traveled to Egypt. He met two friends there, and they sailed down the Nile with "the pleasure of living all day barefoot and only half dressed, and of waking oneself by a header in the river, clambering back by the rudder."[38] A French adventurer later told them to forget the "English routine" of just taking the same way back when they returned, suggesting they "cross the desert and go to Khartoum." So they did just that, crossing the desert on camels in eight days. After Khartoum, Galton traveled on to Beirut and Jerusalem and ultimately to Damascus, where he learned to speak Arabic fluently. He established a residence here, wore Arab clothing, and laughed at his two pet monkeys and his mongoose. But even as he sowed his wild oats during this time, he also reaped the consequences. He wrote to one friend, Montagu Boulton, telling him of a bad tryst he had had. Boulton wrote back, concerned but also amused. "What an unfortunate fellow you are to get laid up in such a serious manner for, as you say, a few moments' amusement." He gave Galton an account of his own attempted conquests, writing how he had planned to purchase a slave girl for himself. But the Ethiopian women paraded before him weren't quite pretty enough. However, the "Han Houris"—the Muslim term for the beautiful virgins of Paradise—"are looking lovelier than ever, the divorced one has been critically examined and pronounced a virgin."[39]

A few moments' amusement. Did he throw away his future? Remembering the wild, adventurous days in Africa, Galton might have been tormented by guilt as he thought of his wife Loui, and what he might not be able to give her now. Did he contract syphilis, and destroy his duty to be a father? Who knew what his carousing could have done to his body. He had been steeped in Christian theology his whole life, including the notion of original sin, and the biblical curse it entailed: the wages of sin is death.

He had met Loui at a Twelfth Night party in Dover in 1851, five years after his oat-sowing in the Middle East. She was the daughter of the Very Reverend George Butler, dean of Peterborough Cathedral, and a distinguished scholar. Galton had just returned from Namibia, another African adventure, and by this time he had made a name for himself as a great explorer. The Royal Geographical Society had just awarded him its gold medal for his maps and vivid accounts, which he published in the Society's *Journal*. He had also just submitted a book called *Tropical South Africa,* a longer account of how he encountered the warring Namaqua and Damara tribes, how he boldly rode an ox into the doorway of the Namaquan chief, and, wearing a pink hunting coat, negotiated a peace. His exciting travel writing had made him somewhat famous, but he was tired, weary, and his health was showing the effects of his vigorous escapades. Though he had enjoyed his popularity, he now "desired to get out of the way of being lionized, which is exceedingly wearisome to the lion after the first excitement and novelty of the process have worn away."[40]

There was something about this woman, Louisa. She was plain, yes, but this seemed to suit him now. She was from a very respectable, wealthy family, and besides, he was thirty years old and ready to settle down. It was time for him to get married and have a family, like any respectable English gentleman.

He proposed to Louisa at the Crystal Palace, the colossal glass edifice in Hyde Park set up to house the Great Exhibition of 1851. The Palace—an iron behemoth with over 1 million square feet of glass—enclosed over 33 million cubic feet of space, and contained over 13,000 exhibits from around the world celebrating the Industrial Revolution, the thrill of progress, and the military and economic superiority of Great Britain. In the midst of this celebration of science and industry, Galton and Louisa agreed to marry. When Galton asked the Reverend Butler for his daughter's hand, he received the reverend's full blessing. After a long engagement, Francis and Louisa were married on August 1, 1853.

That was thirteen years ago. They had since lived a leisurely life together. They had traveled throughout Europe, held lavish dinner parties at Rutland Gate, and became a part of London's social elite. But even as Galton regaled his guests with his bawdy tales, alluding to his wild African adventures, his guilt was tearing at him inside.

His fixations with women were in many ways typical of Victorian ambivalence toward sexuality—especially for a man of his class and breeding. Sex, like everything else, was coming under the "objective light of reason," and conversations about such intimate matters, before not very common in polite society, could now be couched in rational terms, as well as moral, religious, and proper. In this era, such frankness could be shocking at first; but the scientific analysis of human sexuality was creating a space to push the limits of both humor and propriety.[41] Yet much of the focus, as Galton knew, was profoundly negative. The image of the pervert, the masturbator, the homosexual, the hysteric, and the prostitute pervaded discussions of social problems. Science had seemed to reveal each of these types as a distinct sexual species with their own particular constitution. Behind his smirk, Galton must still be haunted by his own few moments' amusement.

But he had just found his life's idea. He had only just begun his disinterested study of hereditary genius, the mystery of propagation, and the majestic evolution of mankind. He had only just begun to proclaim how essential it was for people of good breeding to bear many children, and actively to work to restrain the reproduction of undesirables, those people poor of spirit and breeding. At the same time, the image of the biologically unfit was part of a larger idea sweeping through his era. His friend Herbert Spencer and others were telling the story of a great, epic struggle of biological good and evil, as it were, a "survival of the fittest" where Nature's predestined elect, its chosen people, would ultimately reign triumphant. Were there no gentle ways to make this happen sooner? Could not the undesirables be got rid of and the desirables multiplied?

His own pedigree was impeccable, as was his wife's. But after thirteen years of marriage, it was clear: they could have no children of their own.

WHEN *HEREDITARY GENIUS* was reprinted twenty-five years after first publication, a review in the April 1893 edition of the *Nation* commented that when the book first appeared, "it was commonly believed that the human mind had something supernatural in it" and that "children were born similar in mental ability, subsequent differences being due to surroundings and trainings." But Galton had been proven right. It was now known that "individuals inherit different intellectual capacities . . . and nature limits the powers of the mind as definitely as those of the body. On these points, among thinkers everywhere, the author's opinions have prevailed."[42]

Though it took him almost two years to recover fully from his 1866 breakdown, Galton's scientific work was still fresh and original, and his life's work began to influence sociologists, criminologists, economists, and political reformers, each of whom embraced a profoundly new view of human identity and destiny.

Galton continued to study family pedigrees, and he continued to experience ingenious intuitions. For the first time in the history of science, he came up with the idea to study identical twins, and compare the relative influence of what he alliteratively dubbed "nature and nurture." He published a theory of the physical mechanisms of heredity, claiming the body must contain a host of discrete types of gemmules, which he re-dubbed "germs." The body retained the sum total of every human germ type on something Galton called a "stirp," the Latin word for "root."[43] While Galton was not so much a brilliant systematic scientist as he was a man with an incredibly ingenious imagination, with his notion of a stirp and its code of latent and active germs he had stumbled on an idea that for the first time in history approximated the later discovery of the human genome. He presented his inchoate, undeveloped theory ten years before the German biologist August Weismann developed a more substantial theory of "chromatic loops" and "germplasm," which would evolve into the notion of chromosomes and genes.

But in perhaps his most lasting and influential contribution to science, Galton discovered a new statistical concept: the "coefficient of correlation." Eventually expressed as a number between −1 and 1, it showed how any two statistical traits varied relative to each other. Using the coefficient of correlation, Galton analyzed how arm length relates to height, for example, or how the height of parents relates to the height of their children. Later, this concept would be used to analyze the probable relationship between any two variables, such as rainfall and crop output, inflation and interest, or academic performance and class size. Scientists in a host of disciplines would find this method invaluable as a prediction of probability and risk analysis.

Statistics had been Galton's life; but to the end, long after the blushing, sophomoric quests of his youth, he still maintained his obsession with voluptuous women. A year before he died, Galton wrote a novel he titled *Kantsaywhere,* an at times risqué story describing a scientifically organized Utopia. Stooped with age, his own health failing, he depicts the inhabitants of the land in sexually charged terms. "The girls have the same massive forms, short of heaviness, and seem promising mothers of a noble race," says the hero, Donoghue, an adventuring professor of vital statistics, who discovers a master-race civilization and falls in love with one of its women, Miss Augusta Allfancy. "As for the men, they are well built, practiced both in military drill and athletics, very courteous, but with a resolute look that suggests fighting qualities of a higher order. Both sexes are true to themselves, the women being thoroughly feminine, and I may add, mammalian, and the men being as thoroughly virile."[44] After Galton died, his family ordered the sexually charged novel destroyed. Its explicit sex scenes were embarrassing, they thought, and should never be published. But Galton's niece, who was in charge of destroying the manuscript, saved some parts, thinking they might be valuable for his biographer, his protégé Karl Pearson.

The doctrine of eugenics was beginning to spread throughout the world. The pedigree, the family tree, the fathers and mothers upon the children and the children's children unto the third and fourth generation, had always been an epic source of pride and longing, as well as a biblical image of the curse of sin. Galton made the pedigree an object of scientific study, and helped place the causes of complex human behavior in a new epic story of evolutionary struggle through time. Original sin and hardened hearts were passing away as the ostensible roots of vice and crime, and a new world of heredity, germ-plasm, and genes were beginning to take their place.

The sweeping national policy Francis Galton had proclaimed never really took hold in his homeland, however, and neither had the "negative" side of eugenics. But his vision of a scientifically organized Utopia—an illustrious race of men free from the burdens of unfit, immoral citizens—was resonating across the Atlantic, where American pioneers had long felt they were a people chosen by God to be a beacon to the world, a manifestly destined people who must be set apart from the Old World. Galton the adventurer, obsessed by women and statistics, sought to breed a better race of Englishmen. In the United States, however, a people on a quest for purity embraced his idea of eugenics and applied its methods not on the scaffold or in the battlefield but in the operating rooms of state institutions.

A City Upon a Hill

Morning prayer began in the parlor early, before the sun came up. Charlie Davenport knew that if he was late, he might miss breakfast afterwards, and his father Amzi, face taut and severe, would send him to his room to sit and make no noise. So he woke on time, and he was with the rest of his family as they walked into the room, taking their places on hard, wood chairs. But it was New Year's Day in Brooklyn, 1879, and even as his father began the regular morning ritual, Charlie could barely sit still, fidgeting, suppressing a grin, thinking, "Off! Off! Hurrah! To the city!" He was twelve years old, and nothing made him more excited than the rare times when he could venture out into the busy world around him. The day's planned trip to Manhattan sparked the boy's deep, silent longings.[1]

They lived in a house at 11 Garden Place in Brooklyn: Charlie; his parents, Amzi and Jane; his older brother, Willie, sister, Mary, and little Fanny. In the spring and summer, their father always took them to Connecticut, to Davenport Ridge, the ancient farm the family had owned since the Reverend John Davenport, founder of New Haven and friend of Oliver Cromwell, led a ragged company of five hundred Puritan settlers across the ocean to America in 1638. His father was obsessed with this family pedigree, and constantly told his children about their family's past. In the Connecticut countryside, Charlie felt alive and free, even with his endless chores. But here in Brooklyn, his father kept him close to home, away from bad influences, and this only made the sights of Manhattan more of a thrill: the landscape of tall buildings and church spires, the raucous people with foreign tongues, the throngs of carts and merchants. So, as his father led the prayers, Charlie was anxious to put on his best Sunday clothes and make the trip across the river to the teeming city of New York.

Charlie's extended family had been deeply pious for generations, and morning prayer was a regular routine. When John Davenport and his followers first came to America, they hoped to establish a Christian Utopia, seeing the New World as a new Promised Land, flowing with milk and honey. They saw themselves as a "peculiar people," chosen by God to come to America's wilderness to raise up a new social order, holy and pure. The reverend's direct descendants included a long line of Puritan ministers, some of whom were also farmers, merchants, and patriots engaged in government and politics, men who helped to build the fledgling nation. Charlie's father, Amzi, though not a minister, was also a powerful man in church affairs, serving as an elder and a deacon in three different congregations, his influence unquestioned in the appointment of ministers and the control of treasuries. In Brooklyn, he worked in real estate, buying property and collecting rents, and his commanding voice was just as prominent in civic life. Like his forebears', Amzi's piety was deep and sincere, and he believed the chief end of man was to love and fear God, and obey his precepts always. Foremost among these, he believed, was God's command in Genesis to be fruitful and multiply. He had fathered twelve children himself, though only eight had lived past childhood. With four children still at home, he also had four married sons, and a number of grandchildren. Though Amzi was a very busy man, with his vaulting sense of the importance of his family, he never ceased to pursue his favorite hobby: tracing the Davenport pedigree. He was just now revising his book, *The History and Genealogy of the Davenport Family in England and America,* a 400-page family lineage going back to the time of William the Conqueror. It was a vast family tree, listing the names and great accomplishments of his direct ancestors, and organized with a special system he devised himself.

Charlie was the youngest son, only older than Fanny, who was nine. As he sat through prayer, impatient, watching his stern, inflexible father begin to speak, his earlier excitement was tempered by a slight twinge of resentment. He thought of the harsh discipline, the relentless demands, and his growing sense of alienation.

His father was thin and slight, and his light brown hair, having receded far back on his pate, was combed down straight, like a Roman emperor's. His eyes, through wire spectacles, were blue and piercing, and his thin lips were held taut. He rarely smiled, and, his Bible in his hands, his words were deliberate and distinct. On this morning, Amzi could take time to reflect on the drunken New Year's revelries that took place in the city the night before, and talk to his family about a topic that had consumed him since he was a young man.

"Intemperance is one of the greatest evils that exists in our land," he could say. "Thousands, yes, millions are destroyed by its ravages. Look forth upon our happy country—happy did I say? Yes, thrice happy were it not for this one great evil. . . . What a sad spectacle it is to see our young men on whom the prosperity and thriving condition of our nation is entirely dependent, pressing on carelessly to ruin and blasting every hope of being in any respectable rank in future life."

Happiness. A prosperous and thriving nation. These were God's promises to His peculiar people living in the New World, even now after more than two centuries of carving out a life in America. But, as with the ancient Israelites, God's promise carried with it a threat: if this people failed to keep His precepts, they would lose their blessing and incur His wrath. Amzi's family sat around him, hands folded quietly on their laps, listening to these words, as they did most every day.

"All desire happiness, but many seek it in the wrong way. Some think that they shall obtain it by frequenting the theatre and the ballroom and living a life of ease and luxury, having their table loaded with all the delicacies of life. The poor blind inebriate supposes that he can arrive at pleasure by indulging very freely in strong drink, but he very soon finds that he is deceived and in the lowest stages of degradation. But in my opinion, true pleasure is found in doing one's duty and in the consciousness of having done right. Take for instance such a man as the illustrious Howard, who spent the principal portion of his time in consoling the miseries of the dungeon, relieving the agonies of the dying and imparting assistance to the sick. Did he not enjoy true happiness, I ask? Will any one of experience deny that this is the fact? My inward feelings answer 'No.' Again, on the other hand, take such a man as Bonaparte, who spent his time and energy on gaining victories and making achievements and not preparing for his future existence. He in my opinion was not truly happy. And let us all endeavor to be virtuous and thus be happy. Amen."[2]

When prayers were over, Charlie jumped up for breakfast. They finished, got dressed, and after Willie hitched their horse "Pet" to the carriage, they clambered up and got ready to go. Willie and his father sat up front, and Charlie sat in back with his mother, Mary, and Fanny.

It was cold and brisk, and they were bundled in blankets, but as they jostled along the bumpy road, Charlie's excitement returned as he took in all the sights around him. The carriage clapped through Brooklyn Heights, down Henry Street to Fulton, and then from High to the Catherine Ferry. As they crossed the East River, undulating in the waves, Charlie looked with awe at the sights before him. Massive new stone towers of the Brooklyn Bridge, astonishing them all, held huge steel cables and a framework of half-completed girders. These towers, at 276 feet into the air, were seven times higher than most of the four-story buildings in Brooklyn and New York. On the Manhattan shore, a regimented line of piers held scores of ships and schooners, with unfurled sails and masts the shape of crosses. In the distance, they could see the highest point in the city, the grand spire of Trinity Church, and a few blocks north, near City Hall Park, the massive nine-story headquarters of the *New York Tribune,* built just three years earlier.

When they reached Manhattan, Willie clucked his tongue and Pet pulled the carriage off the boat and onto Catherine Street. They headed west, following beside the East Side Elevated, watching the trains roar by overhead—"much to the terror of Pet, who bolted with his head to see what the thing was!" Charlie later wrote in his diary. They passed near the Bowery slums and the infamous Five Points Mission, a former "pest house of sin" which had been transformed into a Methodist almshouse. Here, waves of impoverished immigrants, Irish, Italian, and Chinese, crammed together in rotting wooden houses and tall brick tenements, came to receive food, clothing, and English classes. When the carriage finally reached Broadway, it was quieter than usual, but even on New Year's Day, hundreds of people were milling about the storefronts and the sidewalks teeming with carts. Amzi pointed out the Manhattan Savings Bank as they passed, the scene of a recent robbery, and Stewart's Department Store, with all its fancy clothes.

Above them was a latticework of wires. Just a year before, the world's first telephone system was woven into the existing telegraph cables in lower Manhattan; one visitor described it as "a perfect maze of wires, crossing and re-crossing each other from the tops of houses. The sky, indeed, is blackened with them; it is as though you were looking through the meshes of a net."[3] More wires would be added, too, as officials sketched the plans to build the world's first commercial power plant on 257 Pearl Street, just blocks from where the Davenports were riding now, inaugurating Thomas Alva Edison's bold new plan to light the city with electricity. But this mesh of wires began to thin as the Davenports left lower Manhattan and headed north on Broadway. After a long drive, they turned onto the boulevard through Central Park, and watched holiday skaters glide over a frozen pond.

They finally arrived at 969 Lexington Avenue, where the rest of the family had assembled. This was the address of John Isaacs Davenport, Amzi's oldest son from his first marriage, who lived here with his wife and children. John's full brother Albert Barnes had come with his family, too. Their younger half brothers, Henry Benedict and James Pierpont, the oldest sons of Amzi and Jane, had also come with their wives and children. As Charlie walked in with the rest of his family, the house was alive with cousins dashing around the three-story house, and three generations of Davenports began to celebrate the New Year.

LIVING SIX MONTHS in the country and six months in the city, experiencing the calm, pastoral pleasures of nature as well as the tumultuous din of urban life, Amzi Davenport's children witnessed more than most an era of profound change. American life up to 1879 had been dominated by the distinct values of rural Yankee Protestants, like the long line of Davenports. Thinkers such as Thomas Jefferson and Hector St. John de Crèvecoeur imagined such people as ideal citizens: yeoman farmers, honest and industrious, producing only what they needed and remaining content. While fiercely independent, these citizens also maintained a spirit of social equality that Alexis de Tocqueville believed was the essence of American greatness. Their rustic egalitarianism, free from European systems of class, he believed, drew men naturally toward the kind of democratic institutions that tempered both radical individualism and tyrannical majorities. For these Americans, all free men were equal before God, and when they engaged their common life together in government affairs, it was to be for the greater good rather than for personal gain. A government of the people merely preserved the moral order of its citizens, while civic life and commerce provided the opportunity to develop what they held most important: a godly character, responsible, restrained, and private.[4]

By the New Year of 1879, however, the Davenport clan was once again in the midst of the shifting plates of culture and history. The Yankee Protestant consensus, dominant for centuries in various forms, was being challenged as never before. Just as New York's great political thinker, Alexander Hamilton, bitter enemy of Jefferson and his vision of a pastoral democracy, had predicted, the future of the country was becoming enmeshed with banks, big business, and the restless energy of the city. Urban industry was beginning to eclipse rural agriculture, and centers of political power were shifting from the countryside to the metropolis. At the same time, waves of immigrants, especially in New York, were bringing new traditions, new languages, and new religious practices to America, unsettling blueblood New England mores.

Even the ostentatious wealth of the new urban entrepreneurs such as Carnegie, Rockefeller, and Harriman troubled those who held the values of the past. Business and commerce now seemed less related to civic responsibility than personal aggrandizement, immodesty, and pride. In New York, palatial mansions were being built along Fifth Avenue, and lavish, frivolous parties seemed to mock sensible virtues. Some of these were epitomized by the formal parties hosted by Mrs. Stuyvesant Fish, the Manhattan socialite and fencing champion, who once threw a gala for her pet, a dog wearing a $15,000 diamond collar. Describing the values of this "gilded" society, Mark Twain wrote in his essay "The Revised Catechism": "What is the chief end of man? —to get rich. In what way?—dishonestly if we can; honestly if we must. Who is God, the one only and true? Money is God. God and Greenbacks and Stock—father, son, and the ghost of same—three persons in one; these are the true and only God, mighty and supreme."[5]

While business entrepreneurs gained enormous wealth, the growing power of foreign immigrants in the city, with their hard drinking and Roman Catholicism, also seemed to be corrupting the virtues of temperance and restraint—as well as the very idea of self-government. The anarchy of the Draft Riots of 1863, when a mob of mostly Irish immigrants looted, burned, and terrorized neighborhoods on the East Side of Manhattan, still lingered in people's memories, and many feared that "carnivals of revenge" would spring again from growing class resentments and labor strife. Political bosses and corrupt "machines" took advantage of the rampant poverty of immigrants, and exchanged government services for votes and power. More and more, Americans with native pedigrees were becoming alarmed by the new urban class, and saw these people as the cause of vice and corruption.

American society seemed to be on the verge of collapse, especially for those who clung to the old Puritan vision of a land of innocence, a New Jerusalem set apart from Europe's dens of iniquity. Impelled by fear, they began to call for changes to thwart these threats. Early temperance crusades and religious revivals flourished throughout the country at this time, and politicians had just begun to formulate new anti-trust legislation. Seeing their country as a land of innocence, many Americans had long clung to the idea of self-purification, attempting to excise that which posed a danger to the social good. So as the calls for reform began to gain momentum, the ideas of Francis Galton, just being formulated across the ocean in England, would soon hold a particular appeal. Eugenics would combine an American self-consciousness with the new and unimpeachable authority of Science.

John Isaacs Davenport, Amzi's oldest son, was a federal election official immersed in these movements of reform. Amzi was fiercely proud of John, who fought for the Union in the Civil War and saw action at Drury's Bluff, the battle of Bermuda Hundred, and the siege of Petersburg. After the war, John became an attorney, and later an editor for the *New York Tribune,* and he had always been deeply involved in New York politics. In 1868, just two years after his half brother Charlie was born, he testified before Congress, describing the rampant voting fraud that took place in New York during the presidential election that year. Two years later, Congress appointed him Chief Supervisor of Elections in New York, "for Promoting the Purity of the Elective Franchise."[6]

This was not an easy job in New York, and Charlie's oldest brother was for years an enemy of Tammany Hall and William M. Tweed, the hulking man who controlled New York's political machine. "Boss" Tweed, describing elective office in the city, once boasted: "The fact is, that New York politics were always dishonest, long before my time. There never was a time when you couldn't buy the Board of Aldermen. A politician coming forward takes things as they are. This population is too hopelessly split up into races and factions to govern it under universal suffrage, except by the bribery of patronage or corruption. . . . I don't think there is ever a fair or honest election in the city of New York."[7] John Davenport's job was to fight this corruption, even after Boss Tweed lost his power, indicted for fraud a year after John took the job as Supervisor of Elections.

Tammany Hall had consolidated its legendary power by welcoming the despised new immigrants, giving ethnic groups a voice in the system—or, rather, in the political machine—and providing them with jobs and shelter. This was a trend throughout the country. While corrupt politicians exploited the poverty of immigrants in the cities, in rural areas, especially in the postbellum South, political machines also exploited the poverty of black Americans, bribing blocs of voters. Ironically, the first poll taxes in the country were often "reform" efforts meant to combat the growing power of corrupt political machines—as well as to disenfranchise non-white American citizens.

In many ways, John Davenport's fight for election reform was part of a complex clash of the old rural Yankee establishment and the new urban immigrant class, the subtle tensions between the values of simplicity, sobriety, and restraint and those of industrial wealth and freewheeling capitalism. At the same time, in scattered intellectual circles in England and America, the specter of Darwinian evolution and natural science was beginning to haunt the old creeds of human nature, providing startling new explanations for the competition between social groups and the causes of mankind's social ills. For families like the Davenports, their long-held vision of America was once again struggling for existence, just as it had when their ancestors first came to this land and endured toil and hardship. The methods of science were beginning to alter the nature of their strivings for innocence and perfection, but the yearning, long burning in the hearts of generations of Davenports, remained much the same.

IN THE MONTHS AFTER the New Year's gathering at his older brother John's, Charlie returned to a monotonous routine he was coming to loathe. He didn't go to school in Brooklyn, since his father preferred to keep him from outside influences and mold the boy's character himself. "Train up a child in the way he should go, and when he is old, he will not depart from it," Amzi liked to quote from Proverbs. So, during the months of fall and winter, Charlie spent most of his days in his father's real estate office, a garden floor, half underground, at 36 Willoughby Street, a few blocks from his home. Sometimes, in the evening, the family played parlor games, but even these events were rare. Charlie spent a lot of time alone.

"O! I want to go to school!" he scribbled in his new diary, which his brother Willie had given him for Christmas. "I hate this office!" On a rainy day in March, he confided, "Pa gone up to the country. (Good!?)"

At the office the greater part of the morning. After feeding Pet and cleaning out the stable I started for the office arriving at that 'Prison House' as I call it, at about 8:15."

His father paid him 25 cents a week to sweep the floors, keep track of receipts and expenditures, and sometimes go out to collect rent from the tenants at the building on Adelphi Street. When he was not doing office chores, Charlie did the lessons prepared by his father, who had been a schoolmaster decades ago. He sat alone at the office desk, sometimes nine hours a day, doing problem sets in arithmetic, grammar, and American history. In the evenings, at home, his father drilled him on these lessons. Sitting next to the fire, cutting clips from newspapers, Amzi rattled off questions, occasionally peering over his spectacles at the boy, as if distracted. But he especially liked to make his son parse verbs, and Charlie later became a famous parser among his peers.[8]

Charlie had fun in solitary ways. He was an avid reader, and during the long hours he spent in the office, he sometimes snuck out to buy "dime novels." Once, after reading a book called *The Balloon Twins,* he tore it up and put it in the fire, for if his father caught him, the consequences would be severe. One of his favorite pastimes, however, was putting together his own magazine, which he dubbed *The Twinkling Star.* He took to this hobby when he was eleven, and since one of his father's tenants had a printing shop, he was able to get professionally made copies printed every month. Charlie wrote and edited the four-page broadsheet himself, though his father sometimes helped him with the layout, encouraging the hobby since it might help him learn an honest, practical trade. *The Twinkling Star* included family news items, essays on Davenport history, and even serial stories told by a character named "Carlos." He also "published" a number of his father's letters, sent during his tour of Europe.[9] In his diary, Charlie carefully listed all his subscribers—family members mostly—and mentioned the money he made peddling *The Twinkling Star* on the street. He was a precocious boy.

Charlie mentioned his father in almost every entry in his diary, and the family faith pervaded his life. He never wrote about his mother. Along with morning prayers each day, he attended three church services each Sunday and one each Wednesday evening. He carefully recorded the biblical texts and sermon topics, usually noting that they were "excellent."

Amzi had instilled in his son the importance of faith and the significance of living a temperate life, but more than anything else, he had impressed on Charlie the privilege of being a Davenport with an illustrious Puritan heritage. As he noticed his father's dedication to updating his massive book on the Davenport pedigree, Charlie too, from an early age, felt this ancient, instinctive pull to record his family tree.

"Hello! Here I am! Charles Benedict Davenport, Esq.!" he wrote in his diary. "Well, old book, how do you do? I've seen you before, but as you see I have not kept up, but I begin again. I am a large well shaped boy, 89½ lb., Fair Complexion—and the editor of The Twinkling Star." Then he wrote a long list of his family, using the biblical rhythm "son of" or "brother to" before each name. For two generations back, he listed his immediate family, including his grandparents and cousins, and for the first time mentioned his mother. "I am: son of, Amzi Benedict Davenport, Mrs. Jane Joralemon Davenport; Brother to, John Isaacs Davenport . . ." and so on for two more pages. Like his father, young Charles was coming to believe it was his heritage, his family history, and something in the past that made him who he was, even as he began to yearn for something of his own.

ON A CHILLY NOVEMBER EVENING in 1849, Miss Jane Dimon felt flattered and confused. On her lap were three letters, in gorgeous cursive handwriting, from a widower she had met just a few weeks before. She lived at 29 Livingston in Brooklyn, the daughter of John Dimon, the powerful Commissioner of the Almshouse. Many in her family thought she should be married by now—she was almost twenty-two—but she was quiet, timid, and shy. Men rarely courted her, and she had never before received letters like those on her lap, full of such passion and love and romance. But she was not sure she could really love the man who wrote them. He was eleven years her senior, with two young sons. Though his family owned land in Connecticut, he was just a schoolmaster, a man who had opened his own private academy when he was nineteen and was still teaching there after thirteen years. The men in her own family were attorneys, wealthy Dutch landowners involved in city politics. Her father not only controlled the tax money allocated for Brooklyn's poor, he also ran his own large farm. Her grandfather, Teunis Joralemon, was one of the most influential judges in the city. A former village trustee, he also owned acres of farmland.

So she wasn't surprised her mother didn't approve of Amzi Davenport when he came calling yesterday. What a funny sight it was! He stood there on the porch, thin and gaunt, with his wire-rimmed spectacles, listening sheepishly as her mother scolded him for coming to the house. Despite her eagerness to find a suitable husband for her daughter, she shook her finger at the frail-looking schoolmaster, and told him his attentions were ruining her daughter's reputation. He tried to protest, and even asked whether Jane were a professing Christian at a church, but her mother just told him to leave.

Two of the letters on her lap she just received today. She had already answered his first, listing the reasons she could never requite his feelings of love. He was persistent, though, and as she read his letters again, she could smile. It was nice to think a man could be so enamored of her.

> My dear Friend,
>   It is with no ordinary emotions I take my pen to address these few lines to you. I take <u>this</u> method thus to express my thoughts, both because it is consonant to my own feelings, & because I believe it will be as agreeable to yours, which I would not in the slightest degree wound, (unless with the arrow of love). . . . First, then, I will say that my feelings towards you are those of respect, esteem, & <u>deep affection.</u> I will not at this time speak of the strength of that affection, or of the causes that may have conduced to it, other than that in the comparatively slight acquaintance I have had with you I have seen that <u>excellence</u> in your person & <u>mind</u> which thus awakened in my heart the emotions to which I have alluded. . . . I know you can have but little knowledge of me, but mine I trust is a transparent heart, & that my own freeness and frankness would fully reveal myself as well as my circumstances to you.[10]

Jane had responded cordially, but she was honest. How could she return the affections of a man with two young boys, one six and the other four? Was she expected to be a mother to them? And even more, how could she abide a man who now lived with his dead wife's parents? In the letter she received today, he responded to these objections and tried to explain. After apologizing for the "embarrassing" incident with her mother, he wrote: "To my two children, I sustain a relation of a solemn & weighty character; & whatever my future relations may be, that responsibility, must be borne, in a greater measure, by <u>myself</u>

alone, and by the blessing of God it will be my endeavor to train them in a way worthy of their long pious ancestry."

He insisted, too, that she should not worry about his in-laws living with him at the moment, because, he told her, he would abide by the principle laid down in Mark 10:7: *For this cause shall a man leave his father and mother, and cleave to his wife.* But even as he tried to reassure her, he told her it was not simply a dutiful wife he desired, or just a mother for his children. He closed with a plea: "The benevolence as well as affection of my nature sometimes leads me to desire that some correspondent heart might share in the many joys of my life, but even these may not be unalloyed. . . . It is not such, but the simple flowing out of an ingenuous heart which had begun to feel a more than ordinary interest in you & which now feels that interest which leads me in another sheet to discuss a subject of ever higher & more momentous interests than that which for the present I dismiss."

The passion and underlined emphasis of his letters gave her pause, but also began to touch her. But though she was a devout Christian herself, active in her church, she wasn't prepared for his third letter, with its "higher and more momentous interests," and written, incredibly, the very same day.

My dear Friend,

In the free interview had with your good Mother on Saturday Evening, the interest I had felt in you, led me to make the inquiry, whether you had publicly professed faith in Christ, or indulged the hope that you were a Christian.

Although you felt that you could not consistently reciprocate the affection of one, who, at the best is but a fellow sinner, I did feel that a few considerations on this most important of all subjects, might not be wholly unacceptable to you from me, & if in any way blessed to your spiritual good, a cause of gratitude to God through Eternity.

Although I had no previous knowledge on this point, I had hoped from your general regard to religious things, & outward performance of religious duty, such as teaching in the Sabbath School, etc., that you had been led, to a personal reception of Christ as your all-sufficient Savior and Redeemer.

Will you not, then, divesting your mind as much as possible, for the time being, of all other topics, give to the subject of personal piety, that calm and prayerful thought, which the interest of the Soul demands?

Whatever else we may do, or not do, unless we have a clear view of

our own unworthiness, & deep sinfulness before God—a deep convic-
tion of our own need, and dependence—<u>an utter renunciation of self &
the world</u> & a child-like Spirit of <u>dependence on Christ</u> & a willingness
to do all those things which shall please him, we cannot be his true &
acceptable Servants. . . .

My dear friend, shall I speak out the deep feelings of my heart, on this
subject?

I had rather you would <u>love Christ</u> & be ready to declare that love to
the world, though you should <u>hate me,</u> than to <u>love me</u> with the
strongest possible affection & not love Christ. May I not then, in the
conscious unworthiness of my own character as a sinner, ask, that you
will not rest till the work of repentance & faith is begun. It is a <u>personal</u>
work. It should be an <u>immediate</u> work. Well do I remember the hour
when these truths came home with power to my own Soul, & I resolved
to cling to these convictions as my best hope; & not only to cling to
them, but yield to them the full assent of my heart; & in this work, O
how greatly was I aided by the counsels of a devotedly pious Mother,
now in heaven. . . .

Christ now <u>offers himself</u> to you. Consider his character. <u>How perfect! How lovely! His terms are repentance & faith.</u> Will you accept the
offers which he makes to you—to <u>be your everlasting Friend?</u> Or will
you reject <u>them</u> & prefer <u>this world,</u> to a hope of heaven?

Accept these few thoughts, spoken in great plainness & yet in love, &
which should they prove in vain, will yet leave me the consciousness of
having sought your good.

> With my best wishes & prayers, I am truly,
> Your Friend,
> A. B. Davenport.

In less than a year, despite her mother's and her own reservations,
Jane married Amzi. On October 30, 1850, in a large ceremony
attended by some of the most eminent families in Brooklyn, Amzi
Benedict Davenport and Jane Joralemon Dimon were joined together
as man and wife.

Jane did indeed become the correspondent heart with whom Amzi
could share his deepest joys. He became her tender friend and lover, an
attentive husband who often brought her gifts. He read to her, and she
to him, and when he traveled, he wrote her almost every day. Once,
after a three-week trip early in their marriage, he began a letter with
the John Donne–like lines: "My dear Jane, Though many a long and
weary mile may lie between us, yet love, like electricity, annihilates
time and space, and awakens those silent, yet sacred communings,

which exists in kindred hearts, well-tried and true. Thus doth my heart commune with thee, my dearest absent one, and though twenty Atlantics were between us, these thoughts would still pervade my breast." Jane treasured his letters, and always kept them safe. But even as she began a new life as a wife and mother of two, drawn into the full expressions of Amzi's earnestness and faith, she remained submissive, shy, and demure.

With the help of Jane's wealthy family, Amzi soon gave up his schoolteaching and opened a real estate office in Brooklyn Heights. He began to manage the finances of local estates and sell insurance policies. He was a tireless worker, and even with the responsibilities of his new business, he also helped to organize new congregations, like the Clinton Avenue Congregational Church in Manhattan or the Plymouth Church in Brooklyn, serving on the board of elders for each. As he gained a reputation for impeccable honesty and reliability, his business began to prosper and his stature grew.

With his new wealth, Amzi built Jane a new summer home at Davenport Ridge in Connecticut. It was a large, white-frame Victorian house, with a huge veranda porch that extended continuously along three of its sides. It was magnificent, Jane thought. The main section was three stories high, and a two-story extension jutted out to the east. A round, six-sided turret with six arched windows extended from the roof, like a small steeple, and provided a panoramic view of the Connecticut countryside. Since the house stood at the top of the ridge, four hundred feet above tidewater, Jane was able to see as far as forty miles over the Long Island Sound to the south. To the north, she could locate the turrets and spires of fifteen churches in the nearby towns of Stamford, Greenwich, and New Canaan. Out in the distance, to the west, she could make out the hills of Westchester County in New York.

Jane came to know that Amzi's deepest joy was his almost obsessive enthusiasm for American history and his Davenport heritage. Her husband was working on an ambitious book, a genealogy of his family going back to 1086, but focusing especially on the American Davenports who branched out from the great patriarch who had founded New Haven two centuries ago. It was only a hobby for him, really, but he worked on it almost every day, constructing this elaborate family tree, organized through a numerical system he had devised himself. To each Davenport he assigned a number, grouping them chronologically with siblings rather than parents.

While doing research for his genealogy, he once took a three-week trip alone to Boston. It would be the first time he saw Plymouth Rock, so the journey became a pilgrimage of sorts. As he traveled, he observed the landscape closely, thinking about the history and culture of his great country, and almost every day he wrote to Jane, describing how he felt about the sights he saw.

When he reached Newport, Amzi was appalled. "Of Newport I might say <u>much,</u> for I <u>saw</u> much," he wrote, with his characteristic underlined emphasis. "Of all the demonstrations of <u>fashion folly and nonsense,</u> three sisters that usually keep company, I never saw aught to exceed what I there witnessed. Of course, the <u>Ladies</u> were <u>beautiful,</u> the elite of the land, and among the number was one an almost perfect picture of yourself."

He then took the train to Plymouth, the "landing place of the Pilgrims," the place he most wanted to see. The experience far exceeded what he ever imagined. His uncle had told him that he would find little there, and would probably be disappointed. But he was wrong. "Seldom have I visited a place & felt more amply rewarded than upon my return from Plymouth," Amzi wrote to Jane. As he took in the place, reflecting on the glorious essence of American history, Amzi was also impressed with the profound differences between Boston and New York. Boston was a town of history, where memory and the past were kept sacred. New York, however, was more a city of engineering and commerce, where man simply imposed his will on the landscape. He described to Jane "this 'city of Notions': Boston rightly merits this appellation, for the Bostonians are 'a peculiar people.' They have their notions about their streets and about theirs houses; every thing ancient must be preserved, be it dwelling, tree, or any thing else. Even the hill must not be leveled, or the valleys filled up. . . . How different from that agrarian Spirit, so rife in Brooklyn and New York, where all things must be brought to a common level, every hill must be brought low, every valley must be filled up, and the rough places be made plain. . . ."

On his journey home, Amzi came across one of the most startling sites he had ever seen. Stopping in Worcester, he was struck by the city's enormous asylum. Even more than that, however, he was awed by the simple spirituality he found there. "The Insane Hospital is one of the largest, if not <u>the</u> largest, buildings I ever saw," he wrote. "Its length must be more than double that of the City Hall N.Y. and contains 450 inmates. My Uncle supplied the chaplaincy there, on the Sabbath, where I attended service, half the day. There was a congregation of between two & three hundred of the inmates, and a more orderly and attentive one I have seldom seen. They have a choir among themselves, which performs the music. . . ."

The sight of orderly and attentive congregations became the norm for Jane and Amzi. During her first two years of marriage, though she had never traveled as far as Boston, Jane accompanied her husband to the many churches he helped run. In Connecticut, they traveled to Stamford and New Canaan during the week to attend services. In Brooklyn, too, they were involved in more than one congregation.

Through the 1850s, they clung to each other still. Their conscious duty to be fruitful and multiply took over a year and a half, and their first child was born in March 1852, a daughter they named after Jane. In two-year intervals, she bore three more children: Henry Benedict in 1854, James Pierpont in 1856, and Margaret Dimon in 1858.

In raising her children, Jane came to rely on a book called *The Mother's Friend: or, familiar directions for forming the mental and moral habits of young children.* In her diary she noted, "It is an English work. . . . It is altogether the best work of the kind I ever saw. Of all instruction, it makes that on religious things the most important, and which to be successful must be from the experience in the heart of the teacher and from the manifestation in her daily life." She was a tender and devoted mother, and her evangelical, heartfelt piety infused the family's daily life.

Jane lost both her daughters. Little Jane died of dysentery when she was three, leaving her mother devastated. Amzi tried to comfort her, and when again away traveling, wrote: "It is when I return to the house that I most realize the truth that she is no more to irradiate it with her presence. 'O no! We shall go to her, but she shall not return again to us.' May the influence of her death so far remain with us that it shall be a golden chain to bind our hearts more closely to Christ and to heaven." Years later, their daughter Margaret succumbed to scarlet fever.

Throughout this time, Amzi and Jane were still reading together and discussing issues of politics and religion. Amzi was one of the founders of Plymouth Church, where they attended services in Brooklyn, and their preacher, Henry Ward Beecher, was one of the most renowned religious orators in the country. His sister, Harriet Beecher Stowe, had written a book just now becoming famous, *Uncle Tom's Cabin,* and Amzi and Jane read it to each other, and attended the anti-slavery rallies at Plymouth Church. At one meeting, they listened attentively to Charles Sumner, the senator from Massachusetts, as he gave a thundering lecture on "The Necessity, Practicability and Dignity of the Anti-slavery Cause."

They were also reading about the storms of controversy in religious matters. A new scientific approach to Scripture and religion was sweeping into many congregations, and "biblical criticism"—as well as Darwin's theory of evolution—was eroding the traditional authority of the Holy Book. Jane noted in her journal the books she and her husband were reading, books with such titles as *A Practical View of the Prevailing Religious System of Professed Christians in the Higher and Middle Classes, Contrasted with Real Christianity,* by William Wilberforce. Another time they read *The Bible Not of Man: or, The Argument for the Divine Origin of the Sacred Scriptures, drawn from the Scriptures themselves,* by Gardiner Spring, pastor of a Manhattan congregation. Jane particularly liked to read about foreign places, since she had never traveled outside New York or Connecticut. "I have been listening to reading aloud by my husband, who has this evening finished volume 1st of 'Recollection of a Lifetime, or, Men and Things I Have Seen,' by an American minister at Paris." As they read these kinds of books, both Jane and Amzi were troubled by the subtle heresies that were threatening traditional Protestant faith.[11]

When the Civil War began, Amzi's oldest son, John, answered Lincoln's call for volunteers. During the war, Jane had more children. She bore twin sons, but lost one of them within a year. Mary came in 1864, and Charles, the youngest son, in 1866. Four years later, she had another daughter, Frances.

Throughout the turmoil of this decade, the Davenports continued to live half the year in their large Victorian house in Connecticut and the other half in their house on Garden Place in Brooklyn. As parents, Amzi became more stern and severe while Jane remained gentle and kind. But both were striving to teach their children in the way they should go, to live pious and devout Christian lives in a tumultuous and confusing world, so that when they were older, they would not depart from it.

IN THE FALL OF 1880, Charlie Davenport's wish finally came true. His father was letting him go to school, to the Brooklyn Collegiate and

Polytechnic Institute, a school for boys just down the street on Livingston. Once again, he could barely contain his excitement.

Plunging into a new life, he was elated by the throng of stamping shoes and voices in the halls. He described the smallest details in his journal—the floor plan of the building, the daily routines, the hymns sung in the morning. The Polytechnic was also a new type of educational institution, focusing less on subjects like Latin and the classics and more on mathematics, engineering, and practical science.

After months in school, Charles became fascinated with natural history and biology. These were never subjects his father had taught him, even though, while living out on Davenport Ridge, he had always liked to spend time observing bugs and worms and birds. But now he was learning such fascinating and interesting ideas about their habits, discovering a new way to see and understand this world he thought he knew so well. It was, somehow, liberating.

Discovering the joys of natural science, Charles soon decided to become involved with the Agassiz Association, a natural history society named for the great American paleontologist and bitter foe of Darwin's theories. Charles's enthusiasm got him elected president of the local chapter in Brooklyn, but he resigned in just a few months, feeling its members showed little passion for the subject. So, with a friend, he formed a new science club, and throughout his years at the Polytechnic a number of boys became active in it, even writing papers to present at the meetings. Enthusiastic and energetic, Charles established contacts with other organizations, corresponding with the Smithsonian Institution in Washington, D.C., sending his observations of birds to the American Ornithological Union, and keeping track of local weather for the Signal Corps.[12] Charles was becoming a master at organizing and planning, even though he still tended to keep to himself, still shy after so many years spent mostly alone.

His passion for natural science was beginning to shape his view of his own place in the world. In an essay he wrote at school, he celebrated "the privilege of adding to human knowledge by studying the stars, by investigating the lives of animals and plants, by revealing their secrets." But if school and the discovery of nature's "secrets" were liberating for him, Charles still felt the heavy hand of his father in his life. He still worked at the real estate office while in school, and with morning prayer and household chores, he had to get up even earlier to fit in an hour or two of study. During his summer breaks, he still went with his family to Davenport Ridge to work on the farm. Charles, like his father, was developing a tireless work ethic.

For over six years, this was his routine, in the country and the city. But just before he turned nineteen, in March 1885, he wrote a long letter to his father, even though they were living in the same house. Charles had always found it difficult to confront his father, and writing was much easier—a more orderly way to convey his thoughts. He wanted his father to know his life's ambition. He wanted to propose something new for his summer work at Davenport Ridge. He knew Amzi was somewhat suspicious of scientific theory—theory detached from practical purposes—but this time Charles wanted to do what was best for himself, for his own desires, and for his own future. He was a man now, after all. Charles had been going to Davenport Ridge his entire life to work, and he knew what was expected. But in this letter he presented a painfully detailed plan for more scientifically oriented tasks. Surely his father would see how important this was to him. So, with all the logic he could muster, on March 18, 1885, he sat down to write one of the most important things he had ever told his father.

My Dear Father,

As the summer is now rapidly drawing upon us, I have begun to look about to find how I may best employ myself during the vacation, and it seems to me that some employment by which I could put in practice those facts which I have learned in the Polytechnic and which would have some bearing upon my future intended occupation would best and most fully carry out your ideas of educating me, which your kindness in sending me to such a good Institute indicate. For the opportunity I cannot thank you too much, and hope I may be able to repay you by a life spent in the service of humanity.

The course of studies in the Polytechnic has taken me through Geology, Zoology, and to some extent, Farm Surveying. In addition to what has been taught me in school, I have become profoundly interested in the sciences of Meteorology, the migration of Birds, the Growth of Plants, the Value of Manures. An extended course in chemistry, Blowpipe Analysis, and Qualitative Analysis has prepared me for the investigation of soils and minerals.

All of the above studies have a direct bearing upon the Science of Agriculture, which I should like to study as my life work, if you approve of it.

The proposition which I wish to put forward is the following: That I shall, after the commencement, proceed directly to Davenport Ridge, that there I shall immediately make preparations for the following Summer's Work. . . .

Charles set out a ten-point plan and outlined it in a lengthy, tedious twenty-page letter. For each point, he first gave a rationale and then an exact description of the work he would do and the time it would take, to the minute. At times obsequious and at times condescending, Charles strove to impress his father with his intellect, longing to show how his youngest son was able to rank first in his class at the Polytechnic. He suggested a methodical way to improve the farm. He explained how he would conduct tests on the soil, and determine how much manure was needed in different areas. But he also wrote how he wanted to edify himself and work on his own research this year. "I hope to do you service no less in my work in science during the greater part of the day, than in those three and a half hours per day, and I hope you will regard the whole idea not as a selfish scheme to get rid of work, but as a proposition to aid science, in which I am particularly interested."

Nearly seven weeks went by before his father responded. Despite the fact they said morning prayers and dined together often, no word of this passed between them. Finally, Amzi communicated with his son, also in a letter, on May 5, 1885.

My dear Son,

Your pleasing letter of March 18th was duly received and had it not reached me in so busy a season of the year would have received an earlier answer—which indeed it did not really require.

The very fact that you were able to write so broad and comprehensive a letter taking so wide a range of theories for engaging the vacation hours of the summer days showed how well you have improved your opportunities for mental development, all of which afforded me much gratification.

But you have failed somewhat in meeting my views of the <u>practical</u> parts of the subject. It had been my intention to have you make a survey of my farm giving as you say its meets and bounds, its division into both, with the location of buildings etc. This I wish you will be able to do. As to spending so much time in looking after the geological character of the place, the nature of the soil, the adaptation of manures and chemical appliances to the improvement of the land etc. I think you misapprehend the nature of my requirements—that you are too theoretical.

I am quite well aware of the nature of the soil of our broad acres and of the manures that soil requires. But the question is how much can you do to improve and cultivate the land already possessed and how far I shall be able to procure manures for the barn yards and from the purchase of wood ashes—than which nothing is better adapted as a fertilizer to the nature of our soil.

I think your object should be, if you tarry with us for your brief vacation period, to make yourself as useful as possible upon the farm in assisting to gather in the crops, in cultivating the land to improve the general condition of the farm, by aiding in the garden, in . . . removing the stones, in repairing fences, etc. all of which will tend to make our place, which in location, in variety of scenery and in magnificent views, one of the most charming places in the country—

As to looking after the birds the insects and the worms and noting their habits their names and customs that must be incidental. It need not consume so much of your time as to call you off greatly from your work.

In *fine* the question of prime importance is how much money can you make for yourself and for me during all your engagements—that is, how much can you produce? For so much of your time as may be free from the engagements to which I have referred I would say, I can be willing to set apart any portion of land not otherwise engaged and offer you a liberal proportion of all that you can make off of it.

Hoping that you will see the reasonableness and wisdom of these views and give them your cordial assent and consent, I remain,

Your affectionate Father,
A. B. Davenport.[13]

Hoeing the garden, removing stones, mending fences. Charles was devastated. His desires were barely acknowledged. That summer, when he and his family left Brooklyn and headed off for Davenport Ridge, he simply worked his normal chores, as he always had, but with bitterness welling up like the billows on Long Island Sound.

ON THE FOURTH OF JULY, 1890, throngs of people crowded the streets of Newport, Rhode Island, having a raucous time. Patriotic music, fireworks, and lots of liquor abounded as citizens hurrahed the nation's birth. But away from the celebrations, alone in his room, Charles Davenport sat appalled.

He could not believe how many stores sold liquor here, and the number of saloons in town. Even the grocery stores were selling alcohol, displaying stacks of bottles in the windows. "On the Fourth," he wrote his mother, "all the saloons were open and I never before saw such a sight of drunkenness and disorderliness as I saw on Thames St. on that day. One could count more reeling men in riding through the length of the street once than he would be apt to see in Brooklyn in a year. This inexpressibly disgusting scene is repeated on a smaller scale on Saturday nights. Perhaps the entire rottenness of the whole thing strikes me the more forcibly that I have lived for so long in Cambridge, where there are no saloons visible."

Reading this letter at Davenport Ridge, his mother might have recalled the note Amzi had written from Newport almost forty years earlier. She knew her son was bitter at his father, even maintaining a silent rage. Charles had been writing Jane often the last few years from Cambridge, but he never mentioned his father, asked about him, or sent him greetings. Having just graduated from Harvard University with a degree in theoretical science, Charles was living in Newport now, doing research on local fauna and tiny marine creatures—the work he had always longed to do. Yet, despite the rupture between them, they were so similar, Jane could think. Charles had the exact same reaction to the revelry of Newport as his father years ago. So when she sat down to respond, she tried to hint how Amzi cared, how proud he was, even if he never showed it.

> To-day I took upon the Boston Post to read, and your father coming in while I was so engaged, I volunteered to read to him the speeches on commencement day. . . . Your father afterwards took up the paper and discovered that you had M.A. following the other letters. We send congratulations. Your father says that generally, he thinks, two years elapse after graduation before M.A. is given. Your father would like another copy of the Post of June 26th as he wants to scrap both sides. Can you send him one? How did you get along during the hot wave?[14]

She was a little surprised, really, that Charles had never explained how he was about to get a master's degree. Why wouldn't he?

Charles had tried to be a civil engineer when he first left home, following the "practical parts of the subject" as his father wanted, but he was miserable, and he eventually followed his desire to be a natural scientist, enrolling in Harvard in 1887. He took every natural history course the college offered, and received an A in each. He was growing to be a gangling young man, six foot and slim, well groomed, with a full beard. But he tended to avoid eye contact in social situations and, in general, preferred to be alone. He studied ferociously, and though he would write his mother, telling her about going to genteel Boston dining clubs, playing croquet and tag with newfound wealthy friends, eating strawberries and cream, he said he would always leave these events early. "I had to return to the grind," he explained.[15]

Jane had been a little worried about her youngest son. There were so many dangerous ideas in Cambridge, and since Charles was alone, without the support of his family and church, she feared he might be led astray. So she tried to keep him informed of life in Brooklyn and on the farm in Davenport Ridge, especially in matters of God, family, and country. She told him how she and Amzi took the open buggy to Norwalk to hear a sermon by the Reverend Charles H. Everest, a former member of their Plymouth Church in Brooklyn, and a noted orator. "He gave a patriotic discourse without a text showing the Providence of God in the developing and establishing of the nation. In some particulars he considered the American nation God's peculiar people of the new dispensation, as the Jews were of the old. . . . Of course, your father spoke with him, as he is personally acquainted." In another letter, during the 1888 presidential election, she described the family arguments about the two hot political topics in the house: suffrage and temperance. "Henry is at boiling heat on the one engrossing subject of the day, your father and James more quiet and dignified. . . . Mary is quite disgusted with Henry's heat. . . . For my part, I am quite rejoiced that I have not the 'right' to vote, for I am sure I should not know what to do on the subject." But she also liked to keep Charles informed on the subject she did know: the importance of the family faith. "Benny—John's son—united with Mr. Scoville's church at the May communion. I rejoice that so many of our dear children are named with God's people and hope and pray that they may walk worthy of their high calling."[16]

Charles tried to do the same. When his mother asked what he did on Sundays, he wrote back and told her about the churches he was attending, the preachers he was going to hear, and how he also attended Friday prayer meetings. She suggested he engage in some sort of "Christian work," perhaps offering his services as a Sunday School teacher.[17]

His mother's fears were not unfounded, however. Cambridge had brought a new lifestyle to the shy young man, and was bringing a dizzying array of new ideas. He was reading not only Darwin and Spencer and the latest theories in natural science but also Ralph Waldo Emerson, commentaries on the life of Galileo, and the science of biblical criticism. The path he was taking was dangerous, he knew, and it might tear at him inside to think he was beginning to fail to walk worthy of the "family's high calling." But somehow it was liberating, too.

After his research in Newport, Charles went back to Cambridge. He began to work on his doctorate, and proposed a thesis, "Observations on Budding in *Paludicella* and Some Other Bryozoa." Despite being shy, he always demonstrated a boyish enthusiasm for his work, which was starting to get noticed by his professors. And he was a tireless worker. In his Harvard labs, as he bent over his microscope, he warned others not to disturb him, inscribing a large eyeshield with the words: "I am deaf, dumb, and blind."[18]

As a former engineer, Charles was taken with exact, quantitative methods. He was soon drawn to the ideas of Francis Galton and his protégé Karl Pearson, who applied statistical methods to the study of biology and zoology. These were methods unknown to most of his professors, so Charles's papers were fresh and original. Soon, Harvard hired him as an instructor and gave him courses to teach in the Department of Zoology. The illustrious college also gave him courses to teach at the "Annex," the women's division of Harvard, later instituted as Radcliffe College. This appointment, it turned out, would change his life.

Gertrude Crotty was a graduate student in biology at the Annex. She was a former zoology teacher, having taught at the University of Kansas, her home state, and she was a small but feisty woman, unlike any other he had known. Far from being demure and soft-spoken—like his mother—Gertrude had a sharp tongue and a no-nonsense demeanor, and like so many other bright scientists in Cambridge, she was highly skeptical of spiritual matters and scoffed at the ideas of traditional religion. Charles was smitten.

They began to spend time together. They walked in Arlington woods, hunting for bugs in the leaves, even in the snow. They read novels together—sometimes Charles reading out loud, sometimes Gertrude. When Charles went to Washington, D.C., he wrote to Gertrude every day, describing the sights he saw. "I have eaten from the same table at which G. Washington <u>may</u> have eaten sauerkraut and schnapps!" he told her, awed as well by the capital's other patriotic shrines.[19]

Again, Charles unknowingly followed the same path as his father: a man in love, filled with wonder at American memorial sites, and writing every day to Gertrude to describe it all. Yet, unlike Amzi, who was bold and confident in his words of love and resolute in his beliefs, Charles was less secure. "Sometimes, I confess, [I wonder] whether you are truly happy in your choice, whether some things in me do not bitterly disappoint you, whether you would not, if I asked you seriously, tell me seriously and frankly the things which you would like to see changed in me, with some hint as to how to change them. As for me, my dear Gertrude, my life is wrapt up in yours, and I long to be brought fully to life by your return."[20]

When Charles asked Gertrude to marry him, she accepted. They worked well together, complementing each other's strengths and weaknesses, and unlike many other couples, they were each other's closest friend. As the date of their wedding approached, it should have been a happy, exciting time. Even the guilt of failing to be what his parents had desired—a pious, industrious Davenport son—was not consuming him as it always had. Then, a month before their wedding, in the middle of May 1894, Charles wrote to ask his father a favor: Since the wedding would be held in Burlington, Kansas, Gertrude's hometown, he hoped to have a reception at his own pastoral home on Davenport Ridge. But when Amzi responded, Charles was filled with rage. His father seemed to be going mad, refusing a simple favor for his wedding. As he read the letter again—and the professionally printed pamphlet it enclosed—he was astonished at how petty and immature it all seemed. Instead of being happy for his son and helping with his marriage plans, Amzi—now stooped with rheumatism, but still as passionate and zealous as ever—ranted about how his children had abandoned him, and alluded to a fight he was having with Fanny, his youngest child, twenty-four years old.

    To Dr. Charles B. Davenport.

    My dear Son,

        I received a few lines from you last week which is the first time, I believe, you have written to your aged father for nearly two years.

        You ask a favor of me which I am both physically and <u>morally</u> unable to grant, but which under other circumstances I might be happy to bestow.

I cannot go into the larger house this summer to suffer as I have done for the past three or four years from the misconduct of my children. I think the family life should be a type of <u>heaven,</u> and it will be where <u>love</u> abounds, otherwise it will more resemble hell. I have sought to bring up my children in the fear and love of God, but they have disregarded my counsels and gone contrary to the letter and Spirit of God's Word.

While during my whole life I have enjoyed the honor esteem and affection of those with whom I have been associated both in the business world and in church relations, it is only in the family for the last twelve or fifteen years that I have failed to receive that respect and filial love which I might justly claim; and thus has been fulfilled the words of the Master "<u>A Man's foes shall bee they of his own household.</u>" And I am often reminded of the words of the Psalmist "<u>I am for peace, but when I speak they are for war.</u>" Now I must have peace and quietness in my last days (if possible) and I know I shall have inward peace and joy while I draw from that divine fountain from which all true happiness flows. After your introductory request you close with cursing your feeble old father— which greatly surprised me. That you may see in what spirit of contempt and cruelty one of whom I once had high hopes bore herself last summer, I send you some of her utterances which I wrote down with great exactness, the <u>correctness</u> of which she did not deny; but of which she would not admit the wickedness. You will see I do not ask her to say she is sorry if she is not; but when she has settled that between herself and God I am sure she will express the sorrow of repentance to me. But I will not enlarge.

> With much love and pity, I remain,
> your affectionate father,
> A. B. Davenport.[21]

His sister's "utterances" were included in a "contract" Amzi had drawn up for Fanny to sign. It was not handwritten or typed; the man had the contract professionally printed on expensive glossy paper! Charles could not believe what he was reading:

I hereby acknowledge to my Father that for some years last past I have not behaved, at all times, towards him in a manner becoming a loving and dutiful daughter; that I have said to him that I did not want to do anything that he asked me to do; and that for the past few vacation weeks I have used the following language, saying repeatedly: 'I will not obey you;' at another time, 'You are a thief and a liar;' 'You have not a friend in the world;' and again, 'You are a crazy fool;' and at last, 'You are a villain, and I will call James to put you out of the house;' and finally, when about to depart to my school, saying, 'There is no respectable person that respects you.' I know I ought to be sorry for these things, and I will try in the future to act in a more filial manner towards one whom I ought to respect and love.

Charles and Gertrude were married on June 24, 1894, in Topeka, Kansas. Exactly one month later, Amzi was dead. A young boy accidentally knocked him to the ground as he was walking down a street near their Brooklyn home, sending him to the hospital. While there he caught pneumonia, and died within a few days. Jane was devastated. She had lost her husband of nearly forty-five years. She could barely speak or write, and after nine months of mourning, she too passed away.

Charles's first year of marriage must have been a tumultuous time. Exactly nine months and six days after his wedding, Gertrude gave birth to their first child, a daughter they named Millia. A month after his marriage, his father died, and a month after the birth of his daughter, his mother died. The grief of losing his parents affected him deeply, but the sudden emotional changes in his life also represented a new beginning. As he resumed his life in Cambridge as a twenty-eight-year-old Harvard instructor, bringing yet another generation of Davenports into the world, he knew his life was now his own. He was a promising young scientist, and with his young family and career, he could begin to devote himself to evolution and eugenics—two subjects that were challenging the Christian faith he long had made his own.

IN THE MIDDLE OF APRIL 1897, Charles Davenport received a letter from London. He had written to Francis Galton, the scientist who had shaped his thinking as much as any other. Charles had heaped his gratitude, explaining how important and influential Galton's work had been to him. He had even asked for a photo of the illustrious man. Now, in a handwritten note, Galton responded.

Dear Sir,
    I am much touched by the extremely kind expressions in your letter, though certain that you ascribe to me are more than I deserve. It is most pleasant to know that one's labours are useful to others at a distance. What perhaps gratified me most was that you perceive a unity in my

work although there is much variety in the subjects. I quite feel it myself, but have often feared that others might think more concentration needed. You ask for a photo. . . . I send in a separate parcel a photo which was made here, an untouched photograph that appeared a year or two ago in an illustrated paper.[22]

Davenport had been focusing his studies on molecular biology and the search for the mechanisms of evolution, especially heredity. He had published papers on "The Germ Layers in Bryozoan Buds" and "Regeneration in Obelia and Its Bearing on Differentiation in the Germ-Plasm," observing how cells react to various stimuli. But

An autographed photograph of
Francis Galton.

his full-length scientific monograph, *Experimental Morphology*, just about to be published by the Macmillan Company, was the first American textbook to explain how to use statistical methods in biology, methods pioneered by Galton and Karl Pearson.

Eugenics suited Davenport. The new science brought together not only his interests in biology, evolution, and engineering, but also the deep-seated devotion to family and country his parents had instilled in him. The use of tobacco and alcohol was still inexpressibly disgusting to him—he abstained his entire life—and he suspected these problems were in many ways caused by poor breeding, in the biological sense. Eugenics not only reaffirmed his Yankee heritage, but with its scientific focus on pedigrees, which Galton pioneered, it was really not very different from the lifelong work of his father, who traced the illustrious family tree of the Puritan Davenport clan. And like Galton and Pearson, he was enthralled by the miracle of statistics. After poring over Pearson's improvements of Galton's methods, Davenport began to formulate another book, *Statistical Methods with Special Reference to Biological Variation,* a text which first brought their eugenic ideas to the United States in 1899. When the two Englishmen launched a new eugenic journal called *Biometrika,* they asked Davenport to be the third co-editor, and he accepted.

Charles was following almost the exact same path as his father—not as a Puritan per se, but as a scientist and eugenicist obsessed with his family tree. He began to travel to see relatives he had not seen for decades, and after meeting Amzi's sister Harriet during a trip to Niagara Falls, he wrote to Gertrude, "You see I have been trying to learn about my ancestors so that I can tell Millia—and so that we can know how to interpret her, by knowing her ancestry—for, no doubt, her mental life will be largely the resultant of the various ancestral qualities— we may expect her to be conscientious, stubborn, with a sense of humor, a love of nature, generous but not a spend thrift, without a great talent for any art but with a love of knowledge. Honest and with common sense above the average. You see I am turning 'Fortune Teller.' "[23]

Millia, it turned out, would grow to be none of these. But in the years to come, as Charles Davenport strove to integrate Galton's eugenic ideas with the American vision of a land free from impurity and sin, he continued to struggle with the specter of his parents' Christian heritage. The memories of a son who had failed to live up to his parents' pious hopes—especially those of his mother—began to evolve. Beyond his own understanding, perhaps, he was compelled to reshape his past.

When he filled out Galton's *Record of Family Faculties* for himself, this was how he described his mother, Jane: "gentle temperament; deeply religious; appreciative of humor; timid." Ten years later, in a similar eugenics questionnaire he published himself and entitled *Index to the Germ-Plasm,* he described his mother "devoted to her children, pious, bold in expressing and sticking to opinions." Again, ten years after this, in another revised questionnaire called *Record of Family Traits,* a pamphlet tens of thousands of Americans would later fill out for the sake of science and eugenics, he now described his mother as "not very conservative, inclined to be skeptical in religion, interested in the results of science." By the end of his life, this was the mother he would describe to his colleagues and friends.[24]

Memory, like the arch of life, evolves, and people looking down into the well of history often simply see the reflection of their current selves. Charles Davenport would be America's first respected scholar of genetics, the first scientist to present the rediscovered ideas of Gregor

Mendel to the American university, and the first to proclaim the new gospel of eugenics. But as he preached a vision of the future, a future that would affirm America's greatness as a shining city upon a hill, he would cling to his past by remaking his mother into his wife. He had failed to live up to his mother's pious expectations, he knew, and this may have subconsciously tormented him. But to the end of his life, he really did desire, as his father and mother always had, to be named with God's people, and to walk worthy of the high calling of the Davenports. But he would do so as a scientist, under the modern creed of eugenics.

VII

The Hideous Serpent of
Hopelessly Vicious Protoplasm

Y ou would be amused to hear how general is now the use of your word *Eugenics*!" wrote Karl Pearson to his eighty-five-year-old mentor, Francis Galton, on June 20, 1906. "I hear most respectable middle-class matrons saying if children are weakly, 'Ah, that was not a eugenic marriage!' "[1] Pearson could almost see his mentor's mirth, the intense, piercing blue eyes that always held that glint of amusement, and he knew that even now the world-renowned scientist, stooped with age, would still appreciate the joke. Eugenics was so much more complex than middle-class matrons could ever know, of course, but if even they were thinking about better breeding now, if housewives and mothers were considering the eugenic fitness of their marriages and their offspring, then Galton's vision of a new secular creed might just come into being.

Pearson was writing from his office at the Galton Eugenics Laboratory at the University of London, where he was analyzing thousands of pedigree forms listing generations of family traits. In the decades since his mentor first rooted his science in the concept of the Good, eugenics had finally begun to spread, spurred on by a mingling dance of new ideas and changing social circumstances. The religious and Romantic resistance to evolution by natural selection had waned—more scientific objections were being raised to the theory at this time, in fact—and it was no longer shocking to compare mankind to a monkey or a mongoose. Even though this week London papers were buzzing with debates over a new education bill—a controversy in which Church leaders were clamoring to restore religious instruction to public schools—the battle had long been lost.[2] Many theologians had modified Christian faith in the light of natural science, and secularism had taken hold in public discourse. And as science and evolution raised their victorious banners, the old, traditional forms of piety were slowly being forced into the private places of the heart and home—or out of them altogether.

But the general use of this clanging word "eugenics," now becoming as ubiquitous as a sudden sneeze, was not arising from a single source. Though from the start it had been a theoretical science as well as a social proposal, eugenics was now proving surprisingly fungible, branching off into sometimes unforeseen fields, and utilized by a spectrum of people with varying motives. On the one hand, questions about evolution had become questions about heredity, and younger scientists, turning away from the merely descriptive and speculative methods practiced by the great Darwin himself, were being drawn to analytical, statistical, and experimental modes of research— like eugenics. As Darwinian analogies were applied to other fields, especially the nascent science of sociology, eugenic ideas began to converge with reforms for relief for the poor, who were now seen as the surviving dross of nature. Yet, on the other hand, while professional scientists were creating a climate in which eugenics could flourish, even popular writers and playwrights were beginning to proclaim Galton's epic quest for human perfection. Untrained middle-class matrons and laymen could grasp the powerful analogy of breeding better plants and beasts, and wonder, too, whether the race of men could be similarly improved. Eugenics had a simple logic that was just beginning to have a broad, cultural appeal for the educated elites.

This appeal, however, also arose from deep-seated anxiety. By 1906 in England, the darker effects of the Industrial Revolution were tempering the optimism of an earlier age. For decades, industrial workers had been organizing into unions, and a series of violent strikes had shaken the country's social repose. Out of this struggle, radical socialist movements such as the Fabians and the Independent Labour Party were formed to challenge the established political order. Even now, during this rainy week in June when Pearson wrote his note to Galton, London papers were chronicling the steady decline of the entire Anglo-American world. There were reports of the meatpacking scandals in the United States, particularly in Chicago. There were accounts of the ferocious Zulu revolt in South Africa, making many recall how Britain's empire and military might had been humiliated in the Boer Wars a few years earlier. Indeed, both conservatives and liberals had something to fear. In many ways, the general mood was reflected in the lines of a well-known poem, "The Darkling Thrush," written by Thomas Hardy on the last day of the nineteenth century, and hinting at the real scientific worry of the day:

> . . . *The ancient pulse of germ and birth*
> *Was shrunken hard and dry,*
> *And every spirit upon earth*
> *Seemed fervorless as I.*

There was something mysterious and unseen causing this social decline. The nation's vitality, the ancient pulse of germ and birth, indeed seemed shrunken hard and dry, and people in both England and the United States were beginning to believe their nations were being swamped with incompetent offspring and the inferior blood of foreigners. Evolutionary science and social Darwinism might have undergirded the theoretical explanations for this "racial degeneration," but across the spectrum, the common sentiment drawing people to the seductive notion of better breeding was a subtle, sexually tinged despair.

In England, in fact, the group most enamored of eugenics, and who began to promote it most zealously, were the avant-garde sexual revolutionaries, the political radicals calling for socialistic reform. It was progressive thinkers, intellectuals interested in a sweeping, scientific reorganization of society and its morals, who first embraced the eugenics creed.

Karl Pearson's own efforts revealed a number of the motives driving the new science. He was a mathematical genius, far more talented than Galton had ever been, and a member of the new generation of scientists excited to probe the mysteries of heredity. For years, he had helped mold Galton's obsessions into a systematic scientific method called "biometrics," or the statistical analysis of biological data. Yet deep within his labors was a grander motivation. Pearson was also a socialist, a political radical himself, and behind his biometric endeavors was a hope for a new society, founded on both the principles of social equality and the unimpeachable authority of science.

Like his mentor, Pearson was drawn to themes of women and statistics. He pondered sex and marriage as well as the vitality of eminent men. Indeed, many within his avant-garde circle were drawn to eugenics because it implied both political and sexual emancipation. Pearson wrote that sexual relations should primarily express "the closest form of friendship between a man and a woman," whether they were married or not. Yet he dallied with prostitutes even while insisting to some female intellectuals in the club that their relationships remain strictly platonic. And though he advocated eugenic marriage to improve the human stock, he also proclaimed that sex should not have to entail the purpose of procreation—a radical notion for the time. He believed hordes of unskilled workers were thwarting the basic right of all to decent labor, so some measures of population control should be imposed. Indeed, by 1906, many progressives shared this view, and birth control crusaders were allying themselves with the logic of eugenics.

Better breeding, under the authority of science, implied a reason to break down the cultural restrictions surrounding the marriage of intelligent, middle-class women. If housewives were now bandying about the word "eugenics," Pearson could look back and see that it was radical women, feminists and suffragists, who were some of the first non-scientists to be drawn to Galton's work over forty years ago. He had been collecting his mentor's letters, preparing a massive four-volume biography of the man he believed to be one of the most original thinkers of his generation, and he was particularly interested in a thirty-five-year-old letter from Emily Shirreff, one of the early pioneers in the cause of women's rights in Victorian Britain. In March 1870, just after she had read Galton's recently published *Hereditary Genius,* Shirreff saw how his ideas could help emancipate women from the irrational shackles of current marriage mores, and immediately wrote to the famous explorer:

> *Your very able and original remarks on some of the causes which tended under the sway of the church to deteriorate the race of men . . . seem to me to bear with great force upon some things going on around us now in this country, and still more I believe in America. I mean the various causes which are combining to turn a large number of the ablest and most active women away from marriage. The luxury of modern life pernicious in so many ways has made marriage more difficult. . . . No one would hail more gladly than I should a rebellion against the miserable*

*system which has driven women to marry for subsistence or position, but unfortu-
nately the feeble will still be content to do so . . . while the abler, the more energetic, the most fit to be the mothers of better generation will revolt against the
injustice of our social arrangements, and struggle singly for an independent
position.*[3]

Eugenic marriages, many radicals had come to believe, would set them
free from the sway of church and class, not only leading to fitter children for the country but also allowing the "ablest" women to have more
independence and more control over their bodies and their lives.
Eugenics for them meant freedom.

At the same time, radical progressives could see individual freedom
as a factor in social injustice. Even they could not ignore the "negative"
side of eugenics, the worry that the ancient pulse of germ and birth was
now on the decline. The same fear, the same sexually tinged anxiety,
affected them as much as their reactionary counterparts, and from the
start there existed a certain paradox in their concern for the conditions
of the working underclass.

Since statistical data was a new phenomenon at the end of the nineteenth century, many people in England and the United States were
shocked by the facts they were beginning to hear. New studies seemed
to show that the numbers of paupers and criminals were increasing
every year, threatening to become an overwhelming economic burden
on hardworking, taxpaying citizens. Dozens of scientific and popular
periodicals, in fact, cited these statistics like prophets portending
doom. In one 1906 article, while Pearson was gloating about the spread
of eugenics, another of Galton's disciples lamented the startling fecundity of the feebleminded poor—as well as the religious traditions that
allowed this trend to continue. "On one day in the United Kingdom
there were 60,721 idiots, imbeciles, and feeble-minded, and of the
number, 18,900 were married or widowed. Here we have—under clerical blessing—a veritable manufactory for degenerates."[4] An influential
textbook on heredity reported that the incidence of "defectives" in England had more than doubled between 1874 and 1896, to almost 12
per 1,000 of the population.[5] As many progressives seized the absolute
authority of secular science, they felt these unsettling trends demanded
new kinds of scientifically based solutions.

"Positive" eugenics might break down social barriers and emancipate the fittest men and women from the tyranny of religious and class

mores, but what could curb this veritable manufactory of fecund degenerates?

So, while attempting to build a political base among the working underclass, many social radicals actually had little concern for the rights and dignity of individuals. They were more interested in keeping them from breeding. Following the conservative Spencer, their primary goal was a society in which the biologically fit could flourish, rather than the welfare of the unfit underclass per se. While they abhorred the class system and railed against the excesses of the undeserving rich—who had inherited most of the nation's wealth and lived like opulent dolts—most progressives, like the conservatives, had no romantic illusions about the poor. In one 1896 tract, *The Difficulties of Individualism,* the Fabian socialist Sidney Webb wrote that capitalism and private property promoted "wrong production, both of commodities and of human beings; the preparation of senseless luxuries whilst there is need for more bread; and the breeding of degenerate hordes of demoralized 'residuum' unfit for social life." The real problem, Webb argued, was that an unregulated capitalist society "ensured the rapid multiplication of the unfit" and created the conditions in which a "horde of semi-barbarians" would grow out of control.[6] The arch conservative Herbert Spencer had abhorred state regulation, of course, and had noted that this "horde" should simply be left to the self-purifying process of Nature's brutal laws, which would usher in perfection on its own. "The process *must* be undergone, and the sufferings *must* be endured," he had written. "No power on earth, no cunningly devised laws of statesmen, no world-rectifying schemes of the humane, no communist panaceas, no reforms that men ever did broach or ever will broach, can diminish them one jot."[7] Yet for social reformers like Webb, careful scientific planning could help prevent this demoralized residuum from running amok. Progressives agreed with Spencer, believing the old Elizabethan Poor Laws should be swept away, but they also wanted to replace these with proactive, scientific methods of prevention. "The superficially sympathetic man flings a coin to the beggar," wrote Havelock Ellis, another Fabian socialist. "The more deeply sympathetic man builds an almshouse for him so that he need no longer beg; but perhaps the most radically sympathetic of all is the man who arranges that the beggar shall not be born."[8]

In the United States, where radical socialism never quite took hold, the Progressive movement also found success by appealing to biological laws and the need for a scientifically organized state. The suffragist Victoria Woodhull, who had fought for such causes as the eight-hour workday and the graduated income tax, and who, with her persistent, boundless energy, was the first female candidate for the U.S. presidency and the first registered female stockbroker on Wall Street, also feared this growing class of semi-barbaric hordes. Despite her fight for social equality, in her influential tract, *The Rapid Multiplication of the Unfit* (1891), she proclaimed: "The best minds of today have accepted the fact that if superior people are desired, they must be bred; and if imbeciles, criminals, and paupers, and otherwise unfit are undesirable citizens, they must not be bred."[9] Indeed, as eugenic ideas began to spread in the United States, they often found their greatest grassroots allies among these types of American women—activists who were just beginning to form their own societies to fight for temperance and suffrage, and who were also actively involved in the reforms of organized charity.

The general use of Galton's word, in fact, really began in the American conferences for charity and corrections. Americans had always been a pragmatic people, given to notions of self-improvement, so while intellectuals in England discussed eugenic notions, American reformers were anxious to move the theory to practice. Spencer had always been more popular in the United States than in his homeland, so his devastating critique of Christian altruism and traditional Poor Laws, buttressed by these shocking new statistics, prompted many Americans to seek concrete, practical methods for preventing the biological causes of poverty and crime. The criminal theories of Cesare Lombroso, an Italian contemporary of Galton who believed crime was caused by a physical and biological "atavism" in certain human beings, had also become enormously influential with many American social thinkers. So, as notions of social Darwinism and theories of hereditary deviance began to take hold among the best minds of the day, they prepared the way for the rough-and-ready solutions of eugenics.

Within this changing social milieu, however, it was a family saga describing six generations of "degenerates" that brought the complex drone of social theory into stark relief. A book about the economic burdens placed on society by one pseudonymous New York family first began to galvanize widespread political support for eugenic reform in the United States—and prepare the path toward involuntary sterilization. The family was dubbed "the Jukes," and their story became a rallying cry for eugenicists and a model for research to come.

In 1885, a man named Richard Dugdale, an expatriate Englishman living in New York City, published *The Jukes: A Study in Crime, Pauperism, Disease, and Heredity.* Not a trained scientist, Dugdale came from a family of wealthy manufacturers, and he devoted much of his time to the art of sculpting and the leisurely life of a Greenwich Village intellectual. Interested in scientifically based reforms as a means to improve society, he also became an active member of a host of New York civic organizations. He was secretary of the New York Association for the Advancement of Science and the Arts, secretary of the New York Social Sciences Society, as well as co-founder and secretary of the New York Sociological Club, which met at his home. He also found time to be the treasurer of the New York Liberal Club and vice president of the Society for the Prevention of Street Accidents.[10]

The idea for his book arose out of his participation in the New York Prison Association. Dr. Elisha E. Harris, registrar of the New York Board of Health and a member of the association's executive committee, had asked Dugdale to assist him in a study on criminal heredity. Harris had been tracing the pedigrees of prisoners, and he thought he had found a criminally prone family which he traced back to a woman he dubbed "Margaret, mother of criminals." Not having the time to complete the study, Harris passed the project to the enthusiastic sculptor, a man with no university training except for a few classes in business and sociology from Cooper Union.

Dugdale began his research with six inmates in an Ulster County prison. Eventually, he found 709 of their relatives and traced them all back six generations to a poor Dutch farmer he called "Max." Max's sons, it turned out, had married into a family of six sisters (one of whom was Harris's "Margaret, mother of criminals"). This unusual fact made it convenient to study the family tree and draw conclusions about heredity. Dugdale dubbed this family "the Jukes" because they never seemed to settle down in a single farm or homestead. The slang word often described shiftless chickens which kept no permanent nests and laid eggs erratically, so a person who "juked" would not stay in one place, but preferred instead to be a wanderer, and probably to parent children along the way.[11] For Dugdale and most Americans, this shiftlessness was not a natural moral state of being; it was like a mark of Cain.

Dugdale did not refer to the work of Galton; neither did he delve into the biology of heredity. In fact, he was far from being a professing eugenicist. But he presented in meticulous detail the economic costs this single family placed on the state. Readers were stunned. He itemized the average costs of almshouse relief, outdoor relief, and prison maintenance for every single family member over the years. He listed the number of prostitutes in the family, the average number of men these Juke women "contaminated with permanent diseases," and the wives these men then contaminated in turn. He also detailed the costs of drugs, the costs of medical treatment, as well as the lost wages due to this immoral behavior. The figures were as devastating as they were unprecedented. As Dugdale explained:

> Over a million and a quarter dollars of loss in 75 years, caused by a single family 1200 strong, without reckoning the cash paid for whiskey, or taking into account the entailment of pauperism and crime of the survivors in succeeding generations, and the incurable disease, idiocy, and insanity growing out of this debauchery, and reaching further than we can calculate. It is getting to be time to ask, do our courts, our laws, our almshouses and our jails deal with the question presented?[12]

Even twenty years after the book was published, social thinkers across the nation were still startled by Dugdale's detailed charts and the line-by-line analysis of the costs this single family placed on taxpayers. Indeed, in 1906, this same question was being asked by more and more policymakers throughout the country. Speakers at the annual national Conferences of Charities and Reforms constantly referred to the Jukes saga, and reformers cited Dugdale's narrative before state legislatures as they lobbied for funds to build new asylums to house the poor.[13] Researchers began studies of their "unfit" populations, and newspapers and periodicals ran stories about "criminal families." The story of the Jukes was becoming part of the popular imagination, feeding into a long-held American fear, a prophecy of doom which threatened that if they did not obey the moral precepts of their God, they would become a story and a byword in the world, and be consumed out of this good land.

WHILE SCIENTISTS LIKE GALTON, Pearson, and Davenport in the lab were conducting statistical research and formulating eugenic theories, it was the American doctors working in institutions for charity and corrections who first moved eugenics from theory to practice. They began to use the general term "feeble-mindedness" to describe the nebulous hereditary malady they believed was the cause of the immoral behavior that brought economic hardship on the state. It was these American doctors who began to look for a practical means to curb the reproduction of a growing degenerate horde. First, they expanded the traditional concept of "indoor relief" to include asylum-like "training schools" and "colonies" for the feebleminded poor. Focusing especially on women of childbearing age, the aim of these new institutions—such as the Virginia Colony for Epileptics and Feeble-minded—was to prevent them from breeding.

The efforts of these American reformers also included new restrictive marriage laws based on the theories of eugenics. The first law had been passed in Connecticut in 1896, and it prohibited the marriage, as well as the sexual activity, of eugenically unfit women under the age of forty-five. Only those who could potentially bear children fell under the legislation, and the minimum penalty for breaking the law was three years' imprisonment. Indiana passed a law in 1905, making illegal the marriage of any person deemed "mentally deficient," any person having a "transmissible disease" (including "feeble-mindedness"), as well as all "habitual drunkards." The Indiana law also declared null and void the marriage of any couple who held their wedding in another state in an effort to avoid the statute. By 1914, at least thirty states would pass eugenic marriage laws.[14] Yet, even in states without laws, eugenic theory was guiding judicial discretion. In a typical courtroom scene, a New York judge told a young man convicted of his second theft that he must not proceed with his impending marriage. If he did, he would be arrested and imprisoned. Despite the tearful pleas of the man's fiancée, the judge insisted: "I have no intention of allowing a marriage that will only result in the breeding of more criminals."[15]

Even so, restrictive marriage laws and judicial injunctions could not prevent extramarital relations. By definition, feebleminded paupers were unable to control their base sexual appetites, and since they were also, by definition, unusually fecund, they often had children out of wedlock. As the image of a proliferating horde of semi-barbarians converged with the sexually tinged anxieties of middle-class Americans, many reformers believed a more efficient and more effective weapon had to be found in the battle against poverty and crime.

In the heartland of America, a number of physicians began to experiment with surgical techniques to "unsex"—or castrate—the unfit. In 1894, Dr. F. Hoyt Pilcher, superintendent of the Kansas Asylum for Idiots and Feeble-minded Youths, announced that he had castrated fourteen girls and forty-four boys to treat the "evil habit of masturbating." But he also suggested the procedure could act as a "prophylactic against a long train of evils, and particularly against the hereditary transmission of vice, disease, and the propensity to crime." When local newspapers reported his actions, however, and denounced them as "cruel," "brutal," and "unjustifiable," the ensuing public outrage cost Pilcher his job.

The medical profession rose immediately to his defense. The authority of science could not be thwarted by the moral squeamishness of the untrained public, especially with so much at stake. With a rhetoric displaying an epic sense of moral urgency, defenders of surgical castration joined the battle against unenlightened ignorance. As one doctor, defending Pilcher, exclaimed in the *Kansas Farmer*:

> *Just as the fury of the war horse is taken out of him by gelding, so the fury of a criminal is. It would be a most beneficent thing for society and the social compact called the state, if all criminals were emasculated. The impulse to crime would be very largely eliminated thereby and the reproduction of criminals would be stopped. Who can measure in money or in words the incalculable benefit to the race if all criminals of both sexes were thus prevented from spawning their ilk in the great stream of human life? Criminal? Nay! Nay! A blessing to the world, a boon to all mankind.*[16]

The urgency in the language of American doctors was in many ways similar to that of the revivalist preachers and temperance crusaders in this era of reform. Each proclaimed the evils of sin and vice, each called for repentance, and each warned that Americans would be consumed out of the good land if they did not change their ways. Another doctor, again defending the need to castrate all types of criminals, used the words of any evangelical revivalist in describing the priestly authority of physicians and their patriotic duty to their country:

> *Looking backward through the vistas of the past to the earliest historical times; yea, farther into the dim twilight of tradition, we see passing before us a stately procession of physicians, who not only ministered to the sufferings of the age in which they lived, but were ever in the front rank in all efforts and undertakings*

to prevent the ravages of disease and death, as well as to raise mankind to a higher physical and moral plane. . . . I believe it is the duty of every patriotic citizen, and especially of every physician, whose training and opportunities give him a broader and more intimate knowledge of all physical, intellectual, and I might say moral deviations or perversions which are insidiously sapping the foundations of our national greatness and endangering our future progress and happiness, to exert every effort in his power, in assisting to neutralize or remove these malign agencies and tendencies to degeneration which are so prevalent and potent at the present time.[17]

Moral disorder was sapping the foundations of America's national greatness, many believed, and some of the most sophisticated thinkers of the era thought that castration was the most viable solution. Condemning the firing of Dr. Pilcher, an editorial in the *Kansas Medical Journal* claimed that those who were outraged by his castrations simply wanted "to fling ordinary political mud" and "denounce proper conduct and enlightened methods." The editorial went on to explain that "these operations are occurring constantly under the trained eye of skilled physicians, and all honor to the devoted man who seeks cure and restoration and gives back to the State a restored citizen, to society an ornament instead of a burden, and who restores him to friends and family clothed in his right mind, robbed only of a beastly and execrated curse."[18]

Yet, despite such arguments, the public, the "unsophisticated masses," were not quite convinced. Castration hardly seemed an "enlightened method" to most nonspecialists. It was brutal; it denied human dignity; it treated human beings like any barnyard beast. When reformers brought a castration bill to the Michigan legislature in 1898, it was easily defeated.

In 1899, however, there was a breakthrough. An Illinois doctor, writing in the *Journal of the American Medical Association,* reported the results of a new, less brutal surgical technique that could achieve the same results. The surgeon-in-chief of St. Mary's Hospital in Chicago, A. J. Ochsner, explained how he had severed the vasa deferentia in certain males, a medical procedure that rendered them sterile without emasculating them. But Ochsner wanted to suggest the treatment could also be used in lieu of castrating habitual criminals. Citing the work of Lombroso, he wrote, "It has been demonstrated beyond a doubt that a very large proportion of all criminals, degenerates and perverts have come from parents similarly afflicted." Recognizing that castration "has met with the strongest possible opposition, because it practically destroys the possibility for the future enjoyment of life," Ochsner explained that in severing the vasa deferentia in males, "it is evidently possible to obtain the same result, so far as sterility is concerned, without in any way interfering with the criminal's possibilities of future enjoyment."[19]

Castration made this new operation seem quite harmless by comparison. If the public deemed "emasculation" too brutal and inhumane, doctors could now claim they had answered this objection: the procedure did nothing to change the person or limit their enjoyment of sex; it only restricted their ability to procreate. It was no more cruel or unusual than imprisonment or institutionalization, and probably less so. After Ochsner's article appeared, Dr. Harry Sharp, the surgeon at the Indiana Reformatory, began to experiment with this new surgical technique. Unlike Ochsner, Sharp was an ardent believer in Galton's tenets, and he had closely followed the discussions of heredity over the years. Working with the feebleminded poor, he immediately found this procedure to be an exciting—and even "heroic"—treatment for society's social ills. Noting the public opposition to castration, Dr. Sharp lamented that it had failed to win widespread support. "It is altogether probable that we, through our spirit of humanity, our broad ideas of liberty and individual right, have gone too far in this direction [protecting criminals]," he wrote.[20] Even so, Ochsner's new procedure could accommodate these "broad ideas" of individual liberty, while still protecting the purity of the race.

In 1905, Pennsylvania doctors were the first to present a bill to allow "men of science and skill" to use new surgical methods to eradicate this most dangerous class. Their bill would give surgeons the discretion to determine which operation—"as shall be decided safest and most effective"—neither prohibiting castration nor specifying sterilization. In a letter urging the assembly to pass the bill, three of the state's top surgeons, including the nationally recognized Dr. Martin Barr, tried to reassure the legislators: "If it seems desirable that the testicles remain for the sake of appearances, the severing of the duct or vas deferens is in itself sufficient to secure sterility." The purpose, they reminded the assembly, was to prevent "offspring that will be necessarily a curse to society." And besides, "A gelding or ox loses nothing but becomes in every respect more docile, more useful, and better fitted for service."[21]

The proposal easily passed both houses of the Pennsylvania legislature, but when it came to the desk of Governor Samuel W. Penny-packer, he immediately vetoed it out of hand, finding both its language and logic justification for an extreme form of cruelty. Resorting to sarcasm, he sent the bill back to the Senate, noting, "It is plain that the safest and most effective method of preventing procreation would be to cut the heads off the inmates, and such authority is given by the bill to this staff of scientific experts." He also gave a rebuttal to the lofty authority of science proclaimed by doctors like Harry Sharp:

> *Scientists, like all other men whose experiences have been limited to one pursuit, and whose minds have been developed in a particular direction, sometimes need to be restrained. Men of high scientific attainments are prone, in their love for technique, to lose sight of broad principles outside their domain of thought. A surgeon may possibly be so eager to advance in skill as to be forgetful of the danger to his patient. Anatomists may be willing to gather information by the infliction of pain and suffering upon helpless creatures, although a higher standard of conduct would teach them that it is far better for humanity to bear its own ills than to escape them by knowledge only secured through cruelty to other creatures. This bill, whatever good might possibly result from it if its provisions should become a law, violates the principles of ethics. . . . To permit such an operation would be to inflict cruelty upon a helpless class in the community which the state has undertaken to protect.*[22]

Yet many doctors wondered why it would be better for humanity "to bear its own ills" if they could be eradicated by such a simple and harmless procedure.

In 1907, the governor of Indiana, J. Frank Hanley, sided with the doctors, and signed into law a bill having almost the exact same language as Pennsylvania's. It was a great victory for the eugenicists, the first law in human history allowing doctors to operate on otherwise healthy citizens against their will, rendering the "unfit" incapable of having children. Harry Sharp, as one of the most influential physicians in the state, had led the battle with his rhetoric of moral urgency, and was able to convince legislators they must institute this "radical" measure if they wanted any hope of quelling the growing horde of paupers. After it was signed by Governor Hanley, the law would become a model for other states.[23]

Again, the law neither expressly prohibited castration nor specified sterilization. But surgery was absolutely necessary. It should be the patriotic duty of men of skill and science, Sharp proclaimed, to perpetuate this known relief for a weakening race. And "the race," for most doctors, was far more important than the individual pauper. By 1917, the leading New York urologist and outspoken birth control advocate Dr. William J. Robinson would state that "it is the acme of stupidity . . . to talk in such cases of individual liberty, of the rights of an individual. Such individuals have no rights. They have no right in the first instance to be born, but having been born, they have no right to propagate their kind."[24]

PEARSON'S NOTE TO GALTON joked about middle-class matrons discussing eugenics, but both men knew that some of the best minds of the day were beginning to embrace their ideas. Yet as men and women used the results of their science in the struggle for social change, the "right" of society to protect itself clashed headlong with the old Enlightenment values of liberal democracy and individual rights. These rights had been limited, for the most part, to wealthy, property-owning males, and had been based on an outdated theological concept of "natural" rights. In the modern world, many believed, natural selection should be the epistemological foundation for political and social theory. The sentiment of better breeding, even if forced upon the unfit, was becoming more widespread, and even the famous George Bernard Shaw, a socialist and ardent follower of Galton's ideas, could now justify the reasoning behind the calls for radical political change:

> We must either breed political capacity or be ruined by Democracy, which was forced on us by the failure of the older alternatives. Yet if Despotism failed only for want of a capable benevolent despot, what chance has Democracy, which requires a whole population of capable voters. . . . Plutocratic inbreeding has produced a weakness of character that is too timid to face the full stringency of a thoroughly competitive struggle for existence and too lazy and petty to organize the commonwealth co-operatively. Being cowards, we defeat natural selection under cover of philanthropy: being sluggards, we neglect artificial selection under cover of delicacy and morality.[25]

Yet it was not in England, where structures of class remained strong, that eugenics could be wedded to legislation, state bureaucracy, and new technology. Galton and Pearson marveled at the successes of their American counterparts, whose practical spirit was infused with a religious zeal quite foreign to most Englishmen.

Their most important colleague, the biologist Charles Davenport, expressed the need to breed "political capacity" in a language that recalled the sermons of Jonathan Edwards rather than the science of Charles Darwin. Combining the biblical cadence of his father with the terms of social science, Davenport boldly proclaimed that "idiots, low imbeciles, incurable and dangerous criminals may under appropriate restrictions be prevented from procreation—either by segregation during the reproductive period or even by sterilization. Society must protect itself; as it claims the right to deprive the murderer of his life so also it may annihilate the hideous serpent of hopelessly vicious protoplasm."[26]

The battle was nothing less than a war against original sin, in which eugenics was the Christ who would crush the serpent's head.

In the summer of 1906, when a girl named Carrie Buck was born in the slums of Charlottesville, Virginia, the quest for artificial selection was just beginning. The science of heredity, for the first time in decades, was also just beginning to explode with revolutionary new discoveries. At this time, reformers were starting to believe the ancient pulse of germ and birth could be rejuvenated by science. But the great challenge was to muster the courage and strength to press past religious niceties, to build a new morality, and to establish tangible social policy. In America, legislators were beginning to pass new laws, and Davenport was organizing a world-class experimental station on Long Island, recruiting an army of researchers to study heredity and promote eugenics. These Americans, who had long had a special affinity for ideas of self-improvement and social purity, and for whom plutocratic inbreeding presented no problem, were just now becoming the first people in the world to embark on this new quest for artificial selection, attempting to take charge of their own evolution and breed a better world.

VIII

But, Oh, Alas for Youthful Pride

It was the end of the summer semester of 1906, and a cool, early evening breeze swept off the waves of Long Island Sound and onto the beach at Cold Spring Harbor. In the sand, dozens of students sang and laughed around a bonfire, celebrating the annual end-of-the-season clambake at the Summer School of the Biological Laboratory, run by the Brooklyn Institute of Arts and Sciences. The clambake had become legendary among the biologists who had studied here over the years, and even the serious, staid director, who sported a neatly trimmed goatee and often dressed in a white three-piece suit, participated in the fun. On this night, however, as he stood in the sand keeping an eye on his two rambunctious daughters, Charles Davenport felt content in a way he hadn't known for his entire life.

He'd been director of the Summer School since 1898, but in years past, the clambake had felt somewhat bittersweet for him, since it always marked the time he had to leave Cold Spring Harbor. As a peripatetic young professor, he had to move his family back to Cambridge, where he taught at Harvard and the Annex. Later, the family had to take the train to Illinois, where Davenport served as an assistant professor at the University of Chicago. Coming here each summer felt like coming home, and he always longed to stay. In the distance, miles across the Sound, he could just make out Davenport Ridge, his ancestral home. Thirty miles to the west, in Brooklyn, stood his old house at 11 Garden Place. But now, after years of financial and professional instability, he could finally call the Harbor home. Two years ago, he had convinced the recently established Carnegie Institution of Washington to fund his idea for a Station for Experimental Evolution, and he had already made it the envy of researchers around the globe. Davenport's 1906 budget, in fact, was $21,000—twice that of Pearson's Eugenics Lab in London—and his $4,000 salary was as high as the best-paid professorships in either country.[1] Though Davenport was still in charge of the Summer School for the Brooklyn Institute, as director of the Carnegie Station he was starting to earn a reputation as one of the most prominent biologists in the United States.

The Station for Experimental Evolution lay on acres of pastoral land which the native Wawepex Indians had once called a "good little water place"—marshes filled with freshwater and saltwater creatures, ideal for biological research. It was bounded on the northeast by the harbor, and to the east by Natchaquatuck Creek. Across the public highway to the south lay the New York State Fish Hatchery, and a private road separated the station from the property to the northwest. Charles, Gertrude, and their daughters Millia and Jane lived in the old homestead, the Victorian house with its sprawling porch much larger than the manor on Davenport Ridge. The new laboratory building was a big two-story structure, designed in Italian Renaissance style. With a farm, barns for animals, and plans for new dormitories, the campus was large and splendid, and scholars came from around the world to conduct research. "Quite an empire for us, isn't it?" Charles once exclaimed to Gertrude.[2]

On this night, as he watched his students and colleagues around the fire, Davenport could smile as they began to sing the traditional clambake anthem, "The Sad Fate of a Youthful Sponge," written years ago by a student and sung to the tune of "John Brown's Body." It was a ridiculous fable written in scientific terms, a *faux* morality tale about a curious ocean blastula, which longed to be free of its mother and feel the pleasures of freedom. Laughing, they began to sing together:

> *There was a little blastula no bigger than a germ,*
> *Who performed evagination from his mother's mesoderm,*
> *And soon his nascent cilia with joy began to squirm*
> *In ecstasy supreme.*
> > *Oh, the joys of locomotion*

Down within the depth of ocean
Oh to feel the great commotion
Within each blastomere.

No protozoan can ever guess the pleasure he did feel
As he felt within his ectoderm a growing gastrocele,
With joy and pride his polar cell at length began to reel
In foolish self content.
    Oh, the joys of locomotion . . .
His gastrocele was filled with pride that comes before a fall
And he felt his mother's ectoderm to be exceeding small
So he freed himself from all restraint by rupturing the wall
And floated out to sea.
    Oh, the joys of locomotion . . .
But, oh, alas for youthful pride as upward he did soar,
He caught the topmost spiculae upon his blastopore
And trying hard to get it off, his ectoderm he tore
A great big ugly rent.
    Oh, the joys of locomotion . . .
"Oh mother, dear," he cried in grief, "Come quickly now and try
To heal my little ectoderm or else I'll have to die."
But his mother dear was sessile and could only sit and cry,
From her excurrent pore.
    Oh, the joys of locomotion . . .
Now every night within the depths his little ghost is found
Lamenting to the annelids that burrow in the ground,
The hydroids wave their tentacles and shudder at the sound,
Of this familiar strain.
    Oh the joys of locomotion . . .[3]

Despite its silliness, the song's jarring juxtaposition of worldviews—a biblical warning of the dangers of sinful pride and pleasure described, ironically, as a biological process of birth and evolution—was somehow appropriate for the successful middle-aged scientist, standing with his family in the sand, free from old restraints.

The research at Cold Spring Harbor was actually quite similar to the detailed proposal Davenport had sent his father twenty-one years earlier, when he was just nineteen and longing to pursue a career in science. Now, instead of mending fences and moving rocks on Davenport Ridge, he was breeding snails and chickens, directing the observation and study of the local ecosystem, and writing scientific papers read around the world. His knack for organization and planning, which had hardly impressed old Amzi, was finally yielding important results.

Charles Davenport, Cold Spring
Harbor, 1941.

Despite Davenport's own evagination, however, he had inherited a
few of his father's old obsessions. Like Amzi, he had an abiding interest
in American family pedigrees and the nation's moral character. These,
more than anything else, drove his scientific aspirations, turning the
initial curiosity of a youthful biologist into the moral fervor of a
eugenicist hoping to shape the course of the nation.

THE FIRST EXPERIMENTS at the station were simple, clumsily seeking the hereditary mechanisms that drove evolution. Long-winged
cinch bugs were crossed with short-winged cinch bugs, and spiders
with two different forms of males were mated with the single form of
female. Goats with abnormal pendants on their necks were crossbred
with normal goats. And in one ambitious (but naïve) experiment,
researchers hoped to simulate the slow crawl of evolution and transform
marine creatures, such as the shore snail *Litorina* and the marsh snail
Melampus, into terrestrial creatures by gradually acclimating them to
drier conditions.[4]

Though few of these early experiments would yield any significant results, Davenport was one of the first scholars in the United States to turn his attention toward the Austrian monk Gregor Mendel, whose forgotten paper on pea plants, now rediscovered, was revolutionizing ideas of heredity. In a 1901 article, "Mendel's Law of Dichotomy in Hybrids," Davenport became the first American to write about Mendel's ideas, and when the station celebrated its formal opening on June 11, 1904, he invited Professor Hugo de Vries, the great botanist from the University of Amsterdam—and the man most responsible for the rediscovery of Mendel's long-forgotten paper—to be the featured speaker. In his address, "The Aims of Experimental Evolution," the Dutch scientist explained that with Mendel's ideas, scientists could find a more exact method to study the puzzles of evolution and heredity. This was the aim of Galton, Pearson, and another of Galton's protégés, William Bateson, of course, but Mendel's study presented a challenge to their overly complex method of biometric analysis. The clash prompted one of the most vituperative debates in the history of science. Since Davenport had been a student of Pearson's biometrics, he had actually been an early skeptic of Mendel. But in the two years since the station's opening ceremony, de Vries's words had proven prophetic. Mendel's Laws, as they were now being called, were not only revolutionizing the study of heredity, they were also giving eugenics yet another burst of momentum.

Mendel assumed that heredity was determined by individual "elements," undefined particles transmitted through the sperm and egg cells of the parents. Like Galton (who was born the same year as Mendel), he did not try to locate and describe these "elements"; he simply inferred their existence by observing the "statistical relations" in his controlled pedigrees. When he crossed green-seed plants with yellow-seed plants, the next generation of plants were always yellow-seed. When he crossed tall-stemmed plants with short-stemmed plants, the next generation of plants were always tall-stemmed. No blending, no intermediate traits. Indeed, he found that for all of his hybrid plants, one trait was always "dominant," and the other always "recessive." In the third generation, however, recessive traits reappeared in certain plants. After analyzing this cycle over generations, Mendel discovered a clear, mathematical pattern with inescapable implications: the statistics implied there must be a hereditary "element" that came in pairs and separated during reproduction.

By the time de Vries gave his address in 1904, studies in other research labs were just starting to confirm the existence of these elements. August Weismann had developed a theory of the "germ-plasm" in 1885, and suggested that the fuzzily seen "chromatic loops" in the nucleus of a cell were the mechanisms causing "hereditary habits or structures."[5] But the real breakthrough came in 1902, when Walter Sutton, a medical student at Columbia University, using microscopes far more advanced than Weismann's, discovered that these "chromosomes" were not loops but pairs, and that they split during the formation of gamete cells, just as Mendel's theory had predicted.[6] A few years later, after various kinds of chromosomes were identified, Sutton's Columbia professor, the brilliant cell biologist Edmund B. Wilson, and his former student Nettie Stevens, a professor at Bryn Mawr, each discovered independently that the sex of an organism was determined by the segregation and independent assortment of the X and Y chromosomes, following Mendel's Laws. In 1905, amid a worldwide flurry of research and discovery, William Bateson, Galton's erstwhile disciple and now the foremost champion of Mendel in England, coined the word "genetics" to describe this fast-growing field of heredity and trait variation. He also developed the influential idea of "unit characters," an expansion of Mendel's concept that would become the foundation for almost all eugenic research. Later, in 1909, the Danish botanist Wilhelm Johannsen would coin the word "gene" to describe these Mendelian "elements," believed to exist somewhere on the chromosomes.

By the summer of 1906, Davenport was leaping into this genetic revolution with the same enthusiasm he had always shown for science. He began his own breeding experiments, and wrote a number of influential papers on the inheritance of poultry and canary traits. But some of his most important work at this time was with his wife, the beloved partner to whom he owed so much of his determination to succeed. With Gertrude as the senior author, they wrote papers together on human eye, skin, and hair color—breakthrough studies that were some of the first Mendelian analyses of human heredity ever conducted. Gertrude was an accomplished scientist in her own right, but in addition to conducting research with her husband, she was running the campus kitchen, minding Millia and Jane, and teaching a course on biology.

Yet, even as the Davenports were among the early pioneers in the science of genetics, Charles was determined to focus more of his time on his expanding role as the nation's leading eugenicist. The idea of working for the national good was becoming his primary obsession, and promoting America's "good genes," as it were, was the work he really wanted to do. The revolution in genetics was only buttressing eugenic logic. If certain aspects of heredity could be reduced to "unit characters," with dominant and recessive traits, Galton's college question, "Could not the undesirables be got rid of and the desirables multiplied?" might have a relatively simple answer. Like flowers and peas, pedigrees could be analyzed and unit pairs deciphered. In theory, scientists could know which people had desirable traits—like intelligence—and should combine their genes. They could also discover who had undesirable traits—like feeblemindedness—and should not.

The question was "How?"

AS AMERICAN STATES experimented with new eugenic laws, the federal government, too, was learning the promises of the new biology. Just a month before the summer clambake, President Theodore Roosevelt's secretary of agriculture, William Hayes, appointed Davenport and other prominent scientists to a national "Heredity Commission," charging them to investigate America's genetic heritage. Its purpose, he said, should be scientific research, but "with the idea of encouraging the increase of families of good blood, and of discouraging the vicious elements in the cross-bred American civilization." The commission should also try to discover whether "a new species of human being may be consciously evolved," even amid the resistance of traditional culture and mores. Davenport was especially excited by Hayes's charge to cooperate with churches and other religious groups—in effect, to unite the religious call for purity with the modern science of eugenics. He could recall the editorial from the *New York Times,* which announced its support for the Heredity Commission:

> *Mr. Hayes would invite science and religion, co-operating with Government efforts, "in an investigation at once conservative, careful, and possibly constructive." There is nothing intrinsically objectionable in the notion of invoking afresh these institutional forces for the purpose of bettering the race. They are already at work to this end. Marriage, sanctified by religion, safeguarded by law, and attended its crises by fostering science; the prescriptions of custom and the legal provisions for educating children, and the ordinary ways and barriers of human intercourse are designed for the uplift and improvement of the human type. But in*

coming years, custom, law, and religion may themselves be profoundly modified by the advances in science.[7]

In some ways, Davenport felt it was men like himself who were the real theologians now, who carried the authority of truth and the secrets of human nature. Though he rarely attended church, he still felt it was an essential, if outdated, institution for the good of the nation.

The idea of a national Heredity Commission had sprung from his own work, actually. Davenport's Mendelian scholarship, as well as his organizational talent, stood at the origin of this government-sponsored organization. For the past few years, he had been an active member in the American Breeders' Association, a group formed in December 1903 to study the impact of Mendel on the older, practical breeding techniques used on farm animals and domestic plants. As an offshoot of the American Association of Agricultural Colleges and Experiment Stations, the association's original mission was to act as a national agricultural society, bringing together scientists and farmers to increase the efficiency and productivity of the nation's farms. With headquarters in Washington, D.C., its first president had been the previous secretary of agriculture, James Wilson. At the urging of Davenport, the Breeders' Association in 1907 agreed to add a new "Eugenics Section." Human breeding was just as important for the nation's interests, Davenport argued, and others soon agreed. The Eugenics Section drew such luminaries as David Starr Jordan, president of Stanford University; Luther Burbank, the world-renowned horticulturalist; Alexander Graham Bell, the great inventor; and a host of others. Since President Roosevelt and members of his cabinet were keenly interested in "racial improvement," it was not surprising that the new Heredity Commission was made up entirely of members of the Eugenics Section of the American Breeders' Association. These were the first government-sponsored organizations to promote a systematic study of better human breeding in the world.

Most of the experimental work of the Heredity Commission and the Eugenics Section was conducted at Cold Spring Harbor and supervised by Davenport. Unlike chickens and canaries, however, human beings could not be bred and studied, so Davenport devised a "Family Records" form—actually a little-revised version of Galton's *Record of Family Faculties*. Sent out to schools, hospitals, churches, and other institutions, these questionnaire pamphlets asked interviewees to describe the traits of three generations: grandparents, parents, as well as self and siblings.

Despite his growing national stature and new sense of contentment, Davenport was still wracked by inner turmoil. When he filled out a Family Records form for his own family, he was particularly harsh on himself. Under "Temperament," he wrote: "nervous, liable to be flustered when excited." His prevailing mood, however, was "optimistic." In contrast, he described Gertrude as "self-controlled in emergency," yet with a prevailing mood that was "pessimistic." Under the blank for "marked mental and moral traits or defects," he wrote of himself: "concentration, parturiacity, selfishness (self-absorption)."[8] For those sharing his New England Protestant heritage, "self-absorption" was a particularly negative moral defect, and Davenport's guilt shaped his particularly reclusive and demure personality. His family tradition had always valued sobriety and restraint, and warned of the dangers of sinful pride and arrogance. Success should never be ostentatious, he knew, so the contentment and satisfaction he sometimes felt about his accomplishments were often accompanied by inner pangs of conscience.

These inner pangs mixed ambivalently with his vaulting ambition. His sense of personal character flaws in many ways helped drive his quest to shape America into a greater nation, free from moral defects. By 1909, when the annual budget for the Carnegie Station reached an astounding $41,000, the director was starting to become restless with the strict parameters of the Carnegie Institution, which had only agreed to fund purely biological investigations. He wanted to accomplish more, to plan more. Yet, even as he worked to organize national societies and participated in conferences around the world, he yearned to flee the outside world and remain within the comfort of his "empire" at Cold Spring Harbor—and in the arms of his only real confidante and friend.

On a trip to a conference in England in the summer of 1909, Davenport expressed this ambivalence just as he was on the cusp of a new direction in his life. He wrote to Gertrude—as he did every day they were apart—describing "one of the most memorable days of my life." He had visited a poultry farm at St. Paul's Crag in Kent, and learned some valuable tips for breeding chickens from the people there. But it was a delightful day, so he decided to walk on to Down House, Charles Darwin's former home. The group he was with was planning to go later, "but I wanted to see it—<u>not</u> in a crowd." The experience was an epiphany:

> *It is a wonderful place and seems to me to give the clue to Darwin's strength— solitary thinking out of doors in the midst of nature. I would give a good deal for such a walk {as the Sandwalk} and have planned such. But the first step is to get the factory corner for the Jones' and Hewletts! Then I would build a brick wall around it, as Darwin did around his garden, so as to shut off from the curious world. Fill it with trees and shrubs and have a walk there. . . . It could be accessible by a tunnel from our garden along the side of the ravine. . . . I know you will laugh at this, but it means success in my work as opposed to failure. I must have a <u>convenient</u>, isolated place for <u>continuous</u> reflection.*[9]

As a boy, Davenport had longed to be free of his father's "prison house" and go to school. He had loved the strange sights and foreign tongues of New York City, and the thrill of urban life. But his romantic longings had since evolved, as it were, from the noisy American vision of Walt Whitman to the solitary musings of Henry David Thoreau—two transcendental writers he often read at Harvard. Now, Davenport's longing for solitude, driven by a subtle fear of the "curious world" and those who might look over him and critique his work, pervaded his persona. He was very sensitive to criticism and never responded well when peers found flaws in his work. Gertrude was the only person, really, with whom he could be himself.

Still, he ran the campus with the same harsh, autocratic hand under which he had bristled with his father. Earlier that year, in April, he became furious when two young women invited a male colleague to their dorm room to share a late night cup of soup. The very suggestion of immodesty appalled him, and he not only forbade this sort of behavior, he also prohibited the consumption of alcohol on campus grounds. Cold Spring Harbor would not be Newport, after all.

But the trip to Down inspired him, and he began to formulate another grand idea. His father had fought for temperance, the emancipation of slaves, and the spreading of the Christian Gospel, and Davenport felt a similar duty to promote the moral purity of the nation. He eventually convinced the Carnegie Institution to purchase "the factory corner," and though he never built a brick wall around it, he did begin to devote himself to continuous reflection, even while presiding over a growing campus. He left the breeding experiments to others, and began to think about a vision for the future, a vision of a breed of Americans free from disease, unhappiness, and the traditional notion of "sin." His Utopia wasn't Kantsaywhere, however, but the Puritan "city upon a hill."

Davenport conceived two new projects to make his ideas a reality. First, with the data he had been gathering from his Family Records forms, he wanted to present his grand vision in a book, which he would later call *Heredity in Relation to Eugenics,* the first full-length monograph on the subject in the United States. Second, he wanted to organize a new Eugenics Record Office, a national "clearing house" that would work with civic groups to identify the defective "germ-plasm" passed on in certain families. "Unfit" family strains could be identified all across the nation, and state governments could work to eliminate the reproduction of defectives in a systematic way.

In September 1909, as he was working on his new book, Davenport seized his opportunity. The legendary railroad magnate Edward Henry Harriman had recently died, leaving his $70 million estate to his widow. Mrs. Mary Harriman, as Davenport knew, had already been planning to become one of the nation's foremost philanthropists, giving not only to hospitals, conservation groups, and institutions for the arts but also to scientific research. Like Carnegie and Rockefeller, she envisioned a great legacy of social improvement through her philanthropy. But unlike these wealthy men, Harriman was not interested in foundations; she wanted to be intimately involved with the projects she was funding. So when her husband died, she was deluged with over six thousand requests, amounting to over $247 million.

Davenport was patient. He knew he had an inside connection. Mary Harriman, the widow's daughter and namesake, had been his student at the Summer School in 1906, so he first sent Mary a letter, reminding her of their work in heredity, before explaining his idea to add another research facility to the Cold Spring Harbor campus. Mary was delighted to help her old professor (she might have recalled singing "The Sad Fate of a Youthful Sponge" at the bonfire) and immediately arranged for Davenport to come to the mansion in New York City and meet her mother in person.

Mrs. Harriman was intrigued by the idea that major social problems could be solved by better breeding in humans. She was an avid horse breeder, so she understood the logic of collecting pedigrees. She agreed almost immediately to fund the project—not through a foundation, however, but through an initial purchase of land and then successive yearly grants. In April 1910, she purchased the eighty-acre Stuart property near the Carnegie Station, along with its magnificent Victorian house, for $50,000. She personally paid for the renovations of the house, and agreed to fund the salaries of a superintendent and six social workers, totaling $20,000 a year, guaranteed for the next five years. By October 1910, the Eugenics Record Office was up and running, supervised by Davenport's protégé, the Missouri schoolteacher named Harry Laughlin.

Amid the excitement at Cold Spring Harbor as this new national clearinghouse began its work, Davenport finished writing *Heredity in Relation to Eugenics*. It was a work of science, explaining the findings of his years of research, yet it was also infused with the same hopes, the same fears, the same Yankee Protestant pride that had defined generations of American Davenports. In both style and substance, *Heredity* stood in a long tradition of American literature, presenting the myth of a chosen "peculiar people," who could recover a lost innocence within this wilderness of Edenic lushness—if it strived to remain holy and pure. If eugenics combined the concept of the Good with the study of genetic origins, Davenport brought a uniquely American perspective to the task. He represented a new stage in the evolving Puritan quest that had shaped the ideas of those who celebrated it, like Winthrop, Emerson, and Thoreau, and even those who criticized it, like Hawthorne and Melville.

Davenport began with a simple outline of the basic eugenics argument, invoking Galton and proclaiming the startling truth that "Man is an organism— an animal; and the laws of improvement of corn and of race horses hold true for him also. Unless people accept this simple truth and let it influence marriage selection, human progress will cease." Though an organism, akin to corn and horses, "the human babies born each year constitute the world's most valuable crop."[10] This led to the leitmotif: statistics indicated this valuable crop was degenerating. Over 8 percent of the children born were "non-productive," while at the same time, the really illustrious strains of the race, which had produced men like Thomas Edison and William James and others who made the nation great, were beginning to decline. The old, hearty pioneer stock, the army of settlers who had pressed westward to build a nation, was losing the battle of survival.

Americans had always imposed their will on the vast wilderness of the New World, but, just like John Winthrop's sermon on the *Arbella,* there always lurked a prophecy of doom. If the promise were to be fulfilled, evil could not be tolerated. Davenport was confident Americans could conquer the inner, genetic flaws that caused these social ills; but this could only happen if they worked to wipe them out, putting their trust in the revealed laws of nature. Eugenics, he wrote, could be "the salvation of the race through heredity."[11]

Davenport explained the methods of Mendelian analysis and the discoveries of chromosome theory. Using Bateson's idea of "unit characters," he illustrated the science of genetics by analyzing over one hundred different types of physical traits, mental abilities, and propensities for disease. He explained how eye, hair, and skin color closely followed a Mendelian pattern, and how defects such as cleft palates, color blindness, and chorea also had genetic roots. And setting the stage for research to come, Davenport reduced a series of complex human behaviors to simple "unit characters." Artistic, musical, and literary ability, for example, could be inherited from an either-or unit trait.

Yet it was not always this simple, even Davenport admitted. In his discussion of pauperism and criminality, he claimed these traits resulted from "a complex of causes"—a combination of a number of discrete unit characters. Paupers normally had a "shiftless" character gene, passed to progeny following Mendel's Laws. And both criminals and paupers, he explained, were usually afflicted with the recessive gene of "feeble-mindedness," which dulled the moral faculty and made a person less likely to live a decent, productive, "God-fearing" life.

Davenport's confidence in the notion of "unit characters" led him to many simplistic and naïve conclusions, but it also yielded some significant and suggestive findings. Diseases such as Huntington's chorea, physical abnormalities such as harelips, and destructive behaviors such as alcoholism, each had hereditary causes, he explained. In his analysis of "narcotism"—another genetic cause of poverty and crime—Davenport was one of the first scientists to attribute the problem of alcoholism to a genetic predisposition. Though he agreed that environment did play a part in the behavior, he claimed that the hereditary drunkard would eventually choose an environment that met his genetic urges. "The bad environment has its result first and chiefly on those individuals with an hereditary predisposition toward narcotics and this hereditary bias is stronger in some families than others, depending on the nature of the family trait, and it occurs in a larger proportion of the cases of some families than others, depending on the nature of the matings that have occurred in that family."[12]

Feebleminded paupers and drunks were not his only concern, however. Immigrants had been flooding into cities like New York since he was a wide-eyed young boy, astonished by foreign tongues and bustling streets. American cities had become like Sodom and Gomorrah, and immigrants were a threat to the nation's social well-being. Though this had been an anxiety for decades, when Davenport wrote about migration and race, he explained the danger in scientific terms, describing the unit characters that distinguished one ethnic group from another.

The Irish, he claimed, without referring to concrete data, brought "on the one hand, alcoholism, considerable mental defectiveness and a tendency to tuberculosis; on the other, sympathy, chastity and leadership of men." As for the Germans, they were "as a rule, thrifty, intelligent, and honest. They have a love of art and music, including that of song birds, and they have formed one of the most desirable classes of our immigrants." Scandinavian traits "include a love of independence in thought and action, chastity, self-control of other sorts, and a love of agricultural pursuits." His longest analysis, however, discussed the impact of Jewish immigrants on the country. According to statistics, he said, the Hebrew crimes were chiefly theft and "offenses against chastity.

"There is no question that, taken as a whole, the hordes of Jews that are now coming to us from Russia and the extreme southeast of Europe, with their intense individualism and ideals of gain at the cost of any interest, represent the opposite extreme of the early English and more recent Scandinavian immigration with their ideals of community life in the open country, advancement by the sweat of the brow, and the uprearing of families in the fear of God and the love of country."[13]

DAVENPORT'S ANTI-SEMITISM was rooted in the threat he thought Jews posed to his image of the ideal American citizen: the rural Yankee Protestant, the yeoman farmer, the person who lived a temperate and chaste life—the same type of citizen celebrated by Jefferson and Crèvecoeur. Italians and Negroes and Poles also fell outside this ideal; but Davenport, like so many others, saw a particular threat from the Jews, who, he believed, while not a violent people, were overly sexual and had an unseemly desire for gain.

As for the "bad blood" already in the country, Davenport called for a new national bureaucracy to coordinate a sweeping scientific study of the population. It could begin with "state eugenic surveys," which would use schoolteachers, say, to interview pupils and their parents. The data could then be employed to construct family pedigrees for every family in the nation. In addition, he called for a "clearing house for hereditary data," which would keep this information on file and allow eugenic experts to analyze it and determine exactly where the nation's bad blood flowed. His own Eugenics Record Office would be perfect for such a task.

But what then? What should be done when this bad blood was located? Davenport, conservative in temperament, was ambivalent. Members of the American Breeders' Association had been discussing a number of possibilities, including segregation, stricter marriage laws, and the new technique of sexual sterilization. Before he began to write his book, Davenport himself expressed the Eugenics Section's consensus when he wrote in its 1909 report that sterilization was one possible way to "annihilate the hideous serpent of hopelessly vicious protoplasm."

But in *Heredity in Relation to Eugenics* Davenport was critical of the procedure, especially as it was applied to males. Despite the alarming danger this hideous serpent posed, he was wary of the promiscuous behavior sterilization could unleash. "But is it a good thing to relieve the sexual act of that responsibility that it ought to carry and of which it has hitherto not been entirely free?" he wondered. "Is not many a man restrained from licentiousness by recognizing the responsibility of possible parentage? Is not the shame of illicit parentage the fortress of female chastity? Is there any danger that the persons operated upon shall become a peculiar menace to the community through unrestrained dissemination of venereal disease?"[14] Actually, it would be much better to castrate them, he thought. This would extinguish "uninhibited lust" and "safeguard female honor." The best solution, he believed at this point, would be to segregate the unfit during childbearing years. Even so, "If we are to build up in America a society worthy of the species *man* then we must take such steps as will prevent the increase or even the perpetuation of animalistic strains."[15] Ironically, though Davenport began his book with the ringing declaration that "Man is an organism—an animal," his traditional moral outlook made him worry about animal licentiousness, which was, for some reason, unworthy of the "species man."

Davenport didn't cling to these views throughout his career, however, and his ambivalence and squeamishness about sexuality often made him waver. His strong commitment to chastity rendered him suspicious of birth control as well—which might promote licentiousness and keep the better human stocks from breeding. Years later, the birth control crusader and fellow eugenicist Margaret Sanger would recall how Davenport objected to contraceptives, and would "lift his eyes reverently and, with his hands upraised as though in supplication, quiver emotionally as he breathed, 'Protoplasm. We want more protoplasm.' "[16]

But Davenport and most other eugenicists also wanted less protoplasm from the unfit, so he eventually acquiesced to the logic of sterilization. In fact, as the ideas of eugenics began to spread to middle-class matrons, methods to revive the ancient pulse of germ and birth focused not on families of "good blood" but on the "vicious elements" of the population. Though David Starr Jordan and Alexander Graham Bell also shared Davenport's initial wariness of the surgical technique, many others at the association believed it to be the most promising breakthrough for a national program for eugenics. The Eugenics Section, in fact, had just organized a "Committee to Study and to Report on the Best Practical Means for Cutting Off the Defective Germ-Plasm in the Human Population," which focused almost exclusively on sterilization. And the most enthusiastic supporter of the procedure was a new face at the association: the recently hired superintendent of the Eugenics Record Office, Harry Hamilton Laughlin. He was the hardest-working member of this committee by far, and though he was not receiving credit for all his efforts quite yet, his research on sterilization—one of his first responsibilities after coming to Cold Spring Harbor—was about to get a worldwide audience.

ON JULY 24, 1912, in the ballroom of the enormous Hotel Cecil overlooking the tree-lined Embankment, some of the world's most renowned leaders and scientists were enjoying a formal dinner gala. The men wore badges on their evening jackets, round patches featuring a profile of a stern-looking Francis Galton, his lips pressed down, suppressing the characteristic smirk. Along the edges of the badge were stitched the words: *First International Eugenics Congress, London, 1912.*

The five-day conference had been organized by the Eugenics Education Society as a tribute to the late Sir Francis Galton, who had died the year before. The tables were filled with a host of luminaries. Among the five hundred guests were Major Leonard Darwin, Charles Darwin's son and the president of the Congress; the Right Honourable Winston Churchill, First Lord of the British Admiralty; and Lord Alverstone, Lord Chief Justice of England. With them were the bishops of Birmingham and Ripon, the Lord Mayor of London and his wife, the foreign ministers of Norway, Greece, France, and a number of others. But these political and religious leaders had come to hear the scientists. It was one of the largest gatherings of illustrious scientists ever assembled, and in the next few days, many of them would be presenting a series of papers on genetics, eugenics, and the possibility of better breeding.

There was a sense of excitement at the Congress. Many of the participants felt this was a historic event, and that humanity was on the cusp of a great era of progress. When Major Darwin gave the President's Address, he lauded the grand hopes and aspirations of eugenics to a cheering audience, and his underlying theme was the moral nature of

Hotel Cecil, London, 1906.

eugenics. The great task before them was an ethical quest that sought the Good to all the world. Better breeding had long been known to farmers, he pointed out, but applying these principles to human beings—Galton's intuition decades ago—now demanded a new kind of moral courage from scientists, one to replace that of a bygone era.

> *We might conclude that though for the moment the most crying need as regards heredity is for more knowledge, yet we must look forward to a time when the difficulties to be encountered will be moral rather than intellectual; and against moral reform the demons of ignorance, prejudice, and fear are certain to raise their heads. {Many in the audience cried "Hear, hear."} But the end we have in view, an improvement in the racial qualities of future generations, is noble enough to give us courage in the fight.* [17]

With Major Darwin's rousing call for moral courage, the scientists did turn to this crying need for more knowledge. The delegations had prepared a hall of exhibits so that, at the break, participants could browse a variety of booths and snappy displays. The Eugenics Education Society had put together a family tree showing the intermarriage of Darwins, Galtons, and Wedgwoods, and the dozens of eminent citizens produced by their interbreeding. The British tended to emphasize the "positive" aspects of eugenics. The American exhibit, however, sponsored by the American Breeders' Association, emphasized its "negative" aspects. Charts and photos featured the work of Davenport's fieldworkers at the Eugenics Record Office. Sixteen pedigrees, as well as statistical tables on the incidence of heredity of defectives, were displayed with a dramatic flair. [18]

The general tenor of the Congress, in fact, focused on the problem of the unfit. And the presentation that stirred the most interest was given on the third day by Bleeker Van Wagenen, a trustee at the Vineland Training School in New Jersey. Not the author, he simply presented a clumsily titled paper: "Preliminary Report to the First International Eugenics Congress of the Committee of the Eugenics Section, American Breeders' Association to Study and Report as to the Best Practical Means for Cutting Off the Defective Germ-Plasm in the Human Population." (The report was actually put together by Harry Laughlin, the secretary of this committee, who was not attending.) In many ways, this paper was the most significant of those presented at the Congress, for it alone discussed a real, "practical means" to cut off the defective genes that caused human problems. While suggesting a number of possibilities—including "life segregation," "restrictive marriage laws," "general environmental betterment," "euthanasia," as well as a Spencerian "laissez faire"—the report focused almost exclusively on a relatively new American technique: surgical sterilization.

No other country had ever attempted anything like this—and certainly none had passed legislation authorizing it. As Van Wagenen read out a short history of American sterilization laws, he explained how eight states had recently approved the procedure, but faced a host of legal challenges and public resistance.

> *Except in Indiana and in California little or nothing has been done to carry out these laws. Their constitutionality is in question. Attorneys-general for the several states do not seem anxious to defend suits and appear to encourage delay in putting the laws into operation, and in Indiana, where for seven or eight years vasectomy was practiced without law and exclusively at the request or with the consent of the person operated upon, and for two years thereafter under the law of 1907 compulsorily, there have been no operations since 1909 except a very few cases at their own request, not ten in all.*[19]

Despite the challenges, the report commended the procedure, and went on to illustrate its possibilities by discussing a number of individual case studies. While information had been hard to obtain, Van Wagenen explained, the operation seemed to have a number of beneficial effects on the patient's behavior. But the larger benefit was to society. "Their sterilization would be an insurance against unworthy progeny, and so eugenically of value."

The report concluded with a list of preliminary findings about sterilization. On the one hand, there was a greater risk associated with female salpingectomy, but the procedure did not seem to increase sexual immorality. And there was the touchy issue of consent versus compulsion. The report also predicted a battle to come: "That the sterilization laws already enacted in the United States will have to undergo vigorous attacks before the highest courts before many more compulsory operations are performed, with the probability that there will eventually be material modifications of them."[20] Indeed, these words would prove prophetic, and the man who wrote them, Harry Laughlin, would devote much of his career to winning the battle for compulsory sterilization.

Most of the participants agreed that this presentation on the "Best Practical Means for Cutting Off the Defective Germ-Plasm" was the most significant at the Congress. On the last day, during the president's Farewell Address, Major Darwin told the participants, "If I judge the tone of the Congress aright, all would place legislation tending to stamp out feeble-mindedness from future generations in a leading place in their programme." And though he could not estimate how much their meetings had assisted in "the ultimate victory" of eugenics, he predicted that "we shall conquer in time."[21] At this, the members of the Congress erupted in cheers.

These great leaders and scientists at the First International Congress of Eugenics were filled with excitement and hope, yet pervading their words was this image of battle. They knew that eugenics was an explosive issue, and that it clashed headlong with a millennia-long consensus of human uniqueness and dignity. In his address, Major Darwin alluded to the vehement reactions to his father's theories in the previous decades, and he warned that the path before eugenicists would also be beset with fear and prejudice:

> *The struggle might be long and the disappointments might be many. But we have seen how the long fight against ignorance ended with the triumphant acceptance of the principle of evolution in the 19th century {Applause and cheers}. Eugenics is but the practical application of that principle, and might we not hope that the 20th century will in like manner be known in the future as the century when the eugenic ideal is accepted as part of the creed of civilization. It is with the object of ensuring the realization of this hope that the Congress was assembled.[22]*

As Charles Darwin's son stepped away from the podium, the illustrious crowd, eminent men from England, France, Italy, Germany, and the United States, rose to give him a standing ovation.

Yes, in the future, years from this day in 1912, they were confident that the twentieth century would be known in history for these ideals of eugenics, this new creed of civilization, this moral commitment to racial purity.

Oh, the Bliss of Being a Mother!

Rain was whipping up against the window in his small, rented
room, upstairs from the old Dutchman and his wife. White
oaks, standing thick around the wood-frame house, swayed
violently amid the pelting rain and wind, making quite a roar. It was
one of those late summer storms that swept down across the Midwest
plains, the kind that sometimes brought tornados. Inside his room,
Harry Laughlin was under the covers of his bed, shivering, rubbing his
hands and feet, and crying as he scribbled a letter to his mother.

He was only sixteen, a thin, slight boy, and a bit of a hypochondriac.
He had just finished his first week as the new schoolmaster here in this
poor, rural community some twenty miles outside Livonia, Missouri,
but already things were not going well. On this first day in September
1896, fifty-one ragged pupils showed up at the tiny one-room schoolhouse. Seven more came in the following day, though there was barely
room for forty. There were none of the expected supplies in the dusty
space: no desks, no chalk, no blackboards, and students had to sit on the
floor. When Harry had first arrived, he found only an old oaken bucket,
a door key, a few tin cups, and a chair.

Getting back and forth to work wasn't easy, either. The isolated
schoolhouse was a mile and a half's trek from the old Dutchman's
house, but since there were no connecting paths, Harry had to walk
through cornfields, skip across a stream, and hike two large hills. Today,
as he walked back in the rain, he had had to wade through the stream,
which welled up past his waist. Then he lost his way near Blognet's

farm, wandering in the storm until he finally found his bearings. The whole time, he later wrote in the letter to his mother, "I was nearly blind with the backache."

Homesick and lonely, Harry had been writing his mother almost every night this week. He described his every ailment—his "bad chill," his constant fever, how his "heart didn't go right," and his "difficulty breathing." His handwriting was a pathetic scrawl, barely legible, and he mostly just complained. The people here in this backwoods community were seventy-five years behind the times, he wrote. His students were dull, and the locals were just starting to pay attention to national affairs. They were like the Jukes, really. But most of all, he could not bear to be without her. "Not five minutes pass without I think of you. Mama, write me a big long letter so I can get it on Saturday. . . . As I said before it is bad up here but I will stick to it if I can just live five months. I think about five months of this will show me how to know the true value of mother, home and my past surroundings." Weeks later, he still complained, still longed for his mother. "I have now told you the dark side of my first experience in life. I would like to tell you the light side but there is not a light side . . . It would be so much easier if I could see you every once and a while."

The following morning, the rain had stopped. Harry got out of bed early and began to get ready to go back to teach. But just before he left, he again jotted a few lines on the messy, scribbled letter on the table near his bed. "A good letter from home would help ever so much. . . . It's now about school time, and I must be off—to jail."[1]

Harry was the seventh of his mother's ten children, but he was somewhat different from the rest. He had always been rather frail, unlike his older brothers, and his mother had paid particular attention to him. She called him "Hi Yi," and for the rest of his life, his brothers and sisters did the same, though not always with the same kindness. Each of his four brothers was studying osteopathic medicine—a new approach to healing that did away with the scalpel—but Harry was the only one of the boys to remain a teacher, like their father. He felt ill often, so there may have been times when his brothers probed his body with their osteopathic hands.[2]

Harry's mother was a wonder. Even as she raised her ten children, she was also one of the most outspoken suffragists and temperance crusaders in the Midwest, and she often gave speeches at churches or wrote articles for local newspapers. Her name was Deborah, and like the great

Hebrew prophet in the Bible, she was a leader among men. Unlike most zealots, however, Deborah was never overbearing or cruel, and her modest demeanor always exuded a calm and dignified strength, especially with her children. She was a small woman, with a plain face; her heavy crinoline dresses always covered her ankles, wrists, and neck, but she loved to laugh. Harry worshipped her; of all the children, he most looked like her, too. She was the single greatest influence in his life, the model of tireless effort to build a better world.

Deborah Ross had always been a modern woman. She had attended Abington College in Illinois just before the Civil War, a time when it was one of only six co-educational colleges in the country. Even though men far outnumbered women, Deborah graduated first in her class in 1862 and gave the valedictorian's address at commencement. Over the years in college, she also developed a warm friendship with the class salutatorian, George Laughlin, who gave the speech preceding hers. Six weeks after the commencement ceremony, the two smartest students at Abington were married.[3]

They were a rare couple for their time. Few other than teachers and ministers studied past the eighth grade, and most of those were men. As a husband and wife with college degrees, Deborah and George were both well read and interested in contemporary ideas. They shared a deep commitment to their Christian faith, and were more devout than most, but they were also part of a new movement in American Protestantism—a progressive, theologically liberal form of Christianity that embraced the findings of biblical criticism and natural science, and consciously modernized traditional faith to fit the times.

Deborah followed her husband as he pursued his career. After three years as a schoolmaster in the common schools of Illinois, George became a minister in the new Christian Church, a fledgling denomination that would later become the Disciples of Christ. Members of this group, often called "Campbellites" after their founders, Thomas Campbell and his son Alexander, emphasized a radical reliance on the Protestant notion of "Scripture alone," rejecting any form of "man-made" creed or catechism. The Campbellites also yearned for greater Christian unity, believing it was dogma alone that caused divisions in the Church. Ironically, though this new denomination was in many ways a conservative movement at the start, while seeking the "purest" form of biblical Christianity, its congregations soon became fertile ground for liberal ideas. As they stripped away all dogmas, Jesus the divine Christ could easily evolve into Jesus the great human Teacher, and Trinitarian orthodoxy could be rejected as simply an extra-biblical, Catholic creed. Like many in the new Christian Church, the Laughlins were among the first to begin to adhere to a "social gospel," preferring the simple aphorisms of the Sermon on the Mount to the complex Christology of the apostle Paul. And the essence of the social gospel was progress, a belief that Christians themselves could usher in the kingdom of God by working for social reform.

Deborah's faith was characterized by a liberal Protestant romanticism that in many ways had more in common with the transcendental musings of Emerson than the old Puritan creeds. Like her husband, she had read Spencer and Darwin, and the exciting ideas of the modern naturalists gave her a particular interest in biology and botany. She was very pious; prayer was part of her life every day, but her God was not a transcendent Other who intervened in human history and offered the miracle of salvation. This was the message of the many revivalist preachers, the enormously popular itinerants who traveled throughout the region at this time. For Deborah, God was more akin to the slow-moving process of nature, performing a mission of love from age to age. Ministering to the wants of the world did not mean simply almsgiving or charity, but attacking the very causes of sin and misery. When the Bible was interpreted in the light of science, without dogmatic creeds, the gospel message of salvation could strip away the man-made social obstacles to progress.

Deborah began to write and lecture early in her marriage, but more than anything else, both she and George longed to have a family with many children. Just after their first wedding anniversary, in May 1863, they had their first son, Nimrod, and over the next nine years they had four more children: Laura, William, Mary Ella, and George Mark. Financially, the young family was struggling, and George barely earned enough to support them. During these years he settled in as the principal of the Ralls County Academy in New London, Missouri, also acting as the district superintendent. On most Sundays, he filled the pulpit for local congregations.

George organized the school system and preached his progressive form of Christian faith, while Deborah devoted herself to her children, whom she considered the greatest joy in her life. But at the same time she was also becoming more and more active in the various movements for reform. Her spirituality informed a fierce commitment to bold, progressive politics, especially those that had to do with women and children. Though small of stature, she was an unusually strong public speaker, with a keen intellect, and she began to give speeches before local women's clubs, churches, and even the small suffragist rallies just beginning to spring up in Midwest states.

Deborah was shrewd. She knew how to argue that a woman's identity was more than simply being a mother and homemaker, while at the same time affirming those roles as some of the greatest in all society. She argued that homemakers should be able to participate fully in the nation's political life, since they were essential in building national character. In a speech she often gave at meetings, she maintained:

> *When a young lady has toiled through college, won her well-earned praises and diploma, and chooses to settle in a home, I have often heard the remark: "Yes, what good will her education do her, if she must be tied down at home?" . . .*
>
> *Alumnae ladies may do very much to eradicate the false and pernicious idea, that an educated lady may not take charge of a home and retain her usefulness; for in no other place in the wide world is genuine culture, discipline of heart, will, and mind needed more than in the home. The varied, far-reaching home duties require an activity of body, heart, and brain exceeding every other catalogue of duties under the sun. . . .*
>
> *If on the one hand, there has been a seeming tendency for woman to climb up, out, and away from the duties of home; on the other hand, there has been a much stronger disposition to "clutch" her by the garments and seat her, not so very gently, with the very lofty advice: "Your place is home; stay down there."*
>
> *Both these mistakes are born of ignorance, the veriest misconception of what home is. No woman or man can climb up out of home, nor can they be tied down in a home. Home is not down, it is up, high up! And if our homes are not there, it is time for us to tremble, and gird us with strength to lift them. . . .*
>
> *It is here, if at all, she must renovate social life, church, and state.*[4]

Soon, Deborah was one of the first to wear the white ribbon of the Women's Christian Temperance Union, which had first taken hold in Ohio in the mid-1870s. Their motto suited her well—"Moderation in all things healthful; total abstinence from all things harmful"—and she began to lead marches in front of the local saloons. A reform spirit infused the wider women's movement in the region, and as clubs began to take a particular interest in poverty and crime—the direct results of drinking, they believed—many middle-class matrons thought the real solution began in the home.

This reform spirit was also infused with new ideas of biology and evolution. Though Deborah may have never lectured on eugenics herself, better breeding was one of the most widely discussed ideas in women's organizations across the Midwestern and Western states.[5] The womb could no longer be seen as simply a receptacle, a carrier for the children of men. The hereditary fitness of women was now considered just as important in making good marriages and fitter families. As the mothers of future generations, they had a crucial role to play in the drama of evolution and survival of the fittest. As the feminist writer Charlotte Perkins Gilman stated at the end of the nineteenth century, women had to realize not only "their social responsibility as individuals, but their measureless racial importance as makers of men."[6]

As Deborah spread her message to renovate social life through the power of a woman's moral sensibility, she continued to have more children. The family left New London in 1874 and settled in Iowa, where George accepted the position of chair of Ancient Languages at Oskaloosa College, a small Christian school surrounded by farms. Here, Deborah gave birth to four more children—Annie Elizabeth, Debbie, Harry, and finally Earl in 1882, the year she and George celebrated their twentieth anniversary. George was becoming more successful, too, and after he was appointed president of Oskaloosa College in 1880, it was easier to support his growing family, teeming now with teens, toddlers, and infants.

In 1883, however, the Laughlins were once again on the move. George had stirred up controversy with his liberal ideas, and Deborah's outspoken views on women sometimes caused a hullabaloo in town. Throughout the Midwest at the time, religious fervor was pulsing through everyday life, and controversies raged as emotional, wailing revivalists clashed with a rational, educated elite. Disputes over the nature of salvation and the proper grounds for biblical interpretation competed for the minds of the faithful. As a Campbellite, George spoke out against what he felt was the dangerous passion of the revivals, and his views on Jesus were considered far too unorthodox for a Christian college. So George left Oskaloosa, accepting the presidency of Hiram College in Ohio. Nimrod and Laura took degrees here, and Deborah taught Latin to help offset their tuition. But tragedy struck the family. Deborah had one more child, Marguerite, who died after only six days. Their second oldest daughter, Mary Ella, also died, succumbing to scarlet fever when she was nineteen and just about to get her own college degree. Controversies raged at Hiram, too, and the family left after four years, moving on to Garfield University in Wichita, Kansas, where George again became chairman of Ancient Languages—a significant step down in his career. When Garfield went bankrupt within three years, the Laughlins had to move again, to Kirksville, Missouri, where they settled for good.

The family struggled for money. George's career had come full circle; he took a position as pastor at the Christian Church in town, and later taught English at the Kirksville Normal School, a vocational college for teachers. The older boys also began to teach at the local public schools, making a small wage. Kirksville was the home of Andrew Taylor Still, the visionary healer whose eccentric treatment methods included physical manipulations to improve the body's natural functions. Still opened a college called the American School of Osteopathy in 1892, a year after the Laughlins arrived. Though Nimrod was almost thirty and William and Mark in their mid-twenties, they were soon studying this radical new form of medicine. Harry, who had just turned twelve, attended the normal school and was training to be a teacher. Laura, Elizabeth, and Debbie stayed at home at first, but in the next few years, each married and started families of their own.

In the fall of 1895, George was struck with typhoid fever. Deborah and the children stayed at his side for most of the eleven weeks he fought the illness, but in the middle of November, he died. Deborah was a widow at fifty-three, and her children, who had moved so often, were now beginning to settle into their own lives, dispersing to other parts of the country. Nimrod went to California to practice osteopathy; William to Kansas City; and Mark George, after marrying the daughter of Andrew Taylor Still, became dean of the American School of Osteopathy. Just a year after his father's death, Harry left for the rural outpost near Livonia for his first teaching position, and Earl attended the osteopathic college, later to move to Arkansas. Deborah's daughters followed their husbands, some leaving the small farm town of Kirksville to other regions bustling with greater opportunities.

For Deborah, rearing her children had been a sacred duty, a responsibility on which the future of the nation hung, and she tried to pass her relentless quest to reform society to her children. After most of her children had left the house, Deborah wrote to Mark George, who was still in Kirksville: "I think every day of our boys and girls (papa's and mine), that we have reared and pray that each one shall live a life of useful-ness—such a life as their sainted father can look down upon with smiles of approbation."[7]

Deborah was alone for the first time in her life. Reflecting on the new loneliness she felt, she composed a "Reverie," describing her feelings of anguish, but distancing herself by writing in the third person. It was an American hymn, agony clothed with optimism, and a description of a new way of life:

> She wandered from room to room asking herself: "What shall I do? How can I get breakfast for just myself? And no one to share it with me?" As she walked listlessly around the house, memory carried her back to the days <u>long ago</u>, when the noise of the little pattering feet, and the babble of childish voices filled the house, <u>now so silent.</u> When busy little fingers were getting into mischief, often undoing the work of weary hours, or getting hurt for mother to kiss and tie up. When motherly ministrations filled the days, and often the nights. And she smiled as she thought, "Well no <u>other</u> mother ever enjoyed her children more than this one. . , ,"
>
> She remembered the times when her prayers died upon her lips. When the agony of heart and soul refused to be voiced—when she could do naught but in a sightless, voiceless way, reach out her hand to the "Father" and He would lead her.
>
> Then memory glides to the happy growing up time of the boys and girls. O, the happy patient days. The school days! How often she went over their lessons with them! What a pleasure now to remember how she loved to help and encourage them. And how her heart swelled with pride when they did well!
>
> O, that all mothers could dwell so happily on the past! Why do mothers sometimes complain because of the children? O, the bliss of being a mother! To be a mother is greater than to sit on a throne! Society! What a bauble! "Lord I thank thee for the gift of motherhood!" All the hard places in her life were forgotten, or covered up in her great joy![8]

The bliss of being a mother, like so many other traditional senti-ments, was beginning to be retranslated for a modern age. Deborah represented a new type of woman, one who based traditional moral duty upon a scientific foundation. For many like her, the purity of women was no longer a simple religious obligation commanded by God. The joy of rearing healthy children was no longer just a cause for individual, domestic satisfaction. These female roles were beginning to be seen as biological responsibilities, a social duty to the purity of the human race and its future. And submission to the strength and authority of a husband could be an impediment to these larger responsibilities. The facts of biology and evolution, especially as explained by the science of eugenics, were now seen as causes for liberation.

At the same time, the image of the wanton woman, the sexually impure woman, the woman who didn't guard the biological chastity of her womb, was seen as a greater threat to society. A woman's purity was not for the satisfaction of a man, many now believed, but for a more important purpose: social and racial purity.

Harry, away from his mother's love and tenderness for the first time, alone in his room above the old Dutchman and his wife, might have remembered the schooldays, the motherly ministrations, and the help and encouragement he had received growing up. Now, he sometimes affected his mother's maudlin tone—which gave him a tendency to whine. But it was her voice, her presence that stayed with him, making him determined not to fail. He would live up to her expectations, he thought. He would make her proud.

LONDON'S ASTONISHING CRYSTAL PALACE, where Francis Galton had first proposed to his wife Louisa, had been the world's first international exposition. It not only celebrated the achievements of science and the triumphs of an industrial age; it also initiated a new era of human commemoration. Instead of simply celebrating and remembering the past, expositions looked forward to the future and invited participation and interaction. In this tradition, San Francisco hosted the 1915 Panama Pacific International Exposition, a breathtaking fair celebrating "all the best that man has done," especially the recently opened Panama Canal. At the entrance stood a dramatic 160-foot arch, near the "column of progress." At the top of the arch, which could be "seen at all points in silhouette against the sky," a sculpture entitled *Nations of the West* featured "the mother of tomorrow."

The "mother of tomorrow," however, was symbolized by an American pioneer, a woman standing at the reins of a covered wagon, calmly raising her fist in the air while flanked by long-horned oxen. To her sides—but conspicuously beneath her—were four men on horseback: a French Canadian trapper, South and North American ranchers, and an American Indian in full-feathered headdress. Hovering above the "mother of tomorrow" was a winged, bare-breasted "spirit of enterprise," emphasizing the female power of the scene. The sculpture was as dramatic as it was subtly modern, recognizing the new place of motherhood in the evolutionary saga of human progress. Indeed, the San Francisco Exposition not only included a series of eugenics exhibits, it also set aside a "Race Betterment Week," which emphasized the importance of physically and morally fit women in the quest for better breeding. But the image of the American pioneer was significant. As the exposition's secretary, Franklin K. Lane, explained: "The long journey of this bright figure of the pioneer is at an end; the greatest adventure is before us, the gigantic adventures of an advancing democracy—strong, virile, kindly—and in that advance we shall be true to the indestructible spirit of the American pioneer."[9]

The Laughlins were among the first of this new breed of American pioneers, led by their own "mother of tomorrow." The old restless urge to press forward into the wilderness, the uniquely American impulse to seek landscapes innocent and new, was being transformed for a modern world. For over a century, Americans had rambled west, settling in places free from the burdens of the past. Close-knit families attempted to build their private Utopias, their cities upon a hill, where they could worship their God unhindered. Even at the cusp of the twentieth century, when the Pacific had been reached and technological progress and modern industry were reshaping much of the wilderness into a powerful modern state, this longing to press forward lingered on. The older restive yearnings were now evolving into a new American individualism, with new ideas of moral and political freedom. Families who had set out into the vast plains did not always disperse from each other as the Laughlins now were doing. They had not emphasized higher education and modern ideas, but lived and died together within their communities, generation after generation, maintaining their particular private creed, while still holding on to a larger vision of an American manifest destiny. Yet, just as the old pioneers looked to the future, so did the members of the Laughlin clan, setting out on their own separate paths "to do some good" for the rest of the country and the world.

In 1899, when he was nineteen, Harry Laughlin was back in school. His teaching career did not have an auspicious start, but the trials of living alone in rural Missouri had made him more of a man. His mother's constant encouragement also helped make him a bit more confident in himself, and he could later write to her, "I can teach alright." He planned to make a career of it, in fact, so when he moved back to Kirksville to teach in the local high school, he studied to get his bachelor's degree from the normal school where his father once had taught. Here, he focused on studying history—as well as biology and agriculture, the most practical subjects in this Midwest farming town.

Deborah was glad to have Harry back in the house, and she continued to try to mold him into a modern Christian man. She convinced him to sign the Christian Women's "temperance pledge"—a promise of abstinence Harry would keep for the rest of his life. She also introduced him to some of her favorite poems, written by contemporary women authors. One of Harry's favorites was a verse by the New York historian and suffragist Maud Wilder Goodwin, and he pasted it prominently in his scrapbook of clippings. With his intensely religious upbringing, he was collecting poems and articles that focused mostly on theological themes such as immortality, predestination, and the tumultuous clash of science and religion.

> *Why should we strive when all things are decreed?*
> *As well may planets tug against the sun,*
> *Or rivers, by resolving, cease to run,*
> *As we by our striving rule our word or deed.*
> *All Darwin's science and all Calvin's creed*
> *Tell the same truth: that which is done is done,*
> *And we, elect or damned ere life begun,*
> *Foredoomed to be a flower or a weed.*
> *Upon the plastic wad of infancy*
> *A thousand years of habit set their seal:*
> *Such as our fathers were, for woe or weal,*
> *Strive we or shirk we, such we too must be.*
> *Thus reason speaks, and has talked her fill,*
> *Something within us, answering, says: "I will."*

This theme of biological fate, determined by the "plastic wad of infancy," made an indelible impression on the teenage Harry. As a student of biology, he marveled at how the Calvinistic theology of predestination—though an anathema to most Campbellites—was somewhat similar to the new science of heredity and the Darwinian theory of evolution. All people are foredoomed, somehow, to be a flower or a weed. The resigned tone of the poem, though it belied a defiant human will in the final line, reflected Harry's youthful melancholy. In the margins of his clippings, he jotted a few lines of his own: "Will the Maker destroy his Masterpiece? / So many Gods, / So many Creeds, / So many ways / That wind and wind / While just the art of being Kind / Is all the sad world needs."

The swirling winds of religious controversy brought young Harry the agony of intellectual confusion and despair. In a note to his mother, he wrote: "I want to be the kind of man that you would like to have me be. . . . My reveries extend back only a tenth as far as yours and are only about a tenth as varied—still enough to make me melancholy at times."[10] By temperament, he was not as secure and strong as his mother, and the Campbellite yearning for Christian unity tormented him as he was bombarded with conflicting ideas. He came to fear change—which had defined his childhood, as he moved from place to place—and to hate all things that seemed foreign or different. But the theme of a predestined elect, whether in theology or science, began to shape his thinking more and more, and as he studied history and biology at the normal school, he began to find a simple, uncomplicated solution.

It was around this time, in an essay on the nature of history and politics, that Harry first wrote about an idea that would consume the rest of his life. He titled it "Cosmopolitanism in America," and in it he presented a sweeping (and sophomoric) outline of the progress of political history, explaining how great empires evolved and fell over the last three thousand years, only to reach a final culmination in the new American empire. For Harry, the primary force moving this history was not God but biology. More specifically, the guiding force of progress was the inherent, biological superiority of the "white man."

"It is an established law of Ethnology that races inferior in intellect, morals, and action recede before the expansive march of a superior blood," the nineteen-year-old began his essay. "It is also generally conceded that, since science and invention are preparing every portion of the globe for the happy and prosperous abode of the white man, that by the tendency toward universality, the rapid transit of thought and matter, the general diffusion of knowledge, the growth of free institutions, the liberal thinking, the higher motives and the common destiny of all Caucasian peoples; that eventually the world will be inhabited by an enlightened race, Caucasian in blood, Christian in religion, and free in government."[11]

This ultimate global destiny, Harry believed, had already begun to take root in America. In many ways, his simplistic vision simply translated his mother's progressive politics into something even more "modern," infusing her social gospel with a scientific racism rooted in ideas of biological determinism. It was not enough for him just to study biology, after all. As a Laughlin—and as an American—there must be a higher calling, a higher purpose toward which to strive. There was a hint of his own family history in his words, and a sense of the manifest destiny of a clan who could disperse like missionaries, only to return to a millennial paradise.

By 1900, Harry's career was more successful. He had made $1.50 a day as a teacher, but he was soon promoted to principal of the Kirksville High School, earning $480 a year. He organized school science exhibitions, started a jug band for the students, and his geology field trips became one of the most popular activities of the school year. Harry's innovative teaching methods, as well as his emphasis on science, made him a new type of educator in Missouri, and people started to take notice.

At the same time, he was also noticed by a woman named Pansy Bowen, the daughter of an older teacher at the school. Harry liked to make her laugh, and they soon began a courtship. Sometimes, when lecturing, he'd see Pansy through the window as she walked down the street. He would flirt, waving to her behind his back, all the while continuing to teach his students.[12] By spring of 1902, they were married.

During the fall of that year, Harry took his new bride to Centerville, Iowa, where he accepted the position of principal at the local high school—a job providing twice his Kirksville salary. After three years there, however, he hoped to achieve a higher position, lobbying to be superintendent of the entire Centerville school district—a post that would again double his yearly salary. Once again, in a letter to his mother, there was a wistful longing in his hopes (as well as a bit of whining) as he hoped to make her proud. On March 11, 1905, Harry's twenty-fifth birthday, he wrote to the aging Deborah:

My Dear Mother:—
Your sweet letter came Sunday. I always make new resolutions when I read one of your messages. I have the best mother in the world. If nothing else had any influence over me your love alone would be enough to make me ambitious to attain those ideals that you always set before me.
. . . Well, I'm 25 at last. I have always looked forward to that point in

my life and wondered what it would be like. It is just like 24. But any way it is just a quarter of a hundred and I suppose that I am a man now. By taking care of my health and looking always on the bright side of life I hope to be able to work for a half century more. If I can't be great I certainly can do much good. And I intend to do it.

I do not however start my 26th year under very auspicious circumstances. I was 25 on Saturday, on Monday my candidates for school board were beaten by 8 votes. They were promised to vote for me for city superintendent at $1500 a year. The count degenerated into a disgraceful church fight. There were more "Wesleyites" than "Campbellites," that's all. I didn't want the place anyhow. If I could get it I would take the superintendency at Kirksville for a year or so. . . .

We are both well, affectionate and happy. Pansy sends love, write when the babies [grandchildren] will let you and think of us when they won't let you write.

Your Loving son,
Harry

Despite the sour, manipulative rationalization of his loss, Harry's drive to be great—or at least to do much good—led him back home when the Kirksville school district offered him the superintendent's job. Then, in 1907, the normal school offered him a position to start a new Agriculture Department, which would, for the time being, consist of him alone. Since this was a college post, Harry began to turn more of his intellectual energy to science and academic research. He studied poultry breeding, egg production, color patterns on the hides of cattle, as well as the Mendelian inheritance of human eye color—studies that would change the course of his life.

On February 25, 1907, Harry wrote to the biologist Charles B. Davenport, the nation's foremost expert on Mendelian genetics, to ask for advice for his breeding experiments with Yokohama and Tosa game hens. Davenport was happy to oblige, and the two began to correspond. When Davenport invited Harry to study with him at the Summer School at Cold Spring Harbor, Harry accepted, and the two began a warm friendship. "I consider the six weeks spent under your instruction to be the most profitable six weeks that I ever spent," he later wrote. During their time in New York, Pansy Laughlin and Gertrude Davenport also became warm friends, and they kept in touch afterwards. "She is Mrs. Laughlin's ideal of a helpmate," Harry told his mentor, describing how much Pansy admired Davenport's hardworking and brilliant wife.[13] Harry also took a pile of Family Records forms back to Kirksville, and faithfully sent them back to Davenport's lab.

Over the next three years, the mentor and protégé and their wives grew closer. In January 1909, when the Davenports traveled to attend a meeting of the American Breeders' Association in Columbia, Missouri, Harry insisted they travel a few hours north and stay at their Kirksville home. Later that year, however, the Davenports extended their new friends a far more serious invitation. Davenport told Harry he would like him to be the superintendent of his new Eugenics Record Office. For Davenport, his friend not only fit into the temperate, morally upright lifestyle he demanded at the Cold Spring Harbor campus, but this energetic young man also shared a vision of America's greatness and a commitment to the hard work needed to improve the purity of the race.

In October 1910, Harry and Pansy Laughlin packed their belongings and boarded a train for New York. Harry must have felt ambivalent about leaving his mother again, but like his older brothers, he was setting out to achieve something great. This was what it meant to be a Laughlin, after all. His $2,400 salary would not be much more than what he was making now, but money wasn't the issue. The prospect of working at an internationally renowned research station (and working to better the human race!) was an opportunity that left him feeling numb with excitement—and not a little intimidated, truth be told. His job would be to train fieldworkers, organize a vast research project gathering hereditary data on American families, and seek the "positive" and "negative" traits American parents passed on to their children. The purpose, as Secretary of Agriculture William Hayes had once said, was "encouraging the increase of families of good blood, and of discouraging the vicious elements in the cross-bred American civilization."

By the time the Laughlins came to Cold Spring Harbor, there was already a primary focus on the negative side of eugenics. But Harry had his own motives to emphasize in his work the "vicious elements" of civilization rather than the wellborn. Though he and Pansy had been married almost ten years, and though both had come from the hearty American "pioneer stock" so prized by the nation's elites, they were choosing, for their own private reasons, not to have children of their own.

Citizens of the Wrong Type

It was bitterly cold in January 1913, but on the campus at Cold Spring Harbor, as ice crusted the shores of Oyster Bay, the community of scientists and fieldworkers felt they were doing much for the good of the country. Even Harry Laughlin, still beset by bouts of melancholy, felt he had done more than he ever imagined he would during his first two years here on the coast of Long Island Sound. Many of the eminent men whose work he used to read in Kirksville were now reading his work as well. His paper for the Committee to Cut Off the Defective Germ-Plasm had been one of the most widely discussed at the First International Congress of Eugenics last year; and now, from a letter Charles Davenport had just received, he could read how his work was leaving an impression on one of the great leaders in the land: the ever virile former president of the United States, Theodore Roosevelt.

Laughlin knew that the indomitable "Bull Moose," a politician who had long appreciated Davenport's eugenic work, had just lost his third-party bid for an unprecedented third term, and was still recovering from an assassination attempt a month before the November election. While stumping in Milwaukee, he had been shot in the chest by a man named John Schrank, a psychotic New York saloonkeeper who believed he was avenging the assassination of President William McKinley. After the shooting, Roosevelt was laid up in a Chicago hospital, unable to campaign the final, frenetic weeks preceding the election. The saloonkeeper, never charged for the attempt on the president's life, had been committed to a state-run mental hospital in Wisconsin, where he would remain until his death.

Laughlin believed that men—and women—like John Schrank were beginning to infest the country. Hospital treatment or "enlightened" altruism for the psychotic and feebleminded did not quell the rising tide of social misfits. Davenport had been keeping Roosevelt informed about their work at the Eugenics Record Office, sending him case studies and explaining how their social workers were collecting data from mental hospitals, prisons, and almshouses. The data showed a dangerous class of congenital misfits lurking everywhere. Indeed, it was no surprise Schrank was a saloonkeeper, since alcohol, mental defects, and criminal inclinations were, as Davenport's father Amzi used to say, "three sisters that usually kept company." In fact, the ostensible problem of saloons was one of the most pressing political issues in the country at the time, and both Laughlin and Davenport approved of the burgeoning temperance movement, now gaining a strength it never had before. But the real cause of this "sin," they both believed, was defective genes, not moral debauchery per se, and the long-term solution could only be found in a nationwide program of eugenics and better breeding. Roosevelt himself was so interested in this solution that he took the time to write this letter, expressing stronger feelings than either of them had realized:

> My dear Mr. Davenport:
>   I am greatly interested in the two memoirs you have sent me. They are very instructive, and from the standpoint of our country, very ominous. You say that these people are not themselves responsible, that it is society that is responsible. I agree with you if you mean, as I suppose you do, that society has no business to permit degenerates to reproduce their kind. It is really extraordinary that our people refuse to apply to human beings such elementary knowledge as every successful farmer is obliged to apply to his own stock breeding. Any group of farmers who permitted their best stock not to breed, and let all the increase come from the worst stock, would be treated as fit inmates for an asylum. Yet we fail to understand that such conduct is rational compared to the conduct of a nation which permits unlimited breeding from the worst stocks, physically and morally, while it encourages or connives at the cold selfishness or the twisted sentimentality as a result of which the men and women who ought to marry, and if married have large families, remain celebates [*sic*] or have no children or only one or two. Some day we will realize that the prime duty, the inescapable duty, of the <u>good</u> cit-

izen of the right type is to leave his or her blood behind him in the world; and that we have no business to permit the perpetuation of citizens of the wrong type.

Faithfully yours,
Theodore Roosevelt[1]

This was true patriotism. Laughlin and Davenport both saw Roosevelt as a great American, a leader who understood the relationship between good citizenship and racial purity. To institute a more rational process of human breeding—encouraging large families of the right type to breed while preventing the propagation of the worst—was their life's work, after all. And though Woodrow Wilson, a Democrat, would be inaugurated as the nation's twenty-eighth president in March, Davenport and Laughlin, both Republicans, weren't too worried. President-elect Wilson himself had signed a eugenic sterilization bill when he was governor of New Jersey, and had also expressed his support for preventing "citizens of the wrong type" from having children.

But during this bone-cold week in January, they had another reason to feel a sense of accomplishment. Both the Station for Experimental Evolution and the Eugenics Record Office were becoming recognized as two of the most prestigious scientific institutions in the country. State eugenics boards and "better breeding commissions" were springing up around the country, basing their authority on the work being done at the august Victorian house here at Cold Spring Harbor. And it was about to get another shot of national prestige: Davenport had organized a Board of Scientific Directors to be the public face of the Eugenics Record Office, and as the *New York Times* reported, it would include the great inventor Alexander Graham Bell, professors Irving Fischer from Yale, William H. Welch from Johns Hopkins, and E. E. Southard from Harvard, plus a number of others. The board would meet once a month to discuss the work being done to improve the American family.

This week's Sunday *Times* had also just published a full-length feature story on the Eugenics Record Office. On January 12, 1913, in the pull-out science section, a bold headline stated, "Social Problems Have Proven Basis of Heredity," and a sketch of the Victorian manorhouse covered half the front page. The sympathetic journalist tried to explain in detail the history and purpose of the two-year-old institution, the support from Harriman and Rockefeller, as well as the type of work

Lawrence F. Abbott
PRESIDENT
William B. Howland
TREASURER
Karl V.S. Howland
SECRETARY

The Outlook

287 Fourth Avenue
New York

Lyman Abbott
EDITOR IN CHIEF
Hamilton W. Mabie
ASSOCIATE EDITOR

Theodore Roosevelt
CONTRIBUTING EDITOR

January 3rd 1913.

My dear Mr Davenport;

I am greatly interested in the two memoirs you have sent me. They are very instructive, and from the standpoint of our country, very ominous. You say that these people are not themselves responsible, that it is society that is responsible. I agree with you if you mean, as I suppose you do, that society has no business to permit degenerates to reproduce their kind. It is really extraordinary that our people refuse to apply to human beings such elementary knowledge as every successful farmer is obliged to apply to his own stock breeding. Any group of farmers who permitted their best stock not to breed, and let all the increase come from the worst stock, would be treated as fit inmates for an asylum. Yet we fail to understand that such conduct is rational compared to the conduct of a nation which permits unlimited breeding from the worst stocks, physically and morally, while it encourages or connives at the cold selfishness or the twisted sentimentality as a result of which the men and women who ought to marry, and if married have large families, remain celebates or have no children or only one or two. Some day we will realize that the prime duty, *the inescapable duty,* of the good citizen of the right type is to leave his *or her* blood behind him in the world; and that we have no business to perpetuate *or let* citizens of the wrong type.

Faithfully yours,

Theodore Roosevelt

Charles B. Davenport Esq.,
Cold Spring Harbor, L.I.

Former president Theodore Roosevelt sent this letter to Charles Davenport on January 3, 1913, less than two months after losing his Bull Moose ticket bid for a third term as president. The handwritten corrections at the end read: "Some day we will realize that the prime duty, the inescapable duty, of the *good* citizen of the right type is to leave his or her blood behind him in the world; and that we have no business to permit the perpetuation of citizens of the wrong type."

conducted by fieldworkers. The subheading promised to explain "what the work done in the Eugenics Record Office at Cold Spring Harbor has proved in scientific race investigation." Indeed, as a host of social thinkers used the science of biology to explain complex human behavior, they also used concepts such as "race investigation" and "racial purity." The *Times* article featured an intimate conversation with Charles and Gertrude Davenport, though it didn't mention the superintendent even once. Harry Laughlin might not be great yet—he was only thirty-two and didn't even have a Ph.D.—but he certainly felt he was doing good.

Lurking in the subtext of the *Times* story, however, was a certain public unease with the science of eugenics. Though so many of the nation's educated elites were giving their enthusiastic support, the public still seemed to be worried by nonscientific questions. Would eugenicists do away with love? Were they going to sterilize people willy-nilly? Were they planning to start a farm of human freaks to analyze and study? The idea that biology determined behavior was troubling for many people, really, and the eugenic logic, which used the analogy of breeding better plants and animals to call for the genetic engineering of the human race, seemed to mock human dignity and threaten the integrity of the freedom of the human will.

In many ways, the article simply sought to alleviate these fears. The story mostly quoted Gertrude, who seemed to dominate the interview with her pugnacious convictions and stubbornness—traits her husband had noted in their own Family Records form. As she laughed at a question about farm breeding, she explained:

> *The work of eugenics is not to attempt to "breed people like cattle" as some persons have surmised; it is to strengthen the race by cutting off weakness; to increase the power of what is good by the inhibition of what is bad. To urge the relief of human blood lines of what is worst does not promise the remaking of the entire human race and the cure of all human ills; but neither does it demand that individuality and personal tastes are to be set at naught in the attempted scientific production of a "super-race." In the marriage of the future, as the eugenists are working for it, love and "eugenic principles" will go hand in hand for happier homes, healthier children, and the minimization of imbecility, hereditary disease, pauperism, and crime.*

Perhaps because of the work Laughlin had been doing on sterilization, even Charles now supported the new laws being passed in a number of states, and was coming to believe the surgical procedure just might be the most efficient and effective method to achieve eugenic goals. Gertrude lamented how states like New Jersey and Illinois were not taking full advantage of their laws. "Of course, sterilization is a grave responsibility and a new thing: people hesitate about it." But for Charles, it was the economic efficiency of the procedure that now made sterilization an appealing solution. He had always called for the forced segregation of the unfit, but this would put a huge burden on taxpayers:

> *Many an American citizen inquires whether a system by which philanthropists drain effective persons of their income until they cannot afford to have children, in order to secure funds to be spent in relieving imbecile parents of any expense of parentage, is a good thing for America.*
>
> *What, then, is the alternative? Let the weak die? Not at all. Our modern Christian civilization would not approve of that. Shelter and feed the poor, give happiness to the feeble-minded, protect the insane, cure the tuberculous and the cancerous, but, as you value humanity, keep them from reproducing their kind![2]*

Happier homes, healthier children, and the minimization of the unfit were obsessive themes at Cold Spring Harbor. Indeed, as Charles's father Amzi used to say at morning prayer, "All desire happiness." His son, using a utilitarian calculus for happiness, sought to maximize it through the state control of human reproduction.

Two years earlier, on January 8, 1911, Charles and Gertrude again knew the joys of being parents when their first son, Charles Davenport, Jr., was born. Their oldest daughter, Millia, was sixteen and Jane already fourteen, so the pregnancy came as something of a surprise.

Little Charlie Junior was a curious, talkative, and precocious boy, and by the time he was two, he was quite a character on campus. Social workers and students doted on him, and he learned to speak at a very young age. After little Charlie turned three, Davenport wrote to Jane, who was away in France, "Sonny is growing in every way. He has a keen sense of sound in words. I called Duke [the family dog]—Charlie said, 'Do you say Douke? I say Dook.' " When his son was five, Davenport was helping little Charlie with a magazine called *The Boy.* It was similar to *The Twinkling Star,* his own childhood broadsheet, which his father had once helped him publish.

Though not a very emotional man, Charles was often overcome by pride in his son. He even wrote him letters whenever he was away,

Charles Davenport with Millia (*right*), Jane (*left*), Gertrude, and Charlie Junior, 1914.

using big, block script for the words, so the three-year-old could read them. The tenderness he showed was very different from what Amzi had shown him. Once, just after Charles Junior's third birthday, when Gertrude had taken him to Kansas to visit her mother, Davenport wrote: "My dear son, Take good care of mother. Help Grandma all you can. Write me a little note now and then to tell me how you are. Have you begun on 'The Boy'? Will you not get one out every week? Good bye, dear son. Your loving father, Chas. B. Davenport."[3]

This spirit of "happier homes and healthier children" at Cold Spring Harbor sometimes soothed Harry's melancholy. Even though the Laughlins themselves did not have children after more than ten years of marriage, Harry and Pansy found that the rhythms of the sprawling campus suited them well, and they became eager participants in this small community, suffused with a particular ethos of American family life. They started a Eugenics Record Office Drama Club, and produced public plays for the neighboring Long Island towns. The purpose of the Drama Club, however, was not simply to entertain but to spread the gospel of eugenics. Pansy wrote a play called *Acquired or Inherited? A Eugenical Comedy in Four Acts.* It was about a hilarious web of infatuations, featuring the characters David Reed, a bachelor interested in eugenics; Hannah Perkins, his flirtatious, feebleminded housekeeper; Jean Reed, his niece; and a butler named Sam. When a group of guests come to their house for dinner, the air is charged with people falling in love. These include a happy-go-lucky and greedy highchair peddler named Felix Rosenfeld; a eugenics fieldworker named Eugenia Traveler; a bachelor maid of forty-five, Sophronia Burton; an indolent but wealthy merchant, Jerry Dunbridge; and finally an ambitious electrician, Lester Gordon. Like a lesson from the book of Proverbs, this morality tale warned of the wiles of a wanton woman and the dangers of bad breeding. The denouement was a message on proper eugenic coupling. With its undisguised anti-Semitism (the greedy Jewish peddler was not a healthy match, it turned out), the play became quite popular in the community, and it played for a number of years.[4]

Yet, for all the laughter and familial love at Cold Spring Harbor, the real obsession was science. Its buildings were abuzz with the work to produce the types of family "memoirs" President Roosevelt had found so ominous. Last year, at the First International Congress of Eugenics in London, Major Darwin had predicted that the most important goal of eugenicists would be to stamp out the perpetuation of feebleminded citizens, and Davenport and Laughlin were now bringing to this goal their own type of religious zeal. The same sentimental, romantic longings that had infused their writings were now defining their research. And their first objective—in addition to training a vast army of social workers—was to locate those citizens of the wrong type, whom they believed, along with Roosevelt, society had no business with perpetuating.

THE GOALS OF THE Eugenics Record Office were stunningly ambitious. From the beginning, Laughlin tried to contact every single public institution concerned with insane, feebleminded, epileptic, criminal, incorrigible, or blind individuals throughout the forty-eight states. He wrote hundreds of letters to these institutions, explaining how the Eugenics Record Office would like to train social workers to observe their patients, then go out into their "home territories" and collect a vast database of family histories. These would then be used to determine the lines of defective "germ-plasm" in the American population. In the first year, Davenport and Laughlin planned to focus specifically on the Mendelian patterns of two diseases: tuberculosis and cancer. On the positive side, they also wanted to find the pattern of two talents: mathematical and musical ability. But in the long term, as they cooperated with state institutions, they wanted nothing less than to seek out every single "unfit" family in the country.[5]

Responses came pouring in. Most institutions apologized to Laughlin about the dearth of funding and the difficulty of convincing their state legislatures to pay for the social workers' training. Even so, most expressed enthusiasm about the proposal, and tried to collaborate as much as possible in the research. Many agreed to take Davenport's Family Records forms—first renamed "Index to the Germ-Plasm" and later "Record of Family Traits"—and fill them out as best they could. A few others agreed to take on social workers and procure the funds to help subsidize their work.

The social workers were mainly young, idealistic women. The first class of six trainees included five females, and all but one were single. Indeed, over the next few years, 131 of 156 social workers trained at the Eugenics Record Office were young women. Davenport and Laughlin even made a point to contact colleges for women in the Northeast and advertise the career potential of studying at Cold Spring Harbor. It was a chance to be a part of a new professional class. Rather than languish in the typical, dead-end jobs for college-educated women—usually typist, clerk, or secretary—they could do important work to solve the nation's social ills. Since the time of Galton, many modern-thinking women had been drawn to eugenics and its promise to reform the mores of marriage and family life, so pursuing a career as a eugenic social worker was one of the most interesting options for a young woman who wanted more in life than acting as "ideal helpmate" to a man.

Davenport and Laughlin developed a summer training program that lasted about a month and a half. The two men gave classroom lectures on Darwin's theory of evolution, Mendel's genetics, anthropometry, endocrinology, types of insanity, and other relevant topics. But they wanted to emphasize the practical work of research, so they arranged weekly field trips to various mental institutions in the area, including King's Park Clinic, the Amityville Hospital for the Insane, and the Letchworth Village for Feeble-minded. Students were to observe the patients and take notes; practice filling out the Family Records blanks; and prepare for a final research project, simulating the work they would later do in the field.[6]

Though the field plan was simple, the actual work would be arduous. First, researchers would locate an individual in an institution, and carefully note his or her particular illnesses. Most of the time, they were to focus on a specific trait, such as feeblemindedness, tuberculosis, or cancer. Then they would travel. They would have to go out and find the family members of that individual, interview them in person, and note their traits as well. Whenever possible, they would also ask neighbors to describe the traits of the case family. Compiling as complete a family tree as possible, they would send their research back to the Eugenics Record Office, where Davenport and Laughlin would analyze it and determine the family's genetic pattern, according to the theory of Mendelian "unit characters."

Before they were placed in the field, social workers received a set of directions: a checklist with nine instructions. These explained such mundane details as which expenses would be paid, when their appointments would begin, as well as the logistics they should follow when compiling data. The list instructed workers to mail a postcard to the office every morning, explaining the places they intended to visit and listing the families they had visited the preceding day. They were also to send a progress report every week, so Davenport or Laughlin could monitor their work and offer suggestions. But the checklist also warned the workers to be discreet, advising them how to approach the poverty-stricken families they would find.

> *The details of a visitor's work at any house are determined by circumstances. Certain general rules may be laid down. First of all, unfailing courtesy and regard for and sympathetic (humanistic) attitude toward the persons you are interviewing are essential from every point of view. . . . To get the truth requires great tact. . . . Seek to be invited to call again, and accept the invitation if possibly useful. It is better to get a few correct pedigrees than many that are fragmentary or full of falsehoods.*

They were, as Jesus once told his disciples, to be "wise as serpents and innocent as a dove." The people they were observing were a menace, after all; they were sapping the country's strength. Courtesy and tact were essential if they were going to gather the proof they needed to weed out this degenerate stock of Americans, and thoroughness was essential when discovering the most hard-to-find unfit. The checklist also reminded them to consult town records or any church genealogies. But the last point offered a final word of common sense: "In conversation with persons seek to collect facts concerning other family traits than that which you are at the time studying. Be on the lookout, all the time, for interesting families."[7]

The first class of social workers was sent out to institutions in Massachusetts, New Jersey, New York, and even as far as Chicago. One of the brightest and hardest-working in this group was a Miss Florence H. Danielson, a recent graduate of Brown University. She was first assigned to study epileptics at Monson Hospital in Palmer, Massachusetts, but she was soon traveling throughout New England, tracking down the family members of inmates. Epilepsy—which was closely correlated with feeblemindedness—was believed to be a telltale sign of the types of people who caused poverty and crime, so Miss Danielson was busy writing to prisons and almshouses, looking for a pattern of degeneracy in the families of epileptics. Her tireless efforts to track down degenerate families yielded the first published memoir to come out of the Eugenics Record Office, a thirty-page pamphlet entitled *The Hill Folk: Report on a Rural Community of Hereditary Defectives.*

Another project from the first class of social workers was a follow-up

ABOVE: Workers at the Eugenics Record Office, 1921.

BELOW AND OPPOSITE PAGE: Eugenics fieldworker training class, in which over nine out of ten trainees were young, idealistic women seeking to improve society; they trained social workers to go out to prisons, mental institutions, and rural hamlets to find "defective" bloodlines. Many classes were held outdoors. Here, the 1916 fieldworkers pose for a group portrait, and members of the 1918 class listen to Harry Laughlin explain the "coefficient of contingency."

to Dugdale's famous and influential study, *The Jukes*. The project was given to the only male fieldworker in the first class, Dr. Arthur H. Estabrook, a Ph.D. in psychology from Johns Hopkins University and a man who would work with the Eugenics Record Office for over a decade. Unlike Miss Danielson, who sent the main office handwritten lists of concrete statistics, Estabrook sent hundreds of pages of single-spaced, typewritten field notes filled with dramatic narratives. Indeed, during the course of his ten-year affiliation with the Record Office, he would send thousands of these pages, which described the history, geography, and towns of a region, as well as the "interesting" families he found. Estabrook's fieldwork eventually yielded two books: *The Nam Family, A Study in Cacogenics,* published in 1912, and *The Jukes in 1915,* published during World War I. But instead of offering concrete, mea-surable data, these books—as well as the tens of thousands of field notes coming into the Eugenics Record Office—were filled with subjective and impressionistic musings. Though the field notes did include the "Record of Family Traits" forms, these, too, were often based on inter-views and impressions, rather than verifiable facts.

The work of the Eugenics Record Office, though clumsy and often ill-defined, was one of the first organized attempts in the history of science to trace the family history of illness. The intuition to seek sex-linked traits was also an innovative idea that, when formulated in a more con-trolled way, would revolutionize the study of medicine. But the same problem that had plagued Galton, the difficulty of defining and then identifying a trait like intelligence, continued in eugenic research.

A promising solution was about to sweep across the country. Since opening the Eugenics Record Office, Davenport had been in close contact with Henry Goddard, the head of the research program at the New Jersey Training School for Feeble-minded Girls and Boys at Vineland. Like Davenport and Laughlin, Goddard was conducting a study of defective families in the tradition of *The Jukes*. But he used a different research technique, one that required far fewer fieldworkers and far less effort. When he had returned from studying in France a few years earlier, he brought back an astonishing new tool: the mental test.

When discussing the possibility of a eugenics fieldworker conference at Cold Spring Harbor, Goddard told Davenport about the French tests he had translated:

> *It has occurred to me that perhaps it would be worth while for me to give a little account of these Binet Tests, inasmuch as it is rather important even in other lines, for the workers to be able to detect mental defect whenever they find it, and we are becoming every day more convinced that these tests are amazingly accurate and they would be very easily applied by any field worker without anybody realizing that they were being tested. If you think this would be worth a few minutes discussion at the meeting I will be glad to present it.*[8]

The mental tests were astounding, and they did seem to be remarkably accurate. From the start, Francis Galton had tried to measure the elusive trait of intelligence, and determine the Gaussian distribution (later called the bell curve) of mental ability in a given population. And as Goddard hinted in his letter, social workers could be cunning in administering them, hiding their true motives. Davenport and Laughlin immediately recognized their utility, and began to incorporate mental testing into their own eugenic curriculum.

The advent of mental tests would transform the science of eugenics. First born from this quest for better breeding and the need for more reliable data, the idea of intelligence testing would later change the course of American history, reshaping ideas of equality, merit, and heredity, as well as revolutionizing access to higher education.[9] These tests seemed to reify an elusive human trait and allow it to be quantified and measured. They then could place everyone where they belonged on a great curve of intelligence. The first accomplishment of mental testing, however, was Goddard's rhetorically brilliant memoir of a Revolutionary War hero who was said to have fathered two families. One family was eminent and intelligent, and the other was defective and dull.

HENRY HERBERT GODDARD did not become famous because he was the first football coach at the University of Southern California (USC). He was a Quaker from Maine and a graduate of Haverford College, the small Pennyvania school established by the Society of Friends. He was young, athletic, and smart, but after graduating in 1887, he traveled west to join the nascent faculty at this new university in Los Angeles, taking a position to teach Latin, history, and botany. USC was just a small institution with a few hundred students at the time—far from the gridiron powerhouse it would later become—so in a few years, the football coach went back east, earning a Ph.D. in psychology from Clark University in 1899. Like Laughlin, Goddard then taught at a state normal school in West Chester, Pennsylvania. Soon he began a new career as a researcher, devoting his life to the problems of the "feeble-minded."

In 1906, Goddard was appointed head of the research department at the Training School for Feeble-minded Girls and Boys in Vineland, New Jersey. Here, he attempted to develop new ways to test the youth, identifying those who could learn and those who could not. This eventually prompted him to travel to France to learn about recent innovations in child psychology. In Paris he met Théodore Simon and Alfred Binet and observed their methods for ranking the intelligence of children. In 1908, Goddard brought these mental tests to the United States and translated them into English.

With this new research tool, Goddard tackled the problem of heredity—which had not been the aim of Simon and Binet. The tests yielded a certain "mental age," independent of actual age, so he tried to test both his students and members of their families to see if there were correlations in their relative mental abilities. It seemed there were. In fact, he seemed to find an entire class of adults who tested at a mental age of around ten to twelve years old. With these new tests, Goddard believed he had discovered a way to positively identify the long-elusive feebleminded, the "high-grade" defectives he dubbed "morons."

In February 1913, Goddard conducted one of the first mass aptitude tests in the history of education. A Harvard professor named Paul H.

Hanus had been hired by New York City's Committee on School Inquiry to evaluate the public school system, which seemed to be riddled with problems. Professor Hanus in turn hired a number of experts, including Goddard. For the first time, thousands of students in New York City's public schools sat down to take the Simon-Binet mental test. On February 8, 1913, the *New York Times* announced that Goddard had found that over fifteen thousand pupils were in fact feeble-minded.[10]

During his research at the Vineland Training School, Goddard had come upon an unusually striking story—and plenty of data—about a feebleminded family spanning six generations, just like the Jukes, and living in the Piney Woods of New Jersey. In 1912, he published *The Kallikak Family: A Study in the Heredity of Feeble-mindedness.*

The story was simple and compelling. Martin Kallikak was an upstanding Quaker in New Jersey. He fought in the Revolutionary War against the British, breaking with his family, who were believers in nonviolence. During his days in the Continental Army, he met a young girl who lived in Piney Woods, a backwoods settlement in central New Jersey. A brief tryst produced a son. After the war, Martin settled on a large, wealthy farm less than twenty miles away from Piney Woods, and there he married a respectable young woman with whom he started a legitimate line of offspring. Over the decades, this family produced eminent judges, professors, and other public servants, and became a famous and elite New Jersey clan. The illegitimate line, however, produced notorious drunkards and misfits.

Goddard dubbed this family "the Kallikaks," from the Greek *kalos* (good) and *kakos* (bad). He told their story in a short, 120-page book, written in a simple, dramatic tone. Unlike the Cold Spring Harbor memoirs, he included dozens of photographs comparing the dark, brooding misfits with their wealthy, upstanding cousins. The stark contrast made a devastating impression on readers throughout the country. Whereas *The Jukes* had caused a sensation among professionals, *The Kallikak Family* was written for a lay audience, and it soon became a brisk-selling popular sensation. His publisher, Macmillan, would reprint the book in 1913, 1914, 1916, and 1919, and it would become a classic—and notorious—work of modern psychology.

Like the Eugenics Record Office memoirs, the story began with an institutionalized individual. "Deborah Kallikak," a twenty-two-year-

old woman from this family, had been an inmate at Vineland for fourteen years, and was a perfect case study in the inheritance of feeble-
mindedness. She seemed normal. A copy of her case file noted that she
was "Graceful. Good in drill. Can copy. Knows a number of words. . . .
Good in entertainment work. Memorizes quickly. Can always be relied
upon for either speaking or singing. Marches well. A good captain.
Knows 'Halt,' 'Right,' and 'Left Face' and 'Forward March.' Always in
step. Knows different notes. Plays 'Jesus, Lover of My Soul' nicely.
Plays scale of C and F on cornet." But she also had a hard time paying
attention in class, and was unusually affectionate and flirtatious.

When Deborah was given the Simon-Binet mental tests, she consistently tested at the level of a nine-year-old child. She could answer cor-
rectly all questions up to age seven, but in the reading sections for age
eight and nine, she could not read the selection in the required time,
nor did she seem to remember what she had read. Her definitions of
abstract terms were judged very poor, and she rhymed "storm" with
"spring," and "milk" with "mill," afterwards using "bill," "will," and
"till." Goddard described Deborah's background and gave similar anec-
dotal accounts of her "defective" relatives. But then he also contrasted
these family members with those in the "good" side of the Kallikak
family, showing how they were as upstanding as those in the feeble-
minded side were "bad."

But the most rhetorically powerful section of the book, perhaps, was
the ringing call for reform in the final chapter, entitled "What Is to Be
Done?":

No one interested in the progress of civilization can contemplate the facts presented
in the previous chapters without having the question arise, Why isn't something
done about this? It will be more to the point if we put the question, Why do we
not do something about it? We are thus face to face with the problem in a practical way and we ask ourselves the next question, What can we do? For the low-
grade idiot, the loathsome unfortunate that may be seen in our institutions, some
have proposed the lethal chamber. But humanity is steadily tending away from
the possibility of that method, and there is no probability that it will ever be
practiced. . . .

It is the moron type that makes for us our great problem. And when we face the
question, "What is to be done with them—with such people as make up a large
proportion of the bad side of the Kallikak family?" we realize that we have a
huge problem.

Photographs of Deborah Kallikak, the primary case study in Henry Goddard's influential and popular book, *The Kallikak Family: A Study in the Heredity of Feeble-mindedness*. Deborah was deemed a high-grade "moron," a word invented by Goddard to describe the dangerous class of misfits he claimed was polluting the nation's genetic pool.

The career of Martin Kallikak Sr. is a powerful sermon against sowing wild oats. Martin Kallikak did what unfortunately many a young man like him has done before and since, and which, still more unfortunately, society has too often winked at, as being merely a side step in accordance with a natural instinct, bearing no serious results. It is quite possible that Martin Kallikak himself never gave any serious thought to his act, or if he did, it may have been merely to realize that in his youth he had been indiscreet and had done that for which he was sorry. And being sorry he may have thought it was atoned for, as he never suffered from it any serious consequences.

Even the people of his generation, however much they may have known about the circumstances, could not have begun to realize the evil that had been done. Undoubtedly, it was only looked upon as a sin because it was a violation of the moral law. The real sin of peopling the world with a race of defective degenerates who would probably commit his sin a thousand times over, was doubtless not perceived or realized. It is only after the lapse of six generations that we are able to look back, count up and see the havoc that was wrought by that one thoughtless act.

Not only was this a powerful argument that appealed to traditional mores, it also served to buttress them with a *scientific* rationale. If there were good, scientific reasons to refrain from "sowing wild oats," even apart from traditional moral strictures, the authority of the Bible might have been right all along, even if it wasn't based on a sure foundation. But in many ways, the degenerate, the poor, and the misfits of society were seen as the wages of sin, an "original sin" in the blood. Moral purity, then, must arise out of biology. The "lethal chamber," he seemed to say with a tone of regret, was not an option—at least not at the moment. So, like an evangelist at a tent meeting, Goddard could proclaim the danger of sinners in the hand of an angry God, and what needed to be done for salvation:

Now that the facts are known, let the lesson be learned; let the sermons be preached; let it be impressed upon our young men of good family that they dare not step aside for even a moment. Let all possible use be made of these facts, and something will be accomplished.

But even so the real problem will not be solved. Had Martin Kallikak remained in the paths of virtue, there still remained the nameless feeble-minded girl, and there were other people, other young men, perhaps not of as good a family as Martin, perhaps feeble-minded like herself, capable of the same act and without Martin's respectability, so that the race would have come down even worse if possible than it was, because of having a worse father.

Others will look at the chart and say, "The difficulty began with the nameless feeble-minded girl; had she been taken care of, all of this trouble would have been avoided." This is largely true. Although feeble-mindedness came into this family from other sources in two generations at least, yet nevertheless these sources were other feeble-minded persons. When we conclude that had the nameless girl been segregated in an institution, this defective family would not have existed, we of course do not mean that one single act of precaution, in that case, would have solved the problem, but we mean that all such cases, male and female, must be taken care of, before their propagation will cease. The instant we grasp this thought, we realize that we are facing a problem that presents two great difficulties; in the first place the difficulty of knowing who are the feeble-minded people; and, secondly, the difficulty of taking care of them when they are known.

With this alarming problem in the nation's midst, a problem that even virtue could not contain, Goddard explored the possible solutions. Decades earlier, Galton had proposed segregating the unworthy in monastery-like institutions as a means to keep them from reproducing. This would be desirable, and was already being done, Goddard wrote, but the costs to taxpayers were enormous. How could good, upstanding parents rear large families if they had to support the hordes of feeble-minded? He then mentioned the possibilities of castration, but again, in a tone of regret, explained that society would consider this too brutal a solution. But the argument was leading the reader to the most logical choice today: surgical sterilization.

In recent years surgeons have discovered another method which has many advantages. This is also sometimes incorrectly referred to as asexualization. It is more properly spoken of as sterilization, the distinction being that it does not have any effect on the sex qualities of the man or woman, but does artificially take away the power of procreation by rendering the person sterile. The operation itself is almost as simple in males as having a tooth pulled. In females it is not much more serious. The results are generally permanent and sure.[11]

The Kallikak Family caused as much a sensation with the general public as the Cold Spring Harbor memoirs had with President Roosevelt. Popular periodicals such as *The Dial* and *Popular Science Monthly* praised it as "a remarkable human document . . . with an endless patience sustained by a scientific insight into the value of principle and detail." The *American Journal of Psychology* hailed the book as a "find" and Goddard as having "the training which enables him to utilize the discovery to the utmost."[12] As a best-selling book on science, the simple, dramatic prose of *The Kallikak Family* also appealed to sentiments in both conservative and progressive political types, setting out a fungible logic in its contrast between "good" and "bad." Conservatives could embrace the emphasis on traditional morals and the need for sexual purity, and progressives could embrace the promise of a future free from social ills. If mankind were to take charge of its own evolution, however, and if Americans were to control their God-given manifest destiny, then the unfit and their defective genes would have to be eliminated.

But sterilization, as Gertrude Davenport had told the Sunday *Times,* was a new thing. Though it was practiced in a number of states now, very few really understood the legal and technical aspects of the controversial procedure. There had been no systematic studies or scientific analyses of the nascent programs, or on the benefits of sterilizing those who caused such problems.

Harry Laughlin knew this. He had only been an administrator during his young career, surrounded by world-renowned biologists, people with Ph D.'s. But as he did more research on the procedure, he was finding it to be an area in which he could carve a niche for himself. Somehow, sterilization suited his beliefs and longings, while calming the storm of self-doubt and sometimes even self-loathing that raged inside him.

IN THE SECOND WEEK of January 1914, Laughlin was getting ready for an important moment in his life. He had been invited to speak at the First National Conference on Race Betterment in Battle Creek, Michigan, a meeting of the nation's leading doctors, geneticists, and eugenicists, organized by Dr. John H. Kellogg, a member of the American Breeders' Association and the superintendent of the Battle Creek Sanitorium. Along with his brother William, Dr. Kellogg had made the sanitorium a famous though somewhat eccentric health spa, emphasizing exercise, intensive health education, and a diet of a new type of food, corn flakes. Over four hundred official delegates would participate in one hundred twenty sessions, and since the conference was open to the public, some of these presentations would be made before audiences of up to two thousand professionals. Thronging the spa were doctors, ministers, and educators, as well as representatives from women's clubs, the Young Men's Christian Association, and the

Women's Christian Temperance Union. While there, they enjoyed eugenic exhibits, movies, and banquets every evening. In addition, a number of Battle Creek schoolchildren participated in "Physical and Mental Perfection Contests," competing for awards to be given on the final day. Laughlin was nervous, but as he traveled with Davenport to Michigan, he knew this was the type of event his mother would have attended.[13]

The conference included a host of topics devoted to "race betterment," and its motto was a quote from Herbert Spencer: "To be a good animal is the first requisite to success in life, and to be a Nation of good animals is the first condition of national prosperity." In spite of the reductionist words of the great social Darwinist, each session opened with a word of prayer, and concluded with a time for discussion and debate. The first was on statistical studies; presentations included "Differential Fecundity" and "The Need of Thorough Birth Registration for Race Betterment." Another section was devoted to alcohol and tobacco—two vices common among the unfit. In the sessions on "sex questions," however, men and women were carefully segregated for modesty's sake. Women heard a presentation by Dr. Kellogg on "The Social Evil," and men listened to one on "Venereal Disease." Laughlin was scheduled to speak on the final day of the conference, in the section on "Eugenics and Immigration."

The conference included an eclectic mix of thinkers, and though the vast majority were "hereditarians" and staunch believers in eugenics, there were also a number of "environmentalist" critics. The muckraking journalist and revolutionary photographer Jacob Riis, whose dramatic photographs documented the bleak conditions of the tenements in New York City, gave a presentation on "The Bad Boy." He told his audience, "We have heard friends here talk about heredity. The word has run in my ears until I am sick of it. There is just one heredity in all the world that is ours—we are all children of God, and there is nothing in the whole big world that we can do in His service without it." If most eugenicists based their assumptions on the philosophical pessimism of a Thomas Hobbes or Herbert Spencer, Riis had a more romantic and Rousseauian bent. "There are, dear friends, not any who are deliberately bad, but plenty whom we make bad."[14] The social thinker Booker T. Washington also warned that hereditary data could be misinterpreted, and he argued that American Negroes "are worth saving, are worth making a strong, helpful part of the American body politic." The physiologist Herbert Adolphus Miller, a professor from Olivet College in Michigan, complained about the current "vogue" for eugenics, and the "cocksureness" and "unjustifiable conclusions" that characterized eugenicists' claims. Social classes were not born, he argued, they were made. Like Riis, he believed bad environments led to poverty and crime. "The social problems have nothing whatever to do with biological inheritance."[15]

These represented the minority view, however, and they were drowned out in a welter of statistics and prophetic warnings of doom from biologists from more prestigious universities and organizations. On January 12, Laughlin presented his own solution in a speech entitled "Calculations on the Working Out of a Proposed Program of Sterilization." The presentation was the final section of the American Breeders' Association's "Report for the Committee to Study and Report on the Best Practical Means of Cutting Off the Defective Germ-Plasm in the American Population," which the Eugenics Record Office was just about to publish under Laughlin's name. For two years, Laughlin had been carefully gathering information on every sterilization law in the United States, analyzing the legislative history and legal issues involved, and describing the institutional procedures in each state's program. But Laughlin also presented a plan for a sweeping, more centralized national program.

After Indiana passed the first sterilization law in 1907, eleven other states soon followed its example. But these varied in scope and purpose. The laws in Indiana, Iowa, Kansas, Michigan, New Jersey, New York, and Wisconsin were passed strictly for eugenic reasons—that is, the legislatures were attempting to "better the race," and not to treat or punish the individual patients being sterilized. California and North Dakota also passed laws for eugenic reasons, but the bills included sections for "therapeutic" sterilizations, which were supposed to help alleviate individual "health problems" such as masturbation. In Washington and Nevada, however, sterilization laws were passed for strictly punitive purposes, and focused only on punishing criminals and "sex perverts."

For Laughlin, each of these laws represented a start, but they did not come close to the far-reaching program needed to accomplish a significant reduction of the feebleminded. A standardized national education program, the legal restriction of marriage, as well as the forced segregation of the socially inadequate were essential cogs in the wheel of better breeding, he explained, but compulsory sterilization was by far the most essential element in wiping out the menace of biological "degenerates." In his opening remarks, Laughlin told his audience:

> *To purify the breeding stock of the race at all costs is the slogan of eugenics.* . . . *It is at once evident that, unless this complementary agency, compulsory sterilization of certain degenerates, is made nation-wide in its application, and is consistently followed by most of the states, it cannot greatly reduce, with the ultimate end of practically cutting off the great mass of defectiveness now endangering the conservation of our best human stock, and consequently menacing our national efficiency and happiness.*[16]

Efficiency and happiness were major themes in Laughlin's argument, but for him, happiness seemed to mean the absence of all physical and mental flaws. If science could somehow eliminate these flaws, he seemed to reason, the country could be "pure" and thus "happy." He went on to cite a bevy of statistics, explaining the extent of the "defective germ-plasm" in the American population, and emphasizing the threat it posed. In a roll call of the types of people carrying defective genes, he included the feebleminded, families who had been paupers in successive generations, "criminaloids," epileptics, the hereditarily insane, the "constitutionally weak," everyone predisposed to specific diseases, everyone congenitally deformed, as well as everyone who was deaf, a deaf-mute, or blind.

These were a lot of people to be sterilized; but Laughlin remained undaunted by the task. In fact, he claimed there were even more carriers of defective germ-plasm, "a group of persons who, though themselves normal, constitute a breeding stock which continually produces defectives; they are so interwoven in kinship with the lower levels that they are totally unfitted for parenthood." Indeed, his audience might have been shocked by the scope of his plan: "The portion of the population which it is sought to cut off is the lowest ten percent of the human stock who are so meagerly endowed by Nature that their perpetuation would constitute a social menace."[17]

Ten percent. To cut off the genes of 10 percent of the population— almost 10 million people at the time—would take a while, of course, and Laughlin tried to analyze the variables that would affect the "rate of efficiency" of his proposal. Given the differential fecundity of defectives, the vehement public opposition to the procedure in many states, as well as the constant influx of defective immigrants, it could be decades before America could significantly reduce the percentage of defective citizens. But according to his plan, sterilization would first begin in state-run institutions—which contained less than 1 percent of the population at this time—and then be gradually applied to those in the general population found to be feebleminded. The initial rate of progress would also depend on the number of new institutions built for the "socially inadequate." But if all went according to his projections, Laughlin thought the number of defectives in the United States could be cut from 10 percent to 2.77 percent by 1970, and 1.32 percent by 1985.

But there was a problem with the current state sterilization programs, he said. Sterilization—and castration before it—had mostly focused on males. The reasons for this were understandable. On the one hand, vasectomies were relatively simple procedures, while salpingectomies were a bit more dangerous, intrusive, and complex There had also been the traditional assumption that female sexuality was passive and demure, while that of males was aggressive and bold. But since eugenic research had revealed both the "hyper-fecundity" and the sexual aggressiveness of the feebleminded female, this prejudice had to change, he explained. Besides, if the goal was to reduce the number of children born, not only should more women be sterilized, they should actually be the *primary focus* of eugenic efforts. Laughlin gave a simple analogy to explain the inefficiency of sterilizing only men:

> . . . [In] the case of domestic animals of less value, having mongrel and homeless strains, such as the dog and the cat, the cutting off of their supply is largely effected through the destruction or the unsexing of the females. As a rule the tax on a female dog is two or three times greater than that on a male dog. Such a difference in taxation is not made because of difference in individual menace, but rather because of a more direct responsibility for reproduction. The females of such homeless strains are not protected, and consequently they increase very rapidly. Consorting with equally worthless mates, their progeny are often excessive in numbers, and of worthless, mongrel sort. The castration of one-half of the mongrel male dogs would not effect a substantial reduction in the number of mongrel pups born.[18]

Laughlin's startling analogy reflected a curious callousness in his tone toward those he deemed imperfect and impure. The comparison to animal breeding had always been the first step in the eugenic argu ment, but Laughlin, for some reason, brought a restrained and subtle bitterness to the problem of the unfit. Throughout his presentation, he relentlessly used the word "worthless," while couching it in a tone of detached scientific analysis. Yet the implicit moral urgency of his words seemed to indicate a deep personal investment in the issue. And though his plan might have seemed shocking and extreme to many people, he himself called it a "conservative" and "humane" program, given the gravity of the problem:

> . . . Unless the people of the several states are willing to attack the problem in its entirety, they cannot hope to find in sterilization, as complementing segregation, anything more than a slight palliative for the present condition from which we are seeking relief. A halfway measure will never strike deeply at the roots of the evil. In animal breeding, when any great results are wrought, or when new and superior breeds are made within a few generations, it is the selected one per cent or at most the tenth part that are selected for reproduction rather than the ninety odd per cent which this conservative program calls for.[19]

Near the end, Laughlin sounded a note of caution. Sterilization, if applied only to inmates of state institutions, might be seen to violate the Fourteenth Amendment's guarantee of equal protection of the law and constitute "class legislation." He expected there would be more lit igation, but he was confident his program could withstand judicial review, even from the Supreme Court. But this possible constitutional obstacle would only apply to the initial phase of his plan. If struck down, they must simply apply the law to everyone immediately. Years later, in his first full-length book on eugenic sterilization, Laughlin would also argue that the procedure's legality was no different from compulsory vaccination, which was becoming the rule in most states. The law, he wrote, operated to protect the welfare of society, and the rights of an individual often had to bow to this need.[20] It was an argu ment Oliver Wendell Holmes, Jr., would find compelling when he wrote the decision for *Buck v. Bell*.

The First National Conference on Race Betterment was a debut for Harry Laughlin. He made a number of contracts there, including the Harvard professor Robert DeCourcy Ward, who was very impressed by Laughlin's paper. Professor Ward was a specialist in immigration issues, and in his own presentation on "Race Betterment and Our Immigration

Harry H. Laughlin, superintendent of the Eugenics Record Office, Cold Spring Harbor; president, American Eugenics Society, 1928–29; ca. 1929.

Laws" had argued that inferior Southern European races must be kept from coming into the country. Laughlin and Ward became professional acquaintances, and the relationship eventually helped the young, upcoming eugenicist to become a specialist in immigration as well. For Laughlin, the conference proved a great success.

The religious zeal of Amzi Davenport and Deborah Laughlin hung over both sons' quest for racial purity, like an ancestor's painting in a Victorian hall. When they were young, they might have never known how much they would have in common with the dominant parents in their lives, how deeply they would share their commitment to temperance, moral purity, American greatness, and, yes, even religion.

As Laughlin and Davenport traveled back to Cold Spring Harbor, they could have suspected the eugenics movement was about to enter a new phase. Public opposition to both eugenics and evolution was beginning to galvanize in certain states, and it was being led mostly by traditional Catholics and a new, anti-modernist Protestant movement labeled "fundamentalism." The Eugenics Record Office had always focused on scientific research, but there was a growing need for more popular publications—or eugenic "propaganda," as they called it. A new battle for public opinion loomed, and they knew they must respond to the powerful condemnations of influential, traditional Christian thinkers.

Despite their mutual insecurities, the passions of their parents energized these two men's devotion to their work, and their desire to spread eugenics throughout the world. Even as war once again loomed in the Old World of Europe, Davenport and Laughlin maintained their long-held American sense of optimism.

Explaining his hopes and his enthusiasm for the work of the Eugenics Record Office, Davenport exclaimed to his benefactor Mrs. E. H. Harriman, with underlined emphasis:

"You see what a fire you have kindled! It is going to be a purifying conflagration some day!"[21]

Catechisms Old and New

On August 8, 1915, at the Second National Conference on Race Betterment, in San Francisco, Dr. John Harvey Kellogg, the eccentric physician from the Battle Creek Sanitarium who advocated the health benefits of corn flakes, stood up to give an address on the pressing need for eugenic engineering. "The world needs a new aristocracy," he proclaimed, even as the aristocrats of Europe were leading their nations into a senseless conflagration, "a real aristocracy made up of Apollos and Venuses and their fortunate progeny."

After waiting for the applause to subside, he continued, "Instead of such an aristocracy, we are actually building up an aristocracy of lunatics, idiots, paupers, and criminals. These unfit persons already have reached the proportions of a vast multitude —500,000 lunatics, 80,000 criminals, 100,000 paupers, 90,000 idiots, 90,000 epileptics— and we are supporting these defectives in idleness like real aristocrats, at an expense of $100,000,000 a year. And this mighty host of mental and moral cripples is increasing, due to unrestricted marriage and other degenerative influences, at a more rapid rate than the sounder part of the population. . . . Every one of these lunatics possesses the right to vote even in States where women are not given the right of franchise."[1]

Despite the malapropian image, this was a typical warning from a liberal, progressive eugenicist, and the leitmotif of startling statistics had been sounded many times before. How could lunatic men have the right to vote, while intelligent, middle-class matrons were kept in the home? Dr. Kellogg thundered from the podium these figures, and used them to call for a nationwide "health registry," which would house the results of an annual "health inspection" for every citizen. Such a registry could then help enforce marriage restriction and sterilization laws, and "would be esteemed as more precious than gold." He continued, "A eugenics registry would be the beginning of a new and glorified human race which some time, far down in the future, will have so mastered the forces of nature that disease and degeneracy will have been eliminated. Hospitals and prisons will be no longer needed, and the golden age will have been restored."

Such audacious rhetoric, however, even though it stood within a long tradition that envisioned America as a city upon a hill, may have been starting to hurt the eugenic cause with its very un-American call for a new aristocracy. With their unabashed elitism and contempt for the individual citizen, such speeches were reported widely by the press, sparking opposition from various local newspapers and civic groups. The leitmotif of statistics was starting to lose its rhetorical power in a democratic land, failing to engage ordinary folk and convince them of the need for better breeding. So, at the onset of the Great War, Galton's science had begun to lose a bit of its momentum in the United States, despite the relentless advocacy of eugenicists at Race Betterment conferences and the support of the nation's political and intellectual elites.

Spearheaded mostly by religious organizations, a growing opposition had raised a cry against this movement to seize control of human reproduction. Even though fifteen state sterilization laws had been passed between 1907 and 1915, as the United States prepared to enter the European theater, efforts to pass new bills stalled and many of the existing statutes began to fall under attack. The enthusiasm for negative eugenics among physicians, scientists, and politicians was being thwarted by a host of legal challenges.

The opposition forced some state legislatures to repeal their eugenics legislation, while in other states a conservative and suspicious judiciary took up its mantle as protector of the Bill of Rights. The laws that used sterilization as punishment for sexual crimes were first to fall. In Washington, the State Assembly rescinded its sterilization statute in 1913, even though the state Supreme Court had ruled the procedure was not cruel and unusual punishment for convicted rapists. The Nevada court, however, ruled that punishing "sexual perverts" with sterilization, while not cruel, was yet unusual, and thus prohibited by the state constitution.

But even the eugenic engineering laws were under attack. In 1913, the "sterilization law for human betterment" in New Jersey, signed by Woodrow Wilson, was ruled both cruel and unusual as well as a violation of the Equal Protection Clause of the Fourteenth Amendment, since it focused only on epileptics confined to state facilities. By the end of the war, state courts in New York and Michigan would also declare their laws to be a violation of this constitutional principle. In Indiana, Iowa, and Oregon, however, courts would reject enforced sterilization as a violation of the "due process" clause of the Fourteenth Amendment, ruling that doctors alone should not have so much absolute decision-making power over the bodies of their patients.[2] Laughlin's predictions, as well as his legal acumen, had been acute.

Yet even where sterilization remained legal, many state-run institutions hesitated making extensive use of their new authority. Officially, only a few hundred inmates were sterilized in states such as Indiana, Nebraska, Oregon, Kansas, North Dakota, and Wisconsin. Indeed, as Laughlin had explained earlier in the "Report of the Committee to Study and Report on the Best Practical Means of Cutting Off the Defective Germ-Plasm":

> It is, therefore, easy to understand why little has been actually done. The machinery of administration has to be created. It was a new and untried proposition. Public sentiment demanding action was absent. Law officers of the state were not anxious to undertake the defense of a law the constitutionality of which was questioned. So we must frankly confess that what has sometimes seemed to be, and has been heralded in some quarters as a remarkable development in this movement for race betterment, is, as yet, little more than the hobby of a few groups of people, and does not really indicate the adoption of a settled policy.[3]

Only in California was an administrative machine set up to make sterilization more than just a "hobby." There happened to be a solid consensus among the doctors and superintendents in the state's seven mental hospitals, and together with California's attorney general, they collaborated in a way that administrators in smaller states—which usually had only one or two mental institutions—did not. Operating rooms in California's mental hospitals were busy: by the onset of the war, over 70 percent of the 1,422 citizens officially sterilized in the United States were operated upon in California.[4] Galton's eugenic dream, now proclaimed by Kellogg as "a real aristocracy made up of Apollos and Venuses and their fortunate progeny," seemed to be imperiled.

As Charles Davenport sat in the audience at the Second National Race Betterment Conference that day, listening to Kellogg's speech, he recognized the biblical allusions mixed amid the classical in Kellogg's address: the "glorified" human body, the image of a new heaven and earth where every tear would be wiped away, a faith more precious than gold. But for his own presentation, Davenport wanted to do something different. He had already called for state health registries in his own writings, and of all the people at the conference, none could claim more practical success in eugenics than he. Yet he had been putting together a lecture a bit less flowery and allusive, but far more theologically serious. While Galton had first pronounced eugenics to be a "virile religion," and while many had echoed this sentiment over the decades, they had simply used the term as a rhetorical analogy. No one had ever tried to present the idea in a serious and systematic way—until today.

Davenport entitled his paper "Eugenics as a Religion," and he began with a simple analysis of the function of religion. Taking an anthropological and biological approach to the question—similar to Freud, a thinker he read very little, if at all—he concluded that religion was a social phenomenon arising out of the clash of instincts and community interests. The "mores," he explained, were lists of rules, written or unwritten, which limited individual freedom for the sake of the community and were expressed, usually, in religious creeds.

This reduced religion to social and historical forces, of course, but Davenport thought it was a bit more complex than that. In addition to the social mores created by a community, individual human beings also had a *genetic* social conscience. "Man as a gregarious animal has not only instincts, but also, typically, an inhibitory mechanism for stopping the instinctive reaction whenever it would be injurious to others." In other words, man as an animal had a natural inner conflict, not only between instincts and mores, but also between competing natural instincts.

In the Bible, the apostle Paul had proclaimed a war between the flesh and spirit, a dualistic doctrine that informed much of Puritan theology. "For the flesh lusteth against the Spirit, and the Spirit against the flesh: and these are contrary the one to the other: so that ye cannot do the things that ye would" (Galatians 5:17). By "flesh," the apostle meant a sinful inner nature, not the body itself. Davenport saw a similar war taking place within the genes: instincts set against instincts, natural inclinations set against "inhibition germs," and he saw religion as both an expression and a resolution of this inner conflict. He also went on to say that there were "special pressures," such as taboos, which society created in order to restrain the destructive instincts, as well as strengthen these "inhibition germs." But like those predestined for either salvation or damnation, some children were endowed with vigorous germs and others with feeble germs. The vigorous were the chosen, while the feeble were the damned.

The lecture in many ways set out Davenport's ambivalence about his own upbringing. His father's faith and strict religious rules, as well as his mother's quiet piety, had often made him bristle. But he could not fully reject the Davenport Puritan traditions. His family religion had always inhibited him, had always kept him from expressing his own desires and individual needs. But now he could only justify this oppression in ostensibly scientific terms:

> This then is, I take it, the function of religion, first to train the inhibitions in so far as they can be trained (ethical culture) and, secondly, to supply those whose inhibitions are weak with a different means of control; namely, opposing a stronger, higher instinct by which the other instinct may be combated (religion in the narrower sense). In general, the function of religion is, as I see it, to secure emotional control and, especially, by means of the emotions.

For many people, such a definition of religion as an "emotional control" would be horrifying and oppressive. But for Davenport, this made religion not only a necessary manifestation of social evolution, but also an essential function in its continued growth. This was why eugenics could be a vital and flourishing religion in itself. Recognizing the genetic causes and evolutionary effects of "inhibition germs"—and especially their varying degrees of strength in different racial groups—could be the first step in purifying the community and eliminating the immoral effects of bad genes.

> Most religions fall short in this; they fail to regard the importance for society of inheritable racial traits. Eugenics, on the other hand, recognizes this fact. Its aim is this: to improve social conditions by securing to the next generation the greatest proportion of persons best fitted by nature to carry on each his own share of the world's work. For why, asks the eugenicist, is the human race here and what has it to look forward to? And he finds his answer in this.
>
> It is for man as a social species to develop a social order of the highest, most

effective type—one in which each person born is physically fit, well endowed men-
tally for some kind of useful work, temperamentally calm and cheerful and with
such inhibitions as will enable him to control his instinctive reactions so as to
meet the mores of the community in which he lives.

Despite cheerfully commending the virtues of religious inhibitions, and, like a good Puritan, concluding that one purpose of the human race is simply to engage in the "world's work" or "useful work," Davenport was not expressing a particularly novel idea. Others had reduced religion to a mere expression of biological instincts, and dozens had charted the evolution of religious and social mores. Since it proposed a serious eugenic religion, however, his paper represented quite a leap. Though Galton and Pearson had played with the terms "eugenic creed" and "new religion," these agnostics always seemed to say these things with a smirk and touch of irony. Davenport, however, who knew the congregational drone of reciting creeds in church, closed his lecture by proposing his own solemn eugenic creed. Standing before scientists from Harvard, Yale, Stanford, and other eminent institutions, he announced:

I believe in striving to raise the human race, and more particularly our nation
and community to the highest plane of social organization, of cooperative work
and of effective endeavor. . . .

I believe that to secure to the next generation the smallest burden of defective
development, of physical stigmata, of mental defect, of weak inhibitions, and the
largest proportion of physical, mental and moral fitness, it is necessary to make
careful marriage selection—not on the ground of the qualities of the individual,
merely, but of his or her family traits. . . .

I believe that I am the trustee of the germ plasm that I carry, that this has been
passed on to me through thousands of generations before me; and that I betray the
trust if, (that germ plasm being good) I so act as to jeopardize it, with its excel-
lent possibilities, or, from motives of personal convenience, to unduly limit off-
spring.

I believe that, having made our choice in marriage carefully, we, the married
pair, should seek to have 4 to 6 children in order that our carefully selected germ
plasms shall be reproduced in adequate degree and that this preferred stock shall
not be swamped by that less carefully selected. . . .

I believe in such a selection of immigrants as shall not tend to adulterate our
national germ plasm with socially unfit traits. . . .

I believe in repressing my instincts when to follow them would injure the next
generation.

I believe in doing so for the race.[5]

Davenport's lecture at the Second National Race Betterment Conference was driven mostly by his own psychological needs and intellectual longings, yet it came at a time when the eugenics movement needed to respond in a new way to growing public opposition. Davenport did not intend his paper to be a new method of eugenic propaganda, yet around the country, eugenicists were beginning to recognize that their movement needed a new type of rhetoric, a new "propaganda" that would engage, and not alienate, a public becoming more hostile and skeptical.

Indeed, after the *New York Times* reported Kellogg's address at the conference, other newspapers around the country again began to take a closer look at the types of ideas being expressed by these elite thinkers. By the end of September, Hearst newspapers were publishing sensational but accurate headlines such as "14 Million to Be Sterilized," outlining the ideas of Harry Laughlin and the work at Cold Spring Harbor. Alexander Graham Bell, at his vacation home in Nova Scotia, was so concerned by the bad press that he quickly wrote to Davenport, asking whether these reports were true. Though Laughlin's proposals clearly implied this figure, Davenport assured him they were not, and that Hearst newspapers were not to be believed.[6]

In February 1916, a Hearst reporter from the *New York American* interviewed the Yale economist Irving Fischer, another speaker at the Second National Race Betterment Conference and a member of the Eugenics Record Office's Board of Scientific Directors. She had pressed him on one of his public statements, demanding evidence for one of the eugenics movement's basic assumptions: the particular danger of the fecund, feebleminded female. Fischer had claimed, "Many women of the borderline type of feeble-mindedness, where mental capacity often passes for innocence, possess the qualities of charm felt in children, and are consequently quickly selected in marriage." It was this surreptitious menace that *really* posed the greatest danger for most eugenicists— these women's uninhibited sexual charms, their ability to seduce eugenically fit males, and their tendency to have a lot of unfit children. Fischer was caught off guard, and he could not provide the data the reporter was asking for. After the episode, he wrote Davenport: "I should have turned her loose on you, had I not known your sentiment on reporters especially of the Hearst journals!" But he knew, as many others in the movement were beginning to suspect, that the battle would be won or lost in the public arena, not the laboratory. "Much as I dislike the tone of their articles," he continued, "if we do not help them, they will do us positive injury. . . . In spite of their sensational-ism, we can utilize them to create respect for the eugenics idea in the mind of the public."[7]

The mind of the public. Fischer was right: newspapers, with writers adept at populist appeals, and organized religion, with the authority of preachers at the pulpit, could do positive injury to the science of eugen-ics if its leaders didn't try to work with them. Indeed, Alexander Gra-ham Bell, who had stood behind eugenics for over a decade, could no longer support the direction of the movement, and after all the negative press, he resigned from the Record Office's Board of Scientific Direc-tors.[8] As for Fischer, he knew, as an economist, that even science must sometimes take account of human squeamishness.

Eugenics must adapt, adopt new strategies, and form new types of publications, many proponents began to believe. Even Darwin's theory of evolution was under fierce attack by great political orators like William Jennings Bryan. As the debate focused more and more on issues of religion, many eugenicists, like Davenport and Fischer, wanted to create respect for science and fight for the mind of the public.

THE PUBLIC BATTLE over evolution and eugenics, however, began to wane as the United States joined the fray in Europe. Work at Cold Spring Harbor, too, slowed during the great conflagration as fieldwork-ers and researchers turned to other matters. And since Harry Laughlin had never received even a master's degree in science, he took a leave of absence and headed off to Princeton to do graduate work in biology.

The men on the Princeton campus had often discussed the "new biology," and eugenics was a common topic of debate—and even rev-elry. In 1914, the talented Princeton undergraduate F. Scott Fitzgerald wrote a playful song, "Love or Eugenics," for the university's annual Triangle Club Show, a vaudeville musical put on every year by one of the oldest comedy troupes in the nation. "Men, which would you like to come and pour your tea? / Kisses that set your heart aflame, / Or love from a prophylactic dame?"[9] Would genetic engineering interfere with human passion?

Laughlin hadn't come to study eugenics, however. Living in a small apartment with Pansy, he set himself to a characteristic drive for great-ness. In 1917, he finally earned a Ph.D., just months after Czar

Nicholas II abdicated his throne in Russia and the U.S. Congress declared war on Germany. His dissertation, "On Mitosis in the Root Tip of the Common Onion," represented a departure from his life's work, but it now bestowed on him the authority and status of an Ivy League doctorate. He also studied under the renowned genetic cytologist (and eugenics sympathizer) Edwin Grant Conklin, which made his accomplishment even more prestigious.[10] Dr. Laughlin could now attend international conferences with bona fide credentials.

In Long Island, however, the ironies of nature's providence struck Charles Davenport and his family. While his oldest daughter Millia was a student at Barnard College in New York City, his younger daughter Jane (or Janet, as the family called her) was having problems. Her Latin was atrocious, and she could barely figure out algebra. "Intellectually you seem to be very well equipped, though your mind does not lend itself readily to many scholastic processes," Davenport reasoned to his second child as she struggled in a boarding school in France. "You seem not to have a good visual memory at least for the letters in words as they appear on the printed or written page." He wrote to her a number of times while she was away, and it was clear that although Jane could carry on conversations with tremendous wit and intelligence, she was somewhat slow in school, often reversing the letters and words she saw. Out of habit, Davenport tended to muse about his children's genetic predispositions, and he would mingle cold, awkward jargon with fatherly tenderness when he explained these things to them. "The organism we call Jane Davenport is surely a new combination of traits—Armstrong practicalness (and inability to spell); Crotty humor and philosophy; Dimon and Davenport love of outdoors and nature. Be your own self and remember, in case of doubt as to action, to act in accordance with your highest ideals. Happiness does lie this way, in being pleased with what you do because you haven't done anything that you are ashamed of."

In another letter, after suggesting she do a few visual exercises to help improve her memory and reading skills, he also tried to encourage his daughter and, in his own way, explain to her the other "family traits" in which she did excel. "Emotionally you seem to be provided with a normal set of inhibitors and we don't anticipate that you will disgrace us by conduct opposed to 'the mores.' " Unlike his own father, however, Charles always let his children know his love. "Janet, dear, your mother and I think a lot of you; we love you dearly."[11]

Charles showed all three of his children a great deal of affection, yet his pride in his precocious son Charles Junior was clear to everyone at Cold Spring Harbor. But little Charlie was sickly. He was constantly ill. He caught colds often and had chronic diarrhea. Earlier that year, in February, he had caught malaria while in Kansas with his mother and had battled a fever on and off throughout the rest of the year. Yet, though the youngest and the object of so much attention on the campus, the boy never seemed to take it for granted, and he bore his afflictions with a typical Davenport stoicism. Once, when playing ball with a social worker at Blackford Hall, he missed a catch and was struck in the face. He hit his head against the wall and fell to the floor. But instead of crying out, as most boys his age would, he just rubbed his head and giggled. "I won't forget that soon," the social worker later recounted to Jane. "That same day so many of us were out on the porch shelling peanuts, we had so few nobody ate any, he was watching us, not asking for any because he knew they couldn't be spared. Then someone else entered, uninvited, and helped herself, saying she was hungry. He said, 'But I was hungry, too.' Most children have been spoiled by the attention he received, but he was unspoiled in every way."[12]

Davenport may have felt a tinge of guilt at his children's problems, especially his son's. Why could Janet not see words in the right order on the page? Why was Charlie ill so often? In *Heredity in Relation to Eugenics*—which had become the most successful textbook on eugenics since it was published in 1911—he had explained that illnesses and diseases were not caused by microbes per se, but rather by a congenital, genetic weakness:

> *With few exceptions, the principle that the biological and pathological history of a child is determined both by the nature of the environment and nature of the protoplasm may be applied generally. It is an incomplete statement that the* tubercle bacillus *is the cause of tuberculosis or alcohol the cause of* delirium tremens *or syphilis the cause of paresis. Experience proves it, for not all that harbor the* tubercle bacillus *show the dread symptoms of tuberculosis (else there were little hope of escape for any of us!). . . . Rather, each of these diseases is the specific reaction of the organism to the specific poison. In general, the causes of disease as given in the pathologies are not the real causes. They are due to inciting conditions acting on a susceptible protoplasm. The real cause of death of any person is his inability to cope with the disease germ or other untoward conditions.*[13]

On September 5, 1916, little Charlie was attacked by the *poliomyelitis* disease germ. His body was unable to cope, and he succumbed quickly, dying the same night.

Charles and Gertrude fell prostrate with grief. The family had to be quarantined for a few days, and this only made the madness of their despair worse. Neither of Charlie's parents could continue with their work; on the verge of breakdown, they retired to a sanitarium in New York, where they would remain for many weeks.[14]

THE STENCH OF DEATH hung over Europe. The carnage wrought by the Great War, the horror of body-strewn trenches, torn horses, and blackened trees in desolate fields, stunned an entire generation. The ambivalence at the turn of the century—when great expositions celebrated the astonishing accomplishments of Western science and technology, when poets lamented the dark, Satanic mills which robbed workers of their ancient pulse of germ and birth, when optimism and despair mingled together in an uneasy dance of changing social circumstances—only intensified in the aftermath of this new, modern war, fought with factory-produced chemicals and airborne machines. With over 9 million slaughtered, there were no epic battles to celebrate, no heroes to welcome home, no glories to revel in.

There was a spiritual crisis in the Western world. Evolutionary progress, it now seemed, did not entail a natural, linear progression toward perfection. Science and technology, astonishing as they were, could not guarantee a world less nasty and brutish, and might even make it worse. For many eugenic reformers, however, the horror of the modern battlefield rendered the quest for human betterment even more acute. As Galton's old protégé, Karl Pearson, wrote in the massive biography of his mentor:

> It is as if the Great War had so thoroughly demonstrated the pitiable failure of humanity, that its thinkers and leaders felt that the old man must be replaced by a new-born Apollo, the worn-out creed which had failed him by a more adequate physical faith. . . . With our present acquaintance with the laws of heredity, with our present knowledge of how customs and creeds have changed, can we not hasten the evolutionary process of fitting man to the needs of his present environment? . . . The new creed bids us seek quality and restrict quantity; separate, where race demands it, the scarce controllable instinct of mating from the

parental instinct, and teach nations to pride themselves on the superior type of their citizens, rather than on their material resources. The eugenic dreamer sees in the distant future a rivalry of nations in the task of bringing to greater perfection their human stocks, and this by an intensive study of biological law applied to man, and its incorporation, it may be gradually, but surely, in a revised moral or social code.[15]

Customs and creeds, it was clear, had changed over the centuries. Society's values and mores, including religious beliefs, had evolved over time. A new epistemology based on evolution seemed to make the "worn-out creeds" and their claims for ultimate, unchanging truth dubious, if not hopelessly outdated. The rigors of Enlightenment doubt had culminated in an understanding of a dynamic, ever-changing world. In the face of such change, the world was demanding a new source of hope, a revised moral or social code that could provide comfort for this present despair. Whether creeds were old or new, most agreed mankind did need "rapid elevating." Most believed nations should indeed begin to pride themselves on the "superior types" of their citizens, and not send them off like lambs to the slaughter.

So, despite the pitiful failure of science and technology to lead mankind into a new era of harmony and repose, this spiritual void created a new space for eugenics to spread even further than it had before. The war years had naturally slowed the progress of research and legislation, but they had also stemmed the successes of the opposition. The resistance to better breeding and forced sterilization, at first led by Catholic theologians and fundamentalist preachers, was sidetracked as the nation turned its attention to other matters.

As Europe and the United States buried their dead, however, and as questions of mankind's nature and destiny became as critical as they had ever been, the conflict between creeds old and new returned with even greater fury. It had always been an inevitable clash of *Weltanschauung*s, raging on and off for almost a century. But now, for both sides, the moral stakes were high, for it seemed something must be done to control reproduction, this "most imperious instinct of living beings," in order to save mankind.

Since the dawn of Darwin's theory of evolution, many (though not all) Christian thinkers were shaken by what seemed to be the theory's inescapable assault on human dignity. If mankind was not created in the image of God, they reasoned, and thus not endowed by the Creator with certain inalienable rights, upon what basis could individuals claim to be different from the beasts of the field, and thus protected from the same breeding techniques farmers used on their livestock and crops? Their revulsion sprung not from the efforts to improve society, but from the way many social scientists reduced human beings to mere animals, how they reduced populations to statistics, and how their calm, rational methods took no account of the spiritual dignity and personal desire of individuals they deemed "manifestly unfit from continuing their own kind."

Many anti-evolutionists saw eugenics simply as the logical outgrowth of Darwinian thinking. If mankind had evolved from beasts, why couldn't men be bred like cattle? Popular preachers took advantage of this ostensibly brutal way of thinking, and cleverly appealed to human sentiment. What could science say to the bereaved as they buried their dead? How could the promises of research console the present needs of a people struck prostrate with loss? The famous itinerant evangelist Billy Sunday would thunder:

Let your scientific consolation enter a room where the mother has lost her child. Try your doctrine of the survival of the fittest. And when you have gotten through with your scientific, philosophical, psychological, eugenic, social service, evolution, proto-plasm and fortuitous concurrence of atoms, if she is not crazed by it, I will go to her and after one half hour of prayer and the reading of the Scripture promises, the tears will be wiped away.[16]

The fundamentalist movement arose out of this complex clash of worldviews. Most evangelical Protestants, like Billy Sunday, emphasized a dramatic conversion experience and a deeply emotional relationship with Christ—a relationship Amzi Davenport had once implored his bride-to-be Jane to embrace. It was a uniquely American type of spirituality: intensely individualistic, suspicious of things rational and intellectual.[17] Most evangelicals ferociously fought any attack on the integrity of the simple, literal meaning of the Bible. The pejorative term "fundamentalist" arose from a series of anti-modernist tracts called *The Fundamentals: A Testimony to the Truth,* published between 1909 and 1915, and written by a number of evangelical scholars to combat the liberal theologies that modernized Christian faith in light of historical criticism and natural science.[18] Indeed, in many states, the religious crusade against teaching evolution spilled over into an assault on eugenic reforms.

Catholic thinkers, however, attacked eugenics more from a theological point of view. Many Catholics could accommodate the theory of evolution, and interpret Genesis within a millennia-long tradition that allowed for moral, typological, and allegorical meanings in Scripture. Ultimate reality and ultimate meaning were spiritual, not historical or carnal, so for them, literal interpretations might even distort the Bible's truth. Catholics thus insisted on the essential *spiritual* dignity of every individual. No matter how God created the physical attributes of mankind, no matter if these evolved over time or sprang from lower forms of life, they were still subject to an immutable "natural law," governed by the laws of God. While many fundamentalist Protestants focused on the integrity of the biblical stories, Catholics emphasized the immortal human soul and the spiritual laws to which it was subject— laws not only revealed in Scripture and the teachings of the Church, but embedded in Nature as well.

Catholic organizations often spearheaded the resistance to state eugenic laws. In Colorado, groups such as the Knights of Columbus and the Holy Name Diocesan Union worked tirelessly to defeat sterilization bills introduced to the State Assembly between 1910 and 1930. Even when the legislature passed a law in 1927, they lobbied the governor to veto the bill, which he did.[19] *The Commonweal,* a popular Catholic periodical, published sterilization data and urged its readers to lead the opposition to state laws. Indeed, eugenic journals reported bitterly that the Roman Catholic Church "furnished the main opposition" to bills in New York and Connecticut, and eugenic laws in general.[20]

By 1930, after over two decades of resistance, Pope Pius XI issued the papal encyclical *Casti connubii* (*On Christian Marriage*) and gave the Church's official position on eugenics and sterilization. Marriage and bringing forth children were considered sacred rights or even duties, of course, but the Vatican also made clear its position on the rights of the state. "Public magistrates have no direct power over the bodies of their subjects; therefore, where no crime has taken place and there is no cause present for grave punishment, they can never directly harm, or tamper with the integrity of the body, either for the reasons of eugenics or for any other reason."[21]

In England, one of the most outspoken critics of eugenics was G. K.

Chesterton, a fiction writer, essayist, and self-described "rollicking journalist" who had famously converted to Catholicism in 1922. That same year, he published a collection of essays entitled *Eugenics and Other Evils: An Argument Against the Scientifically Organized Society.* Though a friend to eugenic sympathizers George Bernard Shaw and H. G. Wells, Chesterton was a vociferous opponent of their ideas of better breeding and social engineering. Chesterton was bitterly opposed to the socialist beliefs of his friends. He had strong religious convictions, libertarian leanings, and his hostility toward eugenics sprang from an unshakable belief that indeed public magistrates had no direct power over the bodies of their subjects. The "scientifically organized state," he maintained, was a grave threat to human dignity and liberty. "It is cold anarchy to say that all men are to meddle all men's marriages. It is cold anarchy to say that any doctor may seize and segregate anyone he likes. But it is not anarchy to say that a few great hygienists might enclose or limit the life of all its citizens, as nurses do with a family of children. It is not anarchy, it is tyranny. . . . But as a vision the thing is plausible and even rational. It is rational, and it is wrong."[22]

Chesterton's commonsense attacks on the theories of better breeding had resonated with the public in England before the war, and they helped thwart the kind of legislation eugenicists were calling for afterwards. In the United States, too, these sorts of rhetorical attacks had been rolling back the successes of earlier eugenic victories.

The eugenic dreamers knew this. If they were to replace the old man with a newborn Apollo—Pearson's classical twist of a biblical image from the apostle Paul—they knew they must do more than simply appeal to their long-heralded analogy of breeding better farm animals, and the promise of scientific progress. If they were to win the hearts as well as the minds of a public yearning for a better world, they would have to speak in a language that the public could embrace.

REPEATING "I BELIEVE" is the genre of the creed, a public act of worship, a communal statement of faith. For the uninitiated, however, or the "catechumen" learning the basics of the faith, the creed is taught by a genre of questions and answers, a "catechism," which is like a child's nursery rhyme.

For many evangelical Protestants in America, reciting the catechism was as common as the ABCs. Religious traditions springing from the

Separatists, the pilgrims who had come from the Netherlands to England and thence on the *Mayflower* to America, usually followed the 129 questions of the Heidelberg Catechism, first published in Germany in 1563. Religious traditions evolving from the Puritan Calvinists, who had arrived on American shores from England with John Winthrop and John Davenport, usually recited the 107 questions of the Westminster Shorter Catechism, completed in England in 1647. "What is the chief end of man?" asked Question One of this famous teaching tool, the most common in Sunday Schools across the nation. Answer: "Man's chief end is to glorify God, and to enjoy Him forever."

Eugenics now had a creed from an American Davenport, but it also needed a primer for lay people to learn the basic principles of better breeding. Imitating the forms and rhythms of its Christian models, a pamphlet entitled *A Eugenics Catechism* was sent out to an audience more and more willing to hear. It was published by the American Eugenics Society, a new organization co-founded by Dr. Harry Laughlin in the 1920s with an advisory committee that included some of the nation's foremost thinkers.

"What is eugenics?" asked Question One. Answer: "Eugenics is the study of those agencies under social control which may improve or impair the inborn qualities of future generations of man either physically or mentally." This was Galton's ancient definition, though the pamphlet didn't say. The primer went on to explain what eugenics was *not,* since the attacks by the religious opposition had always been quick to describe it as a dangerous foe of traditional Christian values:

Q. Does eugenics contradict the Bible?
A. The Bible has much to say for eugenics. It tells us that men do not gather grapes from thorns and figs from thistles. . . .
Q. Does eugenics mean less sympathy for the unfortunate?
A. Eugenics does not mean less sympathy for the unfortunate; it does mean fewer unavoidable unfortunates with which to divide a sympathy which should be more fully and effectively expended on the inevitable unfortunates. At the same time that sympathy and remedial treatment are being extended, something effective should be done to prevent a recurrence of such cases where heredity is to blame. This is a true kindness, both to the victims and to society.
Q. Must one who believes in eugenics believe in evolution?
A. Yes, that evolution is a present and a continuing process. It is not necessary to believe that the original or ancestral man evolved from apes. All admit

that there has been an evolution in the differentiation of the races, and from fossil man to modern man. Should we not want more of such evolution?

A Eugenics Catechism was published in 1926, and the tone was conciliatory. The pamphlet addressed some of the major public issues swirling about the clash of science and religion over the previous ten years, but instead of trying to instigate more rancor, the American Eugenics Society attempted to win over those suspicious of scientific better breeding. A year before, the Scopes "Monkey Trial" had transfixed the nation as Clarence Darrow, the brilliant trial lawyer and outspoken liberal agnostic, took on William Jennings Bryan, the three-time presidential candidate and dazzling orator known as the "Great Commoner." Bryan, a devout fundamentalist, had been a famous critic of Darwinism and an ardent defender of the literal interpretation of the Bible, so he lent his name and fame to Tennessee's case against John Scopes, a high school teacher who flaunted the state's ban against teaching evolution. With the media frenzy surrounding the trial, there was much at stake for the eugenics movement in this case. Indeed, the textbook Scopes had used, George Hunter's *A Civic Biology,* included quotes from the work of Charles Davenport, quotes that explained the concept of evolutionary improvement and the importance of "mate selection" to better the race.[23] Davenport, along with Henry Fairfield Osborn and Edwin G. Conklin, the Princeton geneticist and Harry Laughlin's mentor, were three of the primary scientific experts called by the defense. Before the trial, Davenport had even written a syndicated newspaper column called "Evidences for Evolution" to combat the fundamentalist arguments. (He had heeded Fischer's warning, and joined the battle for the mind of the public.) "Fundamentalists accept what they have been told about the accuracy of description of the origin of the universe given in Scripture," Davenport wrote there. "The biologist has his own idea of what is the word of God. He believes it to be the testimony of nature."[24]

Even though the acrimony of the Scopes trial belied a much more complex relationship—and a greater consensus—between believers and secular scientists than the national media made it appear, as newspapers around the country described Bryan's sputtering inability to defend the literal truth of the Bible, evolution became more and more an accepted fact by the majority of the public. Along with this cultural shift, the science of eugenics also gained a greater credibility. Fundamentalist Christianity, in fact, would retreat from the public arena in the aftermath of the trial; it would not become a significant political voice for the next six decades.

The voices of the new eugenic creed, however, were again gaining momentum. Albert E. Wiggam, a journalist and popular speaker at the time, took up the call to spread the gospel with as much energy and zeal as anyone. In addition to dozens of magazine and newspaper articles, he wrote the best-selling *New Decalogue of Science* in 1923. If Davenport proclaimed a rational and Puritan-like creed, Wiggam spoke the language of a tent revivalist, an evangelist like Billy Sunday who engaged the hearts of his hearers. Indeed, in the tradition of sermons like Jonathan Edwards's "Sinners in the Hands of an Angry God," Wiggam began his book with five dire warnings, including the alliterative and rhythmic "That the Golden Rule without Science Will Wreck the Race that Tries It." It was a screed against the evils of welfare for the poor, but it also provided a new eugenic Golden Rule. If Jesus were to return in the 1920s, Wiggam proclaimed, he would give the world a new commandment: "the biological Golden Rule, the completed Golden Rule of Science. *Do unto both the born and the unborn as you would have both the born and the unborn do unto you.*"[25] The second section of the book listed "The Ten Commandments of Science," and these included "The Duty of Preferential Reproduction" and "The Duty of Trusting Intelligence."

The primary driving force behind eugenics propaganda, however, remained the American Eugenics Society. Founded by men like Laughlin and Osborn, it had been organized after the Great War for the specific purpose of influencing public opinion. It had gone through a number of name changes during its first few years, but by 1926 it had evolved into an organization devoted almost exclusively to the propagation of eugenic ideas. As one member described, its purpose was the "dissemination of popular education concerning the facts of eugenics." To accomplish this, in addition to publications like *A Eugenics Catechism,* its members used a number of different strategies. Focusing attention especially on schools and textbooks, they tried to promote "eugenics as an integral part of various appropriate courses throughout the school system, in the elementary grades through high school as well as the encouragement of special courses in colleges and universities."[26]

But the society, devoted to the new eugenics creed, also sponsored such activities as "eugenics sermon contests," and in 1926 it offered considerable prizes of $500, $300, and $200 for the top entries. Judges included the Yale literary critic William Lyon Phelps, a deacon at Calvary Baptist Church in New Haven, as well as Charles Davenport, science's eugenic theologian. Clergymen from around the country submitted almost three hundred sermons—and they were not only from white Protestants. One submission came from Rabbi Harry H. Mayer of Kansas City, Missouri. Even as Jews were seen as inferior stocks by many eugenicists, in one of Mayer's Mother's Day sermons, he declared, "May we do nothing to permit our blood to be adulterated by infusion of blood of inferior grade."[27]

The Reverend Dr. Kenneth C. MacArthur, who won the $300 second prize for his sermon, had already won a silver cup in the Fitter Families Contest sponsored by the American Eugenics Society at the Eastern States Exposition of 1924. The society used these "Fitter Family" competitions as another strategy to promote better breeding. Fostering interest through competition, such contests had families sign up to see which was comprised of the best "human stock." Family members had to submit to a medical examination, which included a "pedigree study" and an intelligence test. "The time has come," said a contest brochure, "when the science of human husbandry must be developed, based on the principles now followed by scientific agriculture, if the better elements of our civilization are to dominate or even survive." Winners received a gold-colored medal, stamped with the image of a mother and a father passing a torch to their naked young son. The medal was inscribed: "Yea, I have a goodly heritage."[28]

Fitter Family Contests were enormously popular at state fairs across the country in the 1920s, and along with a host of flashy eugenic exhibits, they brought the science of better breeding into the vernacular. The American Eugenics Society tapped into the "Better Baby Contests" that had been common in the Midwest prior to the war, fostered by progressive reformers involved in the women's movement. As early as 1911, eugenic workers had taken notice of these popular contests, first organized by Mary T. Watts, director of the Iowa Parent-Teacher Association, and Dr. Florence Brown Sherborn, a child welfare specialist. After hearing about the contests, Charles Davenport sent Watts a postcard and urged her to "give 50 percent to heredity before you begin to score a baby." Like many progressive women, these two had found in eugenics a liberating message which broke down the barriers of traditional marriage mores and provided a scientific buttress to their efforts

Across the country, hundreds of families participated in "Fitter Families for Future Firesides Contests," in which family members submitted to an intelligence test, a Wassermann test, and observations by eugenics experts. This family won first place in the "Large Family" category in the 1925 Fitter Family Contest at the Texas State Fair.

for reform. In 1920, using sponsorship and funds from the American Eugenics Society, they organized the first "Fitter Families for Future Firesides Contest" at the Kansas Free Fair in Topeka. "The object of the fitter family movement is the stimulation of a feeling of family and racial consciousness and responsibility," explained Dr. Sherborn. Later, in a letter to Yale professor Irving Fischer in 1924, she praised her colleague Mary Watts as the key figure in the battle for the mind of the public. "This movement would not have been possible, it would never even have originated when it did, if ever, had it not been for the indispensable link between science and the public provided by Mrs. Watts. She has the vision, and, as she is fond of saying, she 'speaks the vernacular.' . . . Mrs. Watts has the single-minded enthusiasm of a prophet and the self-confidence of a good salesman."[29]

Throughout the 1920s, the prophets and salesmen promoting eugenics by speaking the vernacular met with much success. There were at least seven to ten Fitter Family Contests at state fairs around the country every year. By 1930, over forty separate organizations were contacting the American Eugenics Society every year, asking for help sponsoring local competitions. Local newspapers splashed bold headlines about the winners on their front pages, and in one instance, the *Denver Daily News* even sponsored a Better Baby show: "Hundreds Witness Judging of Babies in Big News Contest! Cooing Prattlers Cling to Mothers Wide-Eyed with Wonder!"[30]

Fitter Families for Future Firesides became wildly popular, despite being quite demeaning. Contestants had to present their family histories, submit to a medical exam—which included a Wassermann test for syphilis—and then take an intelligence test. Few found this humiliating, though, and scores of families thronged the Eugenics Tent at state fairs. Most of these were simply equipped with makeshift desks consisting of two long boards on trestles, or borrowed kitchen tables and folding chairs. Dozens of families with children crowded into the tent, making a ruckus while social workers and volunteers collected their information, and Cho-Cho the Health Clown entertained them, babbling on about the importance of eugenic fitness, or explaining the brightly flashing exhibits around them. One of these traveling exhibits, sponsored by the American Eugenics Society, featured five blinking lights under the heading: "Some people are born to be a burden on the rest." A sign under one flashing light explained: "Every 15 seconds $100 of your money goes for the care of persons with bad heredity. . . ."

Four generations of winners at an unnamed Fitter Families Contest. Each generation holds its Fitter Family trophy, 1923.

Another had a light flashing every forty-eight seconds to indicate that "a person is born in the United States who will never grow up mentally beyond that stage of a normal 8 yr old boy or girl."

Since the judges used the techniques of the Eugenics Record Office, Mary Watts sent Davenport a full account of the hectic process. Contestants were examined by "busy physicians working on hour schedules, a faithful psychologist who cannot possibly handle all the entries although he works every minute, two or three student nurses to assist the doctors, athletic directors from local YMCAs or YWCAs hurrying to cover all the entries in an hour afternoon and morning . . . and a couple of untrained women at the desk trying to fill in names on score cards and keep up the clerical end of this work while the one available historian tries to write a history in twenty minutes that requires two hours to handle properly."[31]

Still, these Fitter Family Contests fed the kind of anticipation that makes organized competition so enjoyable, and families were eager to join. Who would win? Which family would be proclaimed the fittest, the best stock, endowed by evolution with the most superior genes? At the same time, the contests fed the natural sense of superiority any

"Eugenic and Health Exhibit," Fitter Families exhibit and examination building, Kansas State Free Fair, 1920.

group can feel, a condescension toward outsiders and those of different classes and ethnic groups.

Similar to the old Calvinist and Puritan ideas of the chosen and the damned, eugenics thinkers, wielding the banner of evolution, divided the world into the manifestly unfit versus fitter families, degenerate offspring versus better babies, and citizens of the wrong type versus the hearty American stock. In the aftermath of the horrors of the Great War, when progress could no longer be seen as inevitable, these new creeds preached a new Manichean struggle, this time springing from the blood and the genes. "Purifying the race" through forced sterilization now seemed a worthy and necessary goal, and the ideas of eugenics resonated more and more in the pastoral American countryside, where state fairs and church sermons belied the anxieties of a highly nervous age. Indeed, by 1937, a *Fortune* magazine poll found that two out of three Americans favored the forced sterilization of mental defectives.[32]

The idea of "racial purity," buttressed by eugenic science, had entered the mainstream of American life. The country, maintained the Harvard anthropologist E. A. Hooton, had to do some "biological housecleaning."[33] But such "housecleaning" would not stop with the

UNFIT HUMAN TRAITS

SUCH AS FEEBLEMINDEDNESS EPILEPSY, CRIMINALITY, INSANITY, ALCOHOLISM, PAUPERISM AND MANY OTHERS, RUN IN FAMILIES AND ARE INHERITED IN EXACTLY THE SAME WAY AS COLOR IN GUINEA-PIGS. IF **ALL MARRIAGES** WERE **EUGENIC** WE COULD **BREED OUT** MOST OF THIS UNFITNESS IN *THREE GENERATIONS.*

THE TRIANGLE OF LIFE

YOU CAN IMPROVE YOUR *EDUCATION,* AND EVEN CHANGE YOUR *ENVIRONMENT;* BUT WHAT YOU REALLY **ARE** WAS ALL SETTLED WHEN YOUR PARENTS WERE BORN.
SELECTED PARENTS WILL HAVE BETTER CHILDREN. **THIS** IS THE GREAT AIM OF EUGENICS

The American Eugenics Society prepared traveling exhibits like these to explain genetic theories "in the vernacular," hoping to whip up support for racial purity laws, including forced sterilization.

MARRIAGES,- FIT AND UNFIT

1. PURE + PURE:—
   CHILDREN NORMAL
2. ABNORMAL + ABNORMAL:—
   CHILDREN ABNORMAL
3. PURE + ABNORMAL—
   CHILDREN NORMAL BUT TAINTED:
   SOME GRANDCHILDREN ABNORMAL.
4. TAINTED + ABNORMAL:—
   CHILDREN ½ NORMAL BUT TAINTED ½ ABNORMAL
5. TAINTED + PURE :—
   CHILDREN: ½ PURE NORMAL ½ NORMAL BUT TAINTED
6. TAINTED + TAINTED
   CHILDREN: OF EVERY FOUR, 1 ABNORMAL 1 PURE NORMAL AND 2 TAINTED.

PURE- NORMAL AND TRANSMITTING ONLY NORMAL.
TAINTED- NORMAL BUT CAN TRANSMIT ABNORMALITY.
ABNORMAL- SHOWING THE ABNORMALITY.

HOW LONG

ARE WE AMERICANS TO BE SO CAREFUL FOR THE PEDIGREE OF OUR PIGS AND CHICKENS AND CATTLE,— AND THEN LEAVE THE ANCESTRY OF OUR CHILDREN TO CHANCE, OR TO "BLIND" SENTIMENT?

Flashing-light exhibits, prepared by the American Eugenics Society, were popular in state fairs across the country. They stirred up the fear that the "unfit"—criminals, prostitutes, and the poor—were breeding at an unchecked rate. "You can help to correct these conditions," an exhibit proclaims, hoping to convince onlookers to support racial purity laws.

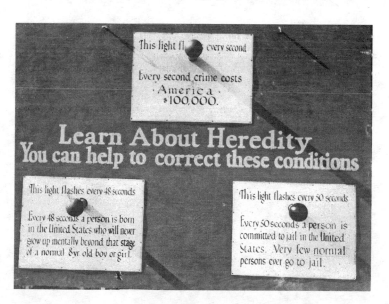

fecund, feebleminded females and criminally prone males. The same intelligence tests created to diagnose this surreptitious menace also seemed to indicate a clear hierarchy of "fitness" among ethnic groups and races. Not only American blacks, the descendants of African slaves, but foreigners within and without were presenting a clear and present danger to the purity of American blood.

Despite the promise of eugenics, the flood of immigration during the first decades of the twentieth century had led some, like Harry Laughlin's old Princeton professor, Edwin Grant Conklin, to despair over the ancient pulse of germ and birth. In this era of rampant immigration, God may have withdrawn His ancient promise to His radiant American bride, His new Israel, since she failed to keep herself pure:

> There was once the supreme chance of breeding here the finest race and nation in the whole history of mankind. . . . The English Puritans thought that "God had sifted the whole nation that he might send choice grain into the wilderness," and this sifting gave us the sturdy, liberty-loving New England stock of our early history. If God had only continued to sift the nations for our benefit, or if our fathers had exercised only reasonable caution . . . we might have had here only the choicest blood and the highest types of culture of all lands, we might have replaced the slow and wasteful methods of natural selection by intelligent selection and thus have enormously advanced and hastened human evolution. That chance has gone forever.[34]

XII

The Making of a Master Race

In her younger and more vulnerable years, Millia Davenport's father gave her some advice that she had been turning over in her mind ever since. Her father, though never one to reserve judgment, had always maintained a measure of infinite hope, a sense that the fundamental decencies parceled out unequally at birth could be genetically engineered, as it were, with desirable traits being multiplied and undesirable traits got rid of. It had always been his habit to explain how character was founded upon the hard rock of a family's genetic "protoplasm," yet he often felt the need to urge Millia to act in accordance with the "mores," those standards of conduct that would keep her from bringing shame to the Davenport name.

The early years of the Roaring Twenties were difficult for Millia and her father. The family had called her Billie since she was a little girl, and she had grown to be a typical tomboy. Far from feminine and graceful, she had dark eyes and slight features, a low, narrow forehead and a round prominent chin that would wrinkle and purse double whenever she closed her mouth. But she was confident and energetic, and very interested in being with boys. To her father's consternation, after she graduated from Barnard College in 1917, Billie moved to Greenwich Village and began to wear tweed slacks, white shirts, and loose ties—a gender-blurring fashion statement common in the Village at the time. She also cut her hair fashionably short into a bob, and wore close-fitting cloche hats. After the Great War, there was a name for these types of young women, a new cliché for such sassy girls who rejected the traditionally demure images of femininity—as Billie's father well knew. In the June 1922 edition of the *Atlantic Monthly*, for example, G. Stanley Hall described this assertive urban girl—who also often drank and smoked—as "a fledgling, yet in the nest, and vainly attempting to fly while its wings have only pinfeathers. . . . [The] genius of 'slanguage' had made the squab the symbol of budding girlhood."[1] Indeed, Millia Davenport was a "flapper."

Her father had just begun his career as a biologist when she was born. Billie arrived after Charles and Gertrude had been married only nine months, during a time when the young Harvard scientist was just starting to develop a particular interest in heredity. A proud new father, Charles told his wife that he was studying his ancestors "so that I can tell Millia—and so that we can know how to interpret her, by knowing her ancestry—for, no doubt, her mental life will be largely the resultant of the various ancestral qualities. We may expect her to be conscientious, stubborn, with a sense of humor, a love of nature, generous but not a spend thrift, without a great talent for any art but with a love of knowledge. Honest and with common sense above the average. You see I am turning 'Fortune Teller.' "[2] Despite the jocular tone, this was more than just a jest. Charles would devote his entire career to the idea.

Now, over twenty-five years later, when he was considered one of the great biologists in the world, Charles could see his daughter's mental life was far from his foretelling.

Billie had little interest in nature, and she spent money without a thought. Secretly, she once charged clothes to her father's account with Wanamaker's, one of the first department store chains. Her father had always been kind and gentle with his children, but when he discovered his daughter's exorbitant spending spree in Manhattan, he was furious, and not a little embarrassed. This was not Davenport behavior. "Dearest Daughter Millia," he wrote,

> I have just received Wanamaker's bill for June. As I had to borrow money to get thru last month I did not pay this bill for May and now the total is $180.10, all of which, I understand, was ordered by you. I am totally unable to pay this, thru lack of funds, and am totally amazed that when it was understood that you should not spend more than $50 in any one month, you would have gone ahead and put the family in such an embarrassing position of running up debts beyond ability to pay.
> I am quite at my wits end what to do. I shall have to borrow from the

bank to meet this bill, and to save the family honor I am writing the following to John Wanamaker:

> Dear Sir: From receipt of this letter please do not charge any goods ordered in name Mr. or Mrs. C.B. Davenport, Jane J. Davenport, unless ordered sent to CSH, or unless on written order, signed by me or Mrs. Davenport and returned with bill.

> It makes me sick to have to write such a letter. Whenever you want anything from Wanamaker's all you have to do is to write or telegraph or phone me and I shall be glad to give you the letter. . . .[3]

To save the family honor—Charles had tried to do this in his own life. His parents had always instilled in him the importance of living in a manner "worthy of the Davenport calling." For Amzi and Jane, this meant living a life of sobriety and restraint, embracing Christ as a personal Lord and Savior, and, in the end, as his mother used to say, "being named with God's chosen people." Charles had demythologized his parents' pious faith, however, reinterpreting the Davenport traditions in light of the discoveries of Darwin and Galton. But it had weighed on him. It felt like a betrayal, especially since he rarely attended church. Yet, even as his psychic needs forced him to forget his mother's simple piety, and as his memory began to remold her into the image of his secular, skeptical wife, he still maintained the same yearning for perfection in his own eugenic creed.

Billie was living a riotous life in Greenwich Village. She went to parties and drank and danced—things strictly forbidden on the campus at Cold Spring Harbor. Her father had once felt the worldly thrills of Manhattan when he was a boy, longing to be free of his "prison house" in Brooklyn. But in the end, he had chosen the pastoral traditions of the American countryside, and the strict, temperate lifestyle of his parents. Billie, however, dove headlong into New York nightlife. "Went to Maurice Becker's on Wednesday night and had an awfully good time," she wrote her more reserved sister, Jane.

> He does those wild things for "the masses." . . . He's a young Russian Jew, twenty-eight or so, but he looks much younger. Dances beautifully, an awfully jolly unaffected sweet sort of chap, very attractive and full of life. . . . I danced a lot with Chamberlain, a big chap who went to Ohio State University once for a month. . . .
>
> One of the Princeton instructors, who labors under the terrible name

of Byron Cassius Gott, fell rather heavily for your little sister. At the end of the second dance he was pleading for my phone number, and I finally did give it to him—with no idea that I should ever see him again. But he called me up only a couple of days later. We went to a place down at Houston Street—had all kinds of fun—danced and argued and drank—well! I never knew there were so many good things to drink which I don't know! Jane, I sure do wish you were here, so I'm on the hunt for a good-looking jolly scout who dances well, so that we can have a nice friendly party altogether once in a while and you would be just right, damn it![4]

As Billie was dancing the Charleston and the Mooch and the Sugar, she was also becoming involved in the burgeoning Village art scene. She had always been something of a flirt at parties for "the masses," but soon after leaving Barnard she met and fell in love with a writer named Arthur Moss. After a brief romance, he asked Billie to marry him, and she accepted.

Given to irony and sarcasm, Arthur Moss was a gregarious wit, an avant-garde Village intellectual, and a ceaseless self-promoter, who had already been married before. He had a grand plan to launch a magazine devoted to Manhattan's bohemian art scene, and he wanted his wife, Billie, to be a part of it. So together with a group of friends, they founded The Quill, a monthly publication featuring poetry, reviews, and simple sketches by local artists. Millia Davenport was the only woman of the eight names on the masthead.

A new type of youthful American spirit jumped from the pages of The Quill. As Arthur wrote in the first Note to Readers: "The editorial policy is neither here nor there. The Quill will not do battle for Causes or Movements, for it does not believe in them. It will not preach. It will not offer any panaceas for the sorrows or joys of life; rather it will serve as a meeting place for the few who are struggling ever and ever for an art that will be truly American. An art that is not hidebound by the deadening influences of a decadent Europe, or the result of intellectual theories evolved by those whose only pleasure in existence is to create laws for others to obey. . . ."[5] This impulse to throw off the strictures of the "Old World," to create a pristine paradise free from the shackles of the past, was merely another version of the Puritan idealism that had formed American self-identity for centuries.

Charles was horrified. He had always held the "creative spirit" in suspicion, and he had little use for the arts in general. The "bohemian" lifestyle embraced by his daughter seemed to border on the moral anarchy he had been trying to fight for most of his life. Here it was, the corrupting influence of the city, which had long been seen as a primary breeding ground for the people who were shattering his and other eugenicists' view of America.

Billie once described her new life as the wife of a Village wit to her aunt Fanny, one of the few family members she really trusted. She had always felt a particular fondness for this Davenport aunt, her father's younger sister. Charles, however, had always been somewhat ambivalent about Fanny, who had never married. She received a Ph.D. in history from the University of Chicago, and was now teaching at Vassar. When Charles filled out his own "Record of Family Traits" form, he described his sister as "the best student in the family . . . high capacity for generalization. Lacks physical courage . . . dislikes all manual work. Somewhat lacking in affection, homosexual motive, probably."[6] Since Fanny wasn't a "traditional" woman, Billie had always been drawn to her, and so she felt comfortable telling her about life in Manhattan.

> I'm very healthy and happy. I wouldn't trade this for Cold Spring Harbor for anything on earth. I can't help it. I don't like frumpy women who know nothing but how to teach biology. That's why I like you, you're an historian—and, yes, you read books and go to the theatre and travel and know when you are eating good food and when you are not. . . . Arthur lived in London for years, in Paris, in Munich, he knows the continent pretty well. He has a tremendous acquaintance among literary and artistic people, both here and abroad. Young as his wife was when she died, she was a musician of some reputation and added to his acquaintance many musicians. I prefer them to biologists. Then there are many charming wealthy dilettantes who give nice dinner parties. And I prefer a dinner for twenty four in the Garden Club at Washington Heights— given by the Tom Vintons—to a Biological Congress at lunch in Blackford Hall.[7]

Such antipathy toward science and the Davenport family values concerned her father. He was also worried about how often Billie was fraternizing with that morally suspect race: the Jews.

Who were these people at *The Quill?* J. C. Rehber, Arthur Gude, Kingsley Moses, Lillie Strunsky, Ida Cohen, Forrest Mann, Perez

Hirschbein, Anton Hellmann? It seemed almost half the contributors had Jewish or Jewish-sounding names. His own eugenic research had shown that, as most everyone had suspected, the Jews were greedy and lascivious. They committed "the greatest proportion of offenses against chastity and in connection with prostitution, the lowest of crimes."[8] And while he often made anti-Semitic jokes, there was truth to them, he thought. When people from the city came out to the country, many would run amok in the clean meadows, ruining the natural scene. Charles once remarked, somewhat tongue-in-cheek:

> *I have spent many a precious Spring Sabbath, sitting near the road with a gun to keep the vandals and automobilists from wrecking a dog-wood grove and gleaning where they have not sowed. As a result of observations in perhaps 50 cases, I have reached the conclusion that at least three quarters of the offenders are obviously Jews who seem not to be able to restrain themselves from getting something for nothing. They gather the most delicate wild flowers which certainly have gone beyond redemption by the time they reach their flats. Nothing has any influence upon them but the presence of a gun. If you learn I have been shot out here you will understand why!*[9]

And now Billie was living among these people, and so shamelessly!

The Quill became quite successful, receiving a rave review from H. L. Mencken and a full write-up in the *Sunday Tribune Magazine*.[10] And as Billie designed many of the magazine's covers, she was becoming an accomplished artist herself. In an issue that featured one of her sketches—it seemed to be a self-portrait, depicting a woman standing at an easel with a palette in her hand—Arthur wrote a sarcastic taunt of her father's science of eugenics. Perhaps Billie had tried to explain to her husband what her father did for a living, prompting Arthur to choose this as the object of his famous wit. In the Heard Round Our Square section, he wrote:

> *After hearing our wife (late Barnard, T.C. and Carnegie Institute) discourse for hours on eugenics, we have come to realize that a constitutional hydropath with a harelip DR + SX = a moran* (or a McGillicudy) with a microcephalic abnormal development of congenital pigments. Further than this we dare not think. (* She says it is a moron.)*[11]

Many people were impressed with her magazine covers, but Billie was also quickly becoming known as the most talented costume designer in all Manhattan. She had learned to sew and make smocks and dresses while living at home on the campus of Cold Spring Harbor, and she decided to put this skill to use. She even placed advertisements in *The Quill*. Next to the ad boxes for the Black Parrot Lounge and the Mouse Trap Tea Room was her own "Gowns to Order, Millia C. Davenport, 133 Washington Square, Telephone Spring 3951."

As her reputation grew, Billie got a job as the designer for the Greenwich Village Theatre at Seventh Avenue and Fourth Street, a new dramatic venture that featured a talented young playwright and director named Eugene O'Neill. Billie helped design and make the costumes for *Anna Christie* and *Desire Under the Elms*, each to critical acclaim. In 1921, a *New York Times* critic wrote of her work: "Millia Davenport's costumes, squeezed evidently from a discreet budget, seemed to set out to prove Goethe's dictum that it is in the working within limits that the master comes out. They were glittering and brilliant and intoxicating clothes as they moved in that charming mirrored room that served to house the shining scenes. They accentuated the success of the whole performance. . . ."[12]

Millia's successes as an artist, however, came as her marriage began to disintegrate. Arthur started an affair with a member of their riotous Village entourage, a dancer named Florence Gilliam. It was not long before he left their new flat at 182 West Fourth Street and moved in with his new lover—despite the fact that he remained married to Billie.

Life in the Village began to change at this time. When Arthur left, Prohibition had just been enshrined in the U.S. Constitution, and the freewheeling bohemian lifestyle seemed to be under attack by powerful lobbies, the self-appointed guardians of American mores, including progressive Christian women and eugenicists. Arthur decided to move his entourage to Paris, where they could live in relative freedom. Describing the famous group of artists' move, a writer for the *New York Times* in October 1921 wrote with the sarcasm of a bona fide Village bohemian:

> *The conquest of Paris by Greenwich Village!*
> *The national and local janitors of our morals, thirst and recreations decreed that Paris should never come to the Village. Well, then the Village shall move to Paris!*
> *It has moved. With bag, baggage, Pegasus and Parnassus, it has planted its far-flung cry of defiance on the slopes and hills of Montparnasse.*

Which is to say that Arthur Moss (who never took a lesson in French, English, dancing or wassailing in his life), together with Florence Gilliam, Allan Ross MacDougall, Wynn Holcombe, Lawrence Vail, Harrison Dowd and Arthur C. Wyman, have founded the super-Greenwich Village in the Latin Quarter of Paris.

Further, this impertinent band of intellectuals have issued within the shadow of the Pantheon and under the very nose of Anatole France, a monthly literary and art magazine, The Gargoyle, written in English by Greenwich Villagers and illustrated by illustrious artists from West Fourth Street and MacDougal Alley. [13]

Billie was left behind. She seemed to take the change serenely, however, and began to date again. Though still legally married to Arthur, she was becoming particularly fond of a man named Bill, even while she was seeing a number of other men. Her father became even more concerned, of course, so he decided to take the train to Manhattan and have a long discussion with his daughter.

He gave her his advice. He explained again the importance of family, the need to live according to the "mores," and to consider carefully who she was going to marry this second time. He explained the heritage of the Davenports, their good breeding, the need to preserve it and hold it more precious than gold. He had given lectures around the world about the importance of eugenic marriage selection, and he was heartbroken at his daughter's unwise choices. He almost begged her not to marry again so soon, and to think of her family, the family honor.

Billie was touched by her father's tenderness. He had always loved her dearly, and had never been remiss to show her affection and warmth. But after hearing once again his views of family and his beliefs in eugenics—which had been drilled into her since she was a girl—she decided to write her father a long letter, explaining how she felt. It may have been the most important thing she had ever told him. She loved him dearly, too, but she could not believe what he believed, and she would not live according to the values taught at Cold Spring Harbor. He must understand this, she thought, understand why the people she knew and loved were not inferior to people like the Davenports.

Dear:
    You absolutely mustn't worry about me, dear dad. To begin with, I won't be able to do anything scatterbrained for another year and if I get married again you'll know all about it, I promise you. Then you must

remember that I may quite well change my mind and, too, that Bill has a great deal to lose in marrying me, and only me to gain, and I'm not at all sure he'd think the price worth paying. . . . We dine together once a week, and in the mean time, I see a lot of men, all the time. <u>Don't worry!</u> I've been thinking of what you said about family.

Dad, take my generation: Harry and Maurice, Roland and Clarence, Pierrepont and his two sisters, Jane & me. Of these nine, three—Harry, Jane and I—are, averaging intelligence, good breeding, good health, physical and moral, the only ones comparable to Bill and Turk. I'd infinitely prefer to introduce either one of them as cousins, to any of the veritable six.

I talk to you of them—Bill's people—I know, as if they were alien creatures, and so, to me, still to some degree, the children of rather more than well-to-do businessmen are. That strangeness kept me from lovely friendships at college. But the world is not made up of college professor's [sic] children, at least, not the dearest part I've found and loved since. Dougie's father had a bagpipe factory in Scotland. Stanislas Szukalski's father was a blacksmith in Poland, and a wonderful person, it appears, but a peasant, and so on and on. . . .

But, you see, I've not the family feeling that you preach, and, fortunately, I imagine, have seldom to practice, so that the theory remains plausible. . . .

Glad I saw you and we had the walk together. Now <u>don't</u> talk this over with the family. I couldn't bear another conference. If we need one next winter, we'll have to bear it, but probably we won't. You know, and we might as well log on, without torturing ourselves, 'till we're faced with a problem, actually.

Love, Billie.[14]

Charles may have long forgotten his own letter to his father. Both he and Amzi had tried to trace so carefully the Davenport family history, each in his own way. But like most who believe they belong to a chosen people, neither had been very keen on complexity or irony. The great movement of history and heredity had forced another clash of values, another clash of human longings, another conflict between parent and child.

The dearest part of the world that Billie had come to find and love was filled not only with artists and intellectuals but immigrants—Irish workmen, Polish blacksmiths, Jewish businessmen from Russia. They were very different from the blueblood Anglo-Saxon Protestants who had first come to the country, and their differences caused people like her father a great deal of anxiety. "The population of the United States will, on account of the recent influx of immigrants from Southeastern Europe, rapidly become darker in pigmentation, smaller in stature, more mercurial, more attached to music and art, more given to crimes of larceny, kidnapping, assault, murder, rape, and sex-immorality and less given to burglary, drunkenness, and vagrancy than were the original English settlers," Charles had once warned in his popular textbook.[15]

It would be ten years before the Pulitzer Prize–winning author James Truslow Adams—a graduate of Brooklyn Polytechnic Institute, like Charles Davenport—coined the cliché the "American Dream" in a book called *The Epic of America.* But Billie was describing to her father what Adams would later call

> *that American dream of a better, richer, and happier life for all our citizens of every rank, which is the greatest contribution we have made to the thought and welfare of the world. That dream or hope has been present from the start. Ever since we became an independent nation, each generation has seen an uprising of ordinary Americans to save that dream from the forces which appeared to be overwhelming it.*[16]

It was the old American dream of a "city upon a hill." Urban immigrants were beginning to see themselves as ordinary citizens, too, and they were embracing the image of America as the new Promised Land—not a land where God demanded moral purity from His "peculiar people," however, but a land in which diligence and hard work would bestow the promise of a richer and happier life. The "American Dream," first understood as a divine calling for restored innocence and social purity (and its corresponding prophecy of doom), had evolved from a heavenly city upon a hill into a land of boundless economic opportunity.

Billie never married Bill. After she got her divorce from Arthur, she continued to live and work as a costume designer in Manhattan, becoming a part of the growing successes of the Greenwich Village Theatre and its director, Eugene O'Neill. By the end of 1923, however, she was writing her family about the new love in her life—another divorcé, a man who had three children. "We ARE going to marry," Billie announced to her sister, Jane, before any of the family had even met him. "And on the 18th! The children are delighted at my marrying their father. And he's a darling, you'll like him."[17]

Charles Davenport, 1938.

On January 18, 1923, Millia C. Davenport married Walter L. Fleisher, a millionaire New York businessman, divorced with three children, and a Jew.[18]

IN THE MORNING HOURS of April 16, 1920, a cold wind buffeted Harry Laughlin as he walked up the steps of Capitol Hill. It was an unseasonably cold day in Washington, D.C.—a "northern disturbance" had swept a blast of Arctic air far down into the Midwest, and it was now flowing up from the South, bringing the possibility of snow at night. Days earlier, not too far away in neighboring Virginia, a woman named Emma Buck had felt the same cold wind as she waited to board a train for Lynchburg and begin her confinement at the Virginia Colony for Epileptics and Feeble-minded.

While the weather was a topic of conversation on Capitol Hill this day, Laughlin knew politicians were also chattering about the fast-changing map of the world, especially in the Middle East. Front-page headlines proclaimed the start of an important international summit in San Remo, Italy, as the Allied powers of Europe, in conjunction with the recently established League of Nations, were meeting to divide the lands of the former Ottoman Empire. As a means to keep the global peace, the League of Nations had helped to organize a new system of international "mandates," which would replace the loosely defined imperialist doctrines of "spheres of influence" or occupied "protectorates." International cooperation, instead of the sheer power of military might, would now determine which Allied nation should control a vanquished region after the Great War. In the most significant regions of this ancient empire, Britain would be given the mandates for Iraq and Palestine, France the mandates for Syria and Lebanon. The United States, however—which had declined to join the League after a majority in the Senate, led by Massachusetts senator Henry Cabot Lodge, felt the country should keep its distance from the affairs of the Old World—was not a party to the conference, and President Woodrow Wilson's hard-line views on Turkey were being politely ignored.[19]

Yet while headlines announced the advent of this historic San Remo Conference, stories also included the status of the fledgling Republic of Armenia, a nation only two years old. In 1915, during a government-sponsored effort to rid eastern Turkey of its Christian inhabitants, Turkish troops had systematically slaughtered hundreds of thousands of Armenians in the region of Anatolia. Over a million more Armenians had been deported to the rocky Caucasus region east of Mount Ararat, near the city of Yerevan. Now, in the aftermath of this mass ethnic deportation and genocide, negotiators at San Remo were trying to determine whether Armenia should receive its own "mandate," and be protected by French or British forces. France opposed such a move; as one newspaper had reported, "Phrases like 'bleeding Armenia' and 'the terrible Turk' leave [the French] cold when they are faced with the practical difficulties of subduing the latter and maintaining the former in a state of freedom and security."[20] The League of Nations eventually agreed, since it seemed both impossible and impractical to protect a landlocked state with no standing army of its own. By the end of 1920, the Red Army would invade Armenia and make it a part of the new Soviet Union, and the first systematic genocide of the twentieth century would be largely forgotten.

Harry Laughlin was very interested in the San Remo story. He had always believed the League of Nations to be one of the most important developments in modern history, and he had been bitterly disappointed when the Senate rejected membership and opted for disengagement. Ever since he was a young man he had dreamed of world unity, and his unease with conflicting cultures and creeds had consumed him as a kind of theological imperative—a Campbellite imperative, which

informed his studies in biology and eugenics. He had long maintained
a vision of a world united under a single government, ruled by superior
"Nordic" Europeans, and he thought the League of Nations could be
the beginning of such a new global order.[21]

On this cold day, however, his thoughts were swirling with more
practical matters, though they stemmed from the same vision of
Nordic (and Christian) supremacy. Today, for the first time, he would
testify before Congress as an expert in eugenics, addressing the question of racial differences and the need for immigration restriction.

The meeting was to begin at 11:10 a.m. Laughlin had been invited
by Albert Johnson, a hard-drinking, cigar-chomping congressman
from the state of Washington, chairman of the Committee on Immigration and Naturalization. They were colleagues at the Eugenics Research
Association, an eclectic group of scholars, politicians, and businessmen
who formally organized to bring the findings of eugenics to practical
legislation. The association had been established at Cold Spring Harbor, the Mecca of eugenic research, and Laughlin had served as its secretary since 1913. He came to speak under the auspices of this group
rather than the Carnegie-funded Eugenics Record Office, which officially discouraged political activities.

Johnson called the committee to order. "We have a gentleman here,
Mr. Laughlin, who came at our request and whom we should hear."

Laughlin sat at a table below the fourteen congressmen. Almost forty
now, he had lost most all his hair, and his appearance was somewhat
striking. His unusually pale skin and balding pate tended to emphasize
the thick lips and upper eyelids, which hung puffed and heavy over his
brown eyes, making him appear to squint. This natural narrowing of
the eyes produced a perpetual scowl, and caused his face to look fierce
and irritated. Getting his notes in order, he began his presentation.

"Mr. Chairman, I want to present the biological and eugenical aspect
of immigration. Some of my remarks will be of a general nature, but I
will support them by special data."

Congressman John Box, a Democrat from Texas, interrupted immediately, asking the chairman, "What particular phase is he discussing?"

"I presume checking immigration," Johnson replied. "We will give
him permission to put his statistics in the record."

Laughlin continued. "The character of a nation is determined primarily by its racial qualities; that is, by the hereditary physical, mental,
and moral or temperamental traits of its people. We have trained field

workers who visit insane hospitals, prisons, and other institutions for the socially inadequate, get in touch with the inmates or patients, find out whether [they are] of native or foreign stock, and then go to their home territories and determine what kind of hereditary material they are made of; in fact, we are trying to solve the problem of the relative influence of heredity and environment in making these degenerate Americans. Since coming under national control our immigration policy has been determined largely upon an economic basis: this was especially true in the earlier years of Federal control, but in later years the sanitary feature quite properly entered, and during the war the element of immediate national safety ruled.

"It is now high time that the eugenical element, that is, the factor of natural hereditary qualities which will determine our future characteristics and safety, receive due consideration."[22]

Fears concerning immigration had been welling up for decades—almost 12 million immigrants had arrived between 1890 and 1910, most from the southern and eastern parts of Europe—and Congress had passed numerous restrictions in the past few years, as members of the committee well knew. The Chinese Exclusion Acts of 1882 and 1902 had severely restricted the inflow of immigrants from the Orient, and in 1907, President Theodore Roosevelt had brokered a so-called Gentleman's Agreement with Japan, in which the Japanese government agreed to curtail the flow of immigrants coming to California to find work. But, as Laughlin pointed out, most arguments for such restrictions emphasized economic fears. The Immigration Restriction Act of 1917, however, did begin to include the "sanitary feature," and covered European immigrants. It broadened the "barred zone" for Asian immigrants even further, but it also banned anyone with any type of mental problem, contagious disease, or history of pauperism. In addition, it banned anarchists, polygamists, and other political agitators influenced by the theories of Karl Marx. The most controversial part of the 1917 Immigration Act was its literacy test, which excluded "aliens over sixteen years of age, physically capable of reading, who cannot read the English language, or some other language or dialect, including Hebrew or Yiddish."[23] It was a provision the old Yankee establishment had been urging for years, though most felt the nation needed even more stringent restrictions.

The welling fears were reaching breaking point, and Laughlin's testimony came as anxiety was particularly rife throughout the country, especially among Anglo-Saxon Protestants. The summer of 1919 had seen the most violent and widespread labor strikes in the history of the country. Race riots raged in American cities as Negro doughboys returned from fighting in Europe and demanded greater rights and dignity. Young women were embracing a "flapper" lifestyle and beginning to drink and smoke. At the same time, the United States endured its first "Red Scare." President Wilson's Attorney General, A. Mitchell Palmer, along with his young assistant J. Edgar Hoover, feared that Communist agents from Russia were planning to try to overthrow the American government. On November 7, 1919, on the second anniversary of the Russian Revolution, they arrested over ten thousand suspected Communists and anarchists and held them without any formal charges. Most of these were released, but almost three hundred were deported. And just a few months before Laughlin gave his testimony before Congress, Palmer and Hoover again arrested over six thousand suspected Communists on January 2, 1920, claiming these radicals were planning a violent revolution to take place on May Day. These "Palmer Raids," as they came to be called, caused mass hysteria, and in New York, five elected Socialists were expelled from the legislature.[24] Most labor unrest and social problems, it seemed, stemmed from foreign immigrants or non-Nordic residents, including Negroes and Jews.

In the midst of such anxiety, Laughlin brought his "special data," his statistics that seemed to show how these crises were caused by the innate, biological defects of non-Nordic immigrants, and the inevitable problems brought about when races intermingled. This social unrest was arising from evolutionary differences in racial groups, he explained, differences rooted in their genes.

After his introduction, Laughlin went on to present a general history of the famous case studies—the Jukes of New York, the Kallikaks of New Jersey, and the Ishmaels of Indiana—and the surreptitious menace now lurking in the country. Seizing on the major theme of these family narratives, Laughlin emphasized to the committee the particular danger of the fecund, feebleminded female. He described the discovery of the "moron"—which drew a number of questions from the congressmen—and explained how these "borderline cases" were very difficult to discern in the population, since they seemed "normal" to the untrained layman, as well as to immigration officers.

"How do you spell that word?" asked Congressman William Vaile of Colorado.

"M-O-R-O-N," replied Laughlin. "It is a Greek word, meaning 'a foolish person.' And those women of the borderline cases are often fertile. If you go to the schools, such as the one at Gainesville, Texas, where we have a field worker now, we find that the moron girl is highly fertile sexually. She has not any sexual inhibitions, as a rule, and her fecundity is limited only by the number of children and coming in contact with men; that is, physiological, not social, conditions limit the fertility of the average female moron that is not placed in an institution and protected.

"Now, a moron can slip through the immigration sieve, as it exists today, pretty easily. And the moron is really a greater menace to our civilization than the idiot. . . . A moron comes before the immigration board, passes the very elementary tests, and is admitted." And, as Laughlin had explained earlier, the danger to the American bloodline would be threatened even by "normal" immigrant women, since not only did they have a higher rate of defective children, but "statistics have shown that immigrant women are more prolific than our American women."[25]

Yes, statistics have shown. Using Galton's old methods, Laughlin pointed out the statistical spikes that seemed to show how prisons and mental institutions were burgeoning with the new immigrants. According to the 1900 Census, he explained, 19.5 percent of the country's inhabitants were foreign-born, and these came not from England or Scandinavia or Germany, but from Italy, Russia, and Austria. And, as Laughlin pointed out, they were "largely Jews."[26] Yet these immigrants were constituting over a third of the nation's insane population. "Now, if that foreign stock was just as good as the stock already here, it ought to have contributed only 19.5 percent," Laughlin reasoned. Add to this the fact that immigrant women were more fecund than the "American stock," and the future of the country was starting to look somewhat bleak.

This was not the only reason, however, to restrict the new immigrants. Laughlin brought up one of the bedrock beliefs of the Eugenics Research Association: the concept of biological and racial purity. When different races live among each other, he said, they tend to intermingle and breed. For Laughlin and other members of the association, this posed a further ominous threat:

"Now, in the matter of race mixture, Mr. Chairman, if I may, I should like to bring up a point: The committee of the Eugenics

Research Association has had the matter in hand, and has failed to find a case in history in which two races have lived side by side for a number of generations and have maintained racial purity. Indeed, you can almost lay it down as an essential principle that race mixture takes place whenever there is racial contact."

Congressman Box interrupted Laughlin. "What do you mean by 'racial purity' as you have used the term?"

"For instance," Laughlin replied, "if the Negroes and the whites live side by side, will all of the children of the next generation be descended only from Negroes in their half and only from whites in their half? . . . The higher races everywhere tend to keep themselves pure on account of the relative chastity out of wedlock of the women of the higher caste, and the lower race tend to mix for exactly the opposite reason. Wherever two races come in contact, it is found that the women of the lower race are not, as a rule, adverse to intercourse with men of the higher. And that has been true throughout history. It is true now."

And it was true, too, for the founder of eugenics. Sir Francis Galton, of course, had obsessed over beauty maps, buxom foreign women, and a master race of "mammalian" females and "virile" males in the land of Kantsaywhere. His dallying with "women of the lower race" probably led to a case of syphilis. His eugenics disciples, too, had obsessed over the "purity of our women," pointing to the feebleminded female or women of a lower race, explaining how they somehow sang an irresistible siren's song and seduced white men of decent standing. (Indeed, in the South, many states were already passing miscegenation laws at this time, forbidding the sexual intermingling of blacks and whites.) Yet no one—neither Laughlin, Davenport, nor any other eugenicist—explained whether this constituted a genetic flaw or constitutional weakness among these Nordic males.

Laughlin also offered the committee a few suggestions on the need to compile eugenic family histories for potential immigrants, the need to screen their genetic fitness before they even began their travels. An "immigration attaché," he said, could supervise this eugenic research in each U.S. Embassy. Yet the primary purpose of his testimony was simply to give a scientific veneer for the severe restrictions Congress was now considering. Laughlin's genetic arguments before the committee only reinforced a decades-long belief in the superiority of the "Nordic" race of Northern Europe.

These sweeping, epic issues had been the focus of Laughlin's life for

over a decade. Stamping out the racial menace, whether through sterilization or immigration restriction, had literally been his goal ever since
coming to Cold Spring Harbor, and his primary desire was to purify the
American land of its social undesirables, its congenitally unfit. Yet even
as his statement before Congress was a great personal and professional
achievement, Harry Laughlin was considering a radical change in his
own life: he was considering moving away from eugenics and devoting
himself to something more pastoral, if safe and mundane. By the end of
spring 1920, Laughlin would take a leave of absence from Cold Spring
Harbor and bring his wife Pansy to the Scripps Institution for Biological Research at La Jolla, California. He had been offered a position to
study the seedlings of date palms and attempt to improve their efficiency. He might even become a partner with one of the commercial
date farms in the San Joachin Valley. He and Pansy could live with a
better salary and lower living expenses, see the stunning ocean views of
La Jolla every day, and experience the gentle weather of Southern California.

Laughlin stayed in California for a year, but soon decided to return to
Cold Spring Harbor. As a cash crop, date palms grew slowly, and the
return on any investment was minimal. The research to improve the
seedlings also proved difficult, if not impossible. And, as his mentor
Charles Davenport warned him, he was risking obscurity from geneticists and men of science if he stayed on the West Coast.[27]

When Congressman Albert Johnson offered him an official position
as the "Expert Eugenics Agent" for the House Committee on Immigration, however, Laughlin was finally convinced he must go back east.
Johnson commissioned him to do a massive study on immigrants in
state institutions, and promised him the franking privilege with official
congressional letterhead. It was another opportunity to do something
really great for America, he thought; so despite the more gentle
lifestyle in California, Harry and Pansy returned to New York.

During his testimony in April 1920, Laughlin had told the committee that the old colonial stock of America, who had long pressed west
and colonized the frontier of America, always bore and reared large
families—just as the new immigrants did now. "Pioneer conditions,"
he said, demanded many children as this hearty stock conquered the
western wilderness. "The premium now is tending toward still smaller
families among the more prosperous and cultivated," Laughlin warned.

"I think that the native stock will have to look to its laurels in this matter."

As he and Pansy returned from their own venture west, however, they remained without children of their own.

THE IDEA OF RACIAL PURITY had a long history in the United States, and Laughlin's genetic arguments before the Committee on Immigration and Naturalization only provided a new eugenic justification for a long-standing belief in the superiority of the "Nordic stocks" which, ostensibly, first colonized the American shores.

Since the end of the nineteenth century, many New England Protestants had been obsessed by the "character of the nation." With the rise of the evolutionary theory of Charles Darwin and the social philosophy of Herbert Spencer, many of the so-called Boston Brahmins—a term coined by Oliver Wendell Holmes, Sr., the father of the great American jurist, to describe the "caste" of the country's most prominent intellectuals and political leaders[28]—began to embrace new theories of ethnic purity and racial supremacy. Earlier Yankees had tended to ignore their Old World ancestry, and emphasize instead the idea that they were a "peculiar people" and, like Abraham of old, chosen by God to leave their homeland behind and live in a new Promised Land in America. Men like Amzi Davenport were most interested in the unique American traits of the hearty Puritan New England caste, and they placed their ideas within a biblical and theological context. But with the creeping rise of rationalism and natural science, this religious self-understanding was evolving into a more secular and scientific form. In universities like Yale and especially Harvard, a theology of divine election was giving way to a theory of race that idealized the ancestry of the Nordics, and in particular the Germanic Anglo-Saxons.

In *fin-de-siècle* Europe, a significant number of intellectuals seized upon the "Teutonic" theory of one French author and diplomat, Comte Joseph-Arthur de Gobineau, who in 1854–55 published a multivolume work entitled *Essai sur l'inégalité des races humaines* (*Essay on the Inequality of the Human Races*). With its deliberate allusion to the French Revolution's call for *liberté, égalité,* and *fraternité,* his essay attacked the very idea of liberal democracy, which, he argued, had stripped power from the natural-born elites of the human race. Tracing the history of the "races," he postulated the existence of an ancient master race called the Aryans. Gobineau had borrowed the idea from the brilliant English philologist Sir William Jones, who in the late eighteenth century discovered the evolutionary tree of language. Jones traced the origin of ancient languages such as Sanskrit and Latin and Greek—the primary parents of many modern European languages—back to an Ur-language in India, which he labeled "Aryan." (In later centuries, Jones's evolutionary idea would still stand, though this ancient language would be called "Proto-Indo-European" instead.) Gobineau, however, claimed that the Aryans were a great and pure and master race, and from them came the nomadic Aryan Teutons who wandered west during the early period of the Roman Empire and settled in Northern Europe. "Civilized" European history, according to Gobineau, began only after the Teutonic invasions, when white-skinned and blue-eyed peoples conquered the darker-skinned, darker-eyed natives. The fall of Napoleon and the present decay of France and Europe, he wrote, were caused by the resurgence of the dark-eyed "Gallo-Roman rabble"—and the Jews.

By the turn of the century, the American intellectual caste began to embrace these "Teutonist" ideas. The Anglo-Saxons, forebears of New England Yankees and so-called Boston Brahmins, were supposed to be an offshoot of these great Germanic Teutons.[29] As millions of Italians, Eastern Europeans, and Jews began to pour into the country, by the spring of 1894, a group of Harvard graduates formed a small but extremely influential club which would become the Immigration Restriction League of Boston.

Its three founders, Prescott Farnsworth Hall, Robert DeCourcy Ward, and Charles Warren, were sons of wealthy Boston Brahmin merchants, and each lived a life of relative ease and leisure, devoting themselves to casual intellectual pursuits—much as Francis Galton had. But their new league soon attracted some of the most serious and eminent scholars in the country. They convinced the famous philosopher and historian John Fiske, foremost champion of Herbert Spencer in the United States, to be the first president, and one of their chief patrons was the Massachusetts senator Henry Cabot Lodge, an isolationist who felt any entanglements with the Old World were dangerous for the country. Within a decade, the Immigration Restriction League's National Committee included A. Lawrence Lowell, president of Harvard; William DeWitt Hyde, president of Bowdoin College; James T.

Young, director of the Wharton School of Finance; and David Starr Jordan, president of Stanford, among many others.[30]

The League's leaders wanted to whip up public opinion to oppose the hordes of non-Nordic immigrants flowing into the country. With their unlimited resources, they published tens of thousands of pamphlets, and sent them out to an impressive mailing list that included over five hundred daily newspapers across the United States.[31] Typically, they invoked the long-standing theological self-understanding of Americans, derived from the Puritan idea of the nation as a "city upon a hill," when they sent out a call to make the country pure for democracy. After all, didn't God command the ancient Israelites to rid the land of its heathen Canaanites? As one pamphlet maintained:

If to this particular nation there has been given the development of a certain part of God's earth for universal purposes; if the world is going to be richer for the development of a larger type of manhood here, then for the world's sake, for the sake of every nation that would pour in upon that which would disturb that development, we have right to stand guard over it.[32]

If the country really was to be as a city upon a hill and develop "a larger type of manhood" for the benefit of the world, then these outsiders should indeed be cast out of the new Promised Land.

For years the League lobbied Congress to keep non-Nordics from coming to the country, with little success. But in May 1911, when Prescott Hall wrote a letter to Charles Davenport, his former classmate at Harvard, and asked him about the work of the new Eugenics Record Office, he found a powerful new ally in the quest to keep the nation pure. Davenport's book, *Heredity in Relation to Eugenics,* had just been published, and had explained the genetic inferiority of certain racial groups. Hall, intrigued with the implications for immigration restriction, hoped to form a greater alliance between the League and this new science.[33] He and Davenport eventually agreed to organize a Committee on Immigration within the Eugenics Section of the American Breeders' Association, which already had strong ties with the federal government and the Departments of Agriculture and Labor.

This new alliance provided the New England Brahmins with a fresh scientific rationale for their calls to curb immigration. Amid a mingling dance of new ideas and changing social circumstances—the rapid shift of power from yeoman farmers to urban industrialists, the rising despair amid the carnage of the Great War, the welling fears from violent labor strife—this new alliance would help lead the United States toward the most restrictive immigration laws ever written.

In 1916, a gregarious member of the Immigration Restriction League wrote a wild-selling book that presented the core beliefs of the Brahmin establishment. It combined the tenets of eugenics with other current anthropological and racial theories, and, borrowing elements of Gobineau's anti-democratic musings on an Aryan master race, warned that the nation would become a story and byword throughout the world if it did not look to its laurels, and preserve its racial purity.

Madison Grant, who would serve as the Immigration Restriction League's president throughout the 1920s, was a Yale man and New York socialite involved in a host of civic activities. A friend and adviser to Theodore Roosevelt, Grant was, like the virile former president, an avid outdoorsman, hunter, and conservationist, appointed to lead the crusade to save the California redwood trees. He was one of the founders of the New York Zoological Society, which had built and now maintained the Bronx Zoo. As a wealthy bachelor who didn't have to work to earn a living, he also became active in over a dozen New York social clubs, and helped create the New York City system of parks.[34] But his popular book, *The Passing of the Great Race,* made him known throughout the world.

Hardly a serious work of science, Grant's book presented the case for Nordic racial supremacy "in the language of the vernacular," outlining dangers posed by the new immigrants. Following the ideas of Gobineau, Grant began his book with a warning of the dangers of democracy—dangers already felt by many New England Brahmins, who had, in truth, become suspicious of elective forms of government. Americans, though traditionally "disclaiming the distinction of a patrician class," nevertheless were endowed with a natural aristocracy that had always supplied the country with its most eminent thinkers and leaders. "In the democratic forms of government," Grant wrote, "the operation of universal suffrage tends toward the selection of the average man for public office rather than the man qualified by birth, education, and integrity. How this scheme of administration will ultimately work out remains to be seen, but from a racial point of view, it will inevitably increase the preponderance of the lower types and cause a corresponding loss of efficiency in the community as a whole."[35]

These ideas were actually quite common among many of the intellectual elite at this time. And more and more, many of them began to embrace the idea that biology determined a natural racial supremacy. A belief in a new Great Chain of Being, based on natural science, was becoming *de rigueur* among some of the most influential intellectuals and politicians in the United States. Indeed, while a small number of scholars in the field of molecular genetics were seriously debating the relative influence of "nature versus nurture," Galton's old phrase, most of the elite had already embraced the idea of hereditary predestination. In his second chapter, "The Physical Basis of Race," Grant gave a simple statement of their faith:

> *Whether we like to admit it or not, the result of the mixture of two races, in the long run, gives us a race reverting to the more ancient, generalized and lower type. The cross between a white man and an Indian is an Indian; the cross between a white man and a negro is a negro; the cross between a white man and a Hindu is a Hindu; and the cross between any of the three European races and a Jew is a Jew.*
>
> *As measured in terms of centuries, unit characters are immutable, and the only benefit to be derived from a changed environment and better food conditions, is the opportunity afforded a race which has lived under adverse conditions, to achieve its maximum development, but the limits of that development are fixed for it by heredity and not by environment.*

The immutability of "unit characters," the bedrock assumption of eugenics, and the Mendelian notion of dominant and recessive traits determined by some chemical mechanism in the chromosomes, provided the scientific justification for strict racial segregation. And these "lower types," Grant believed, which included the feebleminded poor as well as "lower races" such as Negroes and the Jews, should really be annihilated altogether. In his fourth chapter, entitled "The Competition of Races," he explained how other fatuous beliefs, especially the old shibboleths of religion which protected the "sanctity of human life" and the "value" of the unfit, actually deprived the Nordic race of its purity:

> *Mistaken regard for what are believed to be divine laws and a sentimental belief in the sanctity of human life, tend to prevent both the elimination of defective infants and the sterilization of such adults as are themselves of no value to the community. The laws of nature require the obliteration of the unfit, and human life is valuable only when it is of use to the community or race.*[36]

Such obliteration, Grant believed, must begin as soon as possible through strict racial segregation, immigration restriction, and forced sterilization.

The Passing of the Great Race became an immediate best seller, and new editions came out in 1918, 1920, and 1921; it was also translated into German, French, and Norwegian. And as the U.S. Congress began to consider new immigration bills, Grant's arguments, clothed as they were in "science," became the most widely discussed issues on both the House and Senate floors. Not only were "Nordics" physically superior, but it was they who had sprung from the ancient Teutons and their Aryan forebears. "By the process of elimination . . . we are compelled to consider that the strongest claimant for the honor of being the race of the original Aryans, is the tall, blond Nordic."[37] If the country did not try to purify the race, its time would pass, and it would become a byword in the history of the world.

The Passing of the Great Race received glowing reviews from prestigious journals such as *Science,* the *Journal of Heredity,* and the *Saturday Evening Post.* The editor of the *Post,* George Horace Lorimer, later commissioned a series of articles on immigration, and in his editorial for the May 7, 1921, issue, Lorimer wrote, "Two books in particular that every American should read if he wishes to understand the full gravity of our present immigration problem: Mr. Madison Grant's *The Passing of the Great Race* and Dr. Lothrop Stoddard's *The Rising Tide of Color.* . . . These books should do a vast amount of good if they fall into the hands of readers who can face without wincing the impact of new and disturbing ideas."[38] In *Science,* the MIT biologist Frederick Woods wrote that Grant's book was "written both boldly and attractively, and [he] had produced a work of solid merit. . . . The present reviewer accepts, in the main, this racial theory of European historical anthropology."[39] Echoing Grant's ideas that same year, Vice President Calvin Coolidge wrote in *Good Housekeeping* that "America must be kept American. Biological laws show that Nordics deteriorate when mixed with other races."[40] Years later, the *New York Times* would point out that *The Passing of the Great Race,* "besides being a recognized book on anthropology . . . has often been called to Congressional attention in the passage of restrictive immigration laws. . . . [The book] helped frame the Johnson Restriction Act of 1924."[41]

During the early 1920s, no two men worked more tirelessly to convince Congress to restrict "worthless race types" from coming to the

United States than Madison Grant and his colleague Harry Laughlin. But while Grant wrote popular propaganda and used his connections with legislators behind the scenes, Laughlin, as the Expert Eugenics Agent for Congress, was the official voice of the eugenics movement. He was the scientist commissioned to do exhaustive research on the nation's foreign-born inmates in state institutions, and prove Nordic superiority with the hard data of statistics.

In 1922, Laughlin presented the findings of his statistical studies in a paper entitled "Analysis of America's Melting Pot," a survey of 445 state-run mental institutions and over 210,000 inmates. His findings, of course, only reinforced the belief that Southern and Eastern Europeans, the "Alpines" and the "Mediterraneans," were inferior subraces. These tax-funded institutions had a far greater percentage of such types of immigrants than the population as a whole. Though a number of scholars later found flaws in Laughlin's statistical sample and discredited his research (he had not included private hospitals, for example, in which most wealthy Protestants placed their mentally ill, and this skewed his sample irreparably), almost all the House committee members accepted his science as sound. So, in 1923, Congress made Laughlin a Special Immigration Agent to Europe, and commissioned him to spend six months traveling throughout the Continent, gleaning more information on Europe's "defectives" and "socially inadequates." In March 1924, Laughlin again presented his findings in a paper on "Europe as an Emigrant-Exporting Continent and the United States as an Immigrant-Receiving Nation." The findings were the same: Nordics were superior, while Eastern Europeans such as Poles and of course Jews, were not as intelligent. When most pro-restriction experts joined the immigration debate in newspapers, they cited Laughlin's work by name.[42]

In the battle to curb immigration, however, Laughlin and other eugenic reformers also used another statistical weapon, a weapon developed a decade earlier by leading psychologists and used to identify the feebleminded. In the midst of the congressional debates, the tangible, seemingly unimpeachable results of intelligence tests had recently swept across the nation, and scientists and politicians were abuzz about the findings of a massive, government-sponsored program that tested the innate intelligence of over 1.7 million U.S. Army recruits. When the Great War began, the members of the old Eugenics Section of the American Breeders' Association saw an opportunity to examine national intelligence and convinced the federal government to fund a sweeping study. Millions of recruits were pouring into draft centers, which would not only provide an astonishing statistical sample for this new eugenic tool, but would also make the testing process convenient and efficient.

The Army testing program was directed by Robert M. Yerkes, a Harvard professor and president of the American Psychological Association. It also included the indomitable Charles Davenport, as well as Henry Goddard and Lewis Terman, the two pioneers of American intelligence testing. Goddard and Terman collaborated to put together two tests that could be administered in less than an hour. The "Alpha" was a written test, and the "Beta" an oral test, given to illiterates.

The statistical results of this massive study were first published in 1921, though many psychologists had been studying its findings for years. Again, the results seemed to prove a neat racial hierarchy. In one of the most important interpretations of the Army tests, the Princeton professor Carl C. Brigham explained in his book *A Study of American Intelligence*:

> *In a very definite way, the results which we obtain by interpreting the Army data by means of the race hypothesis support Mr. Madison Grant's thesis of the superiority of the Nordic type. Our figures would rather tend to disprove the popular belief that the Jew is highly intelligent. . . . Our results showing the marked intellectual inferiority of the negro are corroborated by practically all of the investigators who have used psychological tests on white and negro groups. . . .*
>
> *According to all evidence available, then, American intelligence is declining, and will proceed with an accelerated rate as the racial admixture becomes more and more extensive.*[43]

Laughlin would refer often to Professor Brigham, a fellow Princeton man, as he testified before Congress. But around the country, in both the popular and academic press, people began to wonder whether American democracy was being threatened by the menace of foreign blood, which was not only increasing the rate of hereditary diseases but also diluting the strength of native intelligence. For democracy to function at all, it seemed, it must be somehow purified.[44]

On May 26, 1924, President Calvin Coolidge signed into law the most restrictive immigration bill in the nation's history. From this time forward, a quota based on only 2 percent of the number of foreign-born residents already living in the United States, per country, could now legally enter. And this strict quota was not based on the most recent census—in which millions of Southern and Eastern Europeans filled the rolls—but on the 1890 Census, before the new residents arrived. This plank was key to eugenic aims: it allowed far more English and Scandinavians to immigrate, while profoundly restricting Eastern Europeans, who had come after 1890. As Representative Robert Allen, a Democrat from West Virginia, openly declared, "The primary reason for the restriction of the alien stream . . . is the necessity for purifying and keeping pure the blood of America."[45]

In 1927, the legal assault on eugenic sterilization came to a resounding halt with the Supreme Court case of *Buck v. Bell,* and eugenicists won another battle in their quest. In stark contrast to the decisions coming down from state courts during the previous decade—decisions which affirmed the rights of citizens to equal protection and due process under the law—the Supreme Court now ruled that those whom Theodore Roosevelt had called "citizens of the wrong type" could be sterilized against their will. "It is better for all the world," wrote Justice Oliver Wendell Holmes in his decision, "if instead of waiting to execute degenerate offspring for crime, or to let them starve for their imbecility, society can prevent those who are manifestly unfit from continuing their kind."

Indeed, around the world, as the United States became the first country to pass eugenic legislation, influential thinkers began to see interwoven social groups pitted against each other in a great, epic struggle. Whether this was the bright and the feebleminded, the proletariat and the bourgeoisie, the Turk and the Armenian, or the Aryan and the Jew, as the twentieth century pressed into its fourth decade, the idea of preventing "those who are manifestly unfit from continuing their kind" was becoming a rallying cry for those who considered themselves the strongest and fittest of the human race. Natural science allowed many to redefine the old, clannish myths of superior ancestry—earlier myths of pride in past accomplishments supported by some god—and change them into the corporeal, biological facts of evolution. Stripped of myth and clothed in scientific facts, the "unfit" were now determined by their genes.

On April 7, 1925, not long after the administrators of the Virginia Colony for Epileptics and Feeble-minded won their first "test case" against Carrie Buck, not long after Congress passed the most restrictive immigration bill in U.S. history, and not long after Millia Davenport married a wealthy Jewish man named Walter Louis Fleisher, Charles Davenport wrote a letter to his friend and colleague Madison Grant:

"Our ancestors drove Baptists from Massachusetts Bay into Rhode Island but we have no place to drive the Jews to. Also, they burned the witches but it seems to be against the mores to burn any considerable part of our population. Meanwhile we have somewhat diminished the immigration of these people."[46]

XIII

Neighborly Love and Beyond

In his office at 94 North Madison, on a warm winter's day in sunny Pasadena, Dr. George Dock, a member of the advisory board for California's Human Betterment Foundation and one of the most highly regarded physicians in the United States, was beginning to change his mind about Adolf Hitler.

The date was January 31, 1934, just one day after the nascent Nazi government had transformed the political landscape of the country where Dock had once studied, instituting the *Gesetz über den Neuaufbau des Reiches*—the "Law Concerning the Reconstruction of the Reich." It was a sweeping statute, a new social contract that eliminated the local political institutions of the German *Länder* and consolidated all power in the hands of the new Reichschancellor, who had been appointed exactly one year earlier. "In accomplishing this task," declared Herr Hitler on January 30, "national socialism has purified democracy, for the new government is only an improved expression of the popular will as against an absolute parliamentary democracy. Abolishing all class opposition, the Nazi State bases itself on the racial foundations of the German people."[1]

A "purified democracy," based on solid "racial foundations," was the same hope behind the Johnson Act of 1924, and an idea quite familiar to American eugenicists like Dr. Dock. A purified democracy was, after all, the primary purpose behind their drive to curb immigration and sterilize the unfit in the United States. But after their great legal and legislative victories almost ten years ago, the country had fallen into financial chaos as a "great depression" engulfed the economies of the entire Western world. Images of rampant unemployment, farm foreclosures and dust bowls, men standing in breadlines, and desperate, destitute mothers clinging to their children had defined these last few years, and little attention was paid to the American quest for racial purity.

Like many Americans, Dr. Dock was a bit concerned about the heavy-handedness of this charismatic German leader. His noisy, militaristic nationalism, while inspiring many in the financially devastated German republic, had made other nations, still distracted by their own financial worries, not a little wary. Some newspapers printed small dispatches describing Nazi brutality: the opening of a concentration camp for political prisoners at Dachau, the steady stream of Jewish refugees fleeing violent persecution, and the military buildup defying the Treaty of Versailles. And the image of the scarlet flag, centered with its white circle and black "swastika"—the Sanskrit term for an ancient Indo-Aryan cross—was already becoming a symbol of Fascist oppression.

But in the past few weeks, Dr. Dock had come upon some startling information, and his concern was slowly turning into admiration. He had been translating for the Human Betterment Foundation another new Nazi statute, the *Gesetz zur Verhütung erbkranken Nachwuchses*—the "Law for the Prevention of Hereditarily Diseased Offspring"—which had also just gone into effect on New Year's Day.

It was a eugenic sterilization law. Hitler had brought eugenics back into the news with a bold new national program for human betterment.

California, of course, had passed its own law back in 1909, introducing by far the world's most aggressive program to better the world through eugenic breeding. It had sterilized over sixteen thousand of its own "hereditarily diseased offspring," which made the state the perfect laboratory to study the effects of such revolutionary genetic engineering. The Human Betterment Foundation, according to its charter, was established in 1928 for "the advancement and betterment of human life, character, and citizenship, particularly in the United States of America, in such manner as shall make for human progress in this life." To help accomplish this epic task, the foundation had funded an ambitious survey to study the long-term effects of forced sterilization in California. Its founder and president, Ezra S. Gosney, and his colleague, a Stanford-educated biologist, Paul Popenoe, had been conducting the survey for years, and had recently published their results in a book, *Sterilization for Human Betterment: A Summary of Results of 6,000 Operations in*

California, 1909–1929. Gosney and Popenoe had asked Dr. Dock for his assistance, since he was the former dean of the medical school at Washington University in St. Louis, and recognized by his peers as "the man who knows more about clinical procedures than anyone in the United States."[2] (Their book also expressed a particular debt to the pioneering work of Harry Laughlin, "who has followed the development of sterilization in the United States more closely than any other student.")[3] Now, as Germany embarked on a similar quest to better the world, eugenicists throughout the United States watched with interest—and sometimes even awe and envy.

Most American reformers already knew that German thinkers had been intrigued by the U.S. legislation—especially the sterilization program on the West Coast of the United States. But as he translated the official proclamation of this far-reaching German law, Dr. Dock did not really expect to discover what he did: immediately after the full text of the law, the Nazi government had singled out the example of California and the statistics provided by the Human Betterment Foundation to justify what it knew would be a controversial national program.

"It seems to me the German law is an excellent one," Dock wrote to Gosney. "Probably some of the details would not fit into our legal modes, but the medical work is good, and the patient seems to be protected in every respect.

"I think the reference to the California work, and the work of the Foundation is a very significant thing. The matter has given me a better opinion of Mr. Hitler than I had before. He may be too impulsive in some matters, but he is sound on the theory and practice of eugenic sterilization."[4]

There were a number of reasons to consider this "a very significant thing." In contrast to the haphazard, hit-or-miss, and rarely enforced statutes in the United States, Germany, after this consolidation of power to the Reich, could become more efficient in its own quest for racial purity, and would no longer have to face opposition from local or religious authorities—a significant problem in the United States. And the Nazi proclamation indicated that the Human Betterment Foundation was doing important, internationally recognized work. Their study, it seemed, had helped shape this new German program, which would be the first to be applied to an entire nation. Indeed, Germany could now be the first country in the world to institute a systematic program for bona fide genetic engineering.

Along with his letter, Dr. Dock sent Gosney translations of the German documents. Hitler had officially announced the legalized practice of eugenic sterilization in the July 25, 1933, issue of the *Reichsgesetzblatt*—the *Imperial Legal Journal*—and had signed the declaration along with the minister of the interior, Wilhelm Frick, and the minister of justice, Franz Gürtner. The official commentary and justification for the law, however, had been prepared by Dr. Arthur J. Gütt, a decorated SS officer who now served as director of the Office for Population Politics and Hereditary Health Teachings. As Dr. Dock translated parts of Gütt's commentary, he found familiar arguments, arguments that had been made by Theodore Roosevelt, Calvin Coolidge, and a host of other American leaders. Indeed, in the first official proclamation of the sterilization statute, Officer Gütt cited the example of the United States as justification:

> *Of interest in this connection is the fact that in the United States of North America, according to the statistics of the Human Betterment Foundation, 16,000 persons have been sterilized—about 7,000 men, and more than 9,000 women, up to January 1, 1933. A pamphlet distributed by the Human Betterment Foundation describes the effect of eugenic sterilization as carried out in California as follows:*
>
> 1. *Only one effect—it prevents descendants.*
> 2. *It does not in any way unsex the patient.*
> 3. *It is a protection, not a punishment; it does not imply any shame or degradation.*
> 4. *It is accepted by the patient to be sterilized.*
> 5. *It is accepted by the relatives and friends of the patients.*
> 6. *It is approved by physicians, social workers, and guardians of the peace. . . .*
> 7. *It permits the return to their homes of many patients who would otherwise have to be forcibly detained in institutions. It aids in the preservation of families and prevents their dissolution.*
> 8. *It prevents the birth of children to be brought up by imbecile or feeble minded parents, or by the State.*
> 9. *It relieves the taxpayers of a great expense, and permits the state to care for many more sick people than otherwise.*
> 10. *A marked diminution of sexual crimes has followed.*
> 11. *It permits marriage to many persons who without sterilization could not lead a normal life.*
> 12. *It is a practical and essential step to prevent racial degeneration.*[5]

It was stunning, really, to see their work quoted by the German government in such a prominent place. These twelve points, summarizing the findings of Gosney and Popenoe's research, were the first pieces of evidence cited in the official Nazi rationale for the law.

Though Dr. Dock did not know it, a week earlier, Officer Gütt had already highlighted the example of the United States in explaining the significance of the German law to a group of foreign correspondents in Berlin. As the *New York Times* reported, he had defended the statute by telling them that "all civilized races stand in imminent danger of degeneration because civilization has turned natural selection, which eliminated the sick and unfit automatically, into 'counterselection,' which not only keeps the unfit alive but also enables them to breed more rapidly than the healthy and ambitious.

"We do not want to abolish the benefits of civilization," Gütt continued. "We will still care for the sick and infirm. But we do want to prevent the hereditarily afflicted from transmitting their afflictions to their children, thereby poisoning the entire bloodstream of the race. We go beyond neighborly love; we extend it to future generations. Therein lies the justification of the law." He then went on to praise "the good example set by the United States, both in Federal immigration laws and in numerous State sterilization laws."

Officer Gütt's premises reflected the long history of social Darwinism in Germany. This complex, multifaceted theory had undergirded eugenic thinking in his country just as it had in the United States and Britain. But whereas a host of British and American thinkers had seized upon the laissez-faire doctrines of Herbert Spencer and begun a powerful tradition of freewheeling, free-market social struggle, German thinkers had shown far less confidence in the inevitable progress of evolution. Landlocked German-speaking nations had never found much success in the struggle for colonies around the world, and for centuries, internal social conflict had led to bloodshed and economic devastation. It was difficult, if not impossible, for them to see evolution as a force "working out the great scheme of perfect happiness," as Spencer had once maintained. Since the Thirty Years' War, perhaps, most Germans could feel "their ancient pulse of germ and birth / Was shrunken hard and dry"—and only worsening after the devastating defeat of the Great War. Many British thinkers, of course, experienced the same darkling clouds of despair, but at the same time their great worldwide empire confirmed their strength and fitness. Americans, too, felt these same anxieties and fears, but even during anxious times, their tradition of optimism could abound amid the limitless promises of a New World.

When Officer Gütt used the term "counterselection" in this chat with foreign correspondents, he was alluding to the ideas of the first and most influential German eugenicist, Dr. Alfred Ploetz, the man who coined the term *Rassenhygiene,* or "racial hygiene." Ploetz had always emphasized the negative aspects of counterselection in his writings, explaining how Germany's bloody wars and revolutions over the centuries had contributed to its failure to succeed—the strong had been slaughtered, while the weak had survived. But even more, Ploetz warned that the tradition of caring for the sick and weak was poisoning the bloodstream of the German race. Weaker Germans were actually protected, allowed to reproduce, and thus causing the inevitable degeneration of a noble people.[6] This was the classic eugenic argument, of course, but with no reason to feel optimistic about the progress of evolution and no reason to feel satisfied with the current state of affairs, German leaders began to embrace the state control of human reproduction in ways their American counterparts could only imagine.

Within the mingling of new ideas and changing social circumstances, eugenic movements in these countries evolved in very different ways. American optimism had always been tempered by a sweeping prophecy of doom, so American eugenicists had found a measure of success with their programs for genetic engineering, despite significant opposition. British eugenicists, however, were never able to pass such "negative" legislation. As an Anglo-American idea, eugenics evolved more slowly in Germany, but its own tradition of despair soon led it to observe, as Officer Gütt now reminded the foreign correspondents, "the good example set by the United States."

Almost a decade earlier, for example, in the fall of 1920, the great German geneticist Erwin Baur, a student of Alfred Ploetz and one of the leading eugenic thinkers in Germany, had written a letter to his colleague Charles Davenport, seeking his expertise in the history of American eugenic legislation. The situation in Baur's war-ravaged country was bleak, he explained:

> *The Medical Division of the Prussian Government has asked me to prepare a review of the eugenical laws and* Vorschriften *which have already been introduced into the differed {sic} States of your country. Of especial interest are the*

marriage certificates {Ehebestimmungen}—certificates of health required for marriage, laws forbidding marriage of hereditarily burdened persons among others—further the experiments made in different states with castration of criminals and insane.

It is at present extraordinarily difficult to gather together the desired material. I am thinking, however, that perhaps in your institute all this material has been already gathered. That, perhaps, there may be some recent printed report on the matter. If my idea is correct I would be exceedingly thankful to you if you could help me with a collection of the material.

The entire work of eugenics is very difficult with us, all children in the cities are entirely insufficiently nourished. Everywhere milk and fat are lacking, and this matter will become yet greater if we now shall give up to France and Belgium the milch cows which they have requisitioned. The entirely unnecessary huge army of occupation eats us poor, but eugenically the worst is what we call the Black Shame, the French negro regiments, which are placed all over Germany and which in the most shameful fashion give free rein to their impulses toward women and children. By force and by money they secure their victims—each French negro soldier has, at our expense, a greater income than a German professor—and the consequence is a frightful increase of syphilis and a mass of mulatto children. Even if all French-Belgian mishandling by German soldiers were true, they have been ten times exceeded by what now in peace!—happens on German soil.

But I have wandered far from my theme. We have under the new government an advisory commission for race hygiene . . . {which} will in the future pass upon all new bills for the eugenical standpoint. It is for this commission that I wish to prepare the Referate {reports} on American eugenic laws.[7]

Davenport quickly wrote to Laughlin, telling him to send a complete set of reprints of any publications on sterilization to Baur's colleague Eugene Fischer, the director of the Kaiser Wilhelm Institut für Anthropologie, Menschliche Erblehre, und Eugenik—a Berlin-based institute funded by the Rockefeller Foundation.[8] Though German scientists had begun to report on the American legislation and propose similar measures in the early 1920s, the German government was never able to adopt any sweeping national bills. The Roman Catholic Church was powerful among the *Länder,* and as German racial hygienists debated such questions over the years, local governments were unable to pass new laws. It was not until the National Socialist regime took control, and Reichschancellor Hitler consolidated political power, that German eugenics could follow the American lead.

In the early 1930s, most American eugenicists were working to tighten U.S. immigration laws even further, and still trying to encourage a more widespread and systematic application of their sterilization programs—now the law in almost thirty states. Most believed the new German experiment, however, represented a significant leap forward for eugenics. As the American press reported the details of the new Nazi program, many began to remember again the profound, epic nature of their quest. Officer Gütt was simply stating the rationale preached for decades in Britain and the United States: the quest for the Good, the ancient enjoinder for real neighborly love.

So now the Nazi government made careful use of the American precedent as it publicly pronounced the details of its sterilization statute. America had been the first, after all, to embark on the quest for racial purity, and for American eugenicists like Dr. Dock and his colleagues Gosney and Popenoe, as well as Davenport and Laughlin and others, it seemed their influence was unmistakable—a very significant thing.

ON THE EAST COAST, at the Eugenics Record Office in Cold Spring Harbor, Harry Laughlin, too, was beginning to feel a certain pride when he saw the details of Hitler's sterilization program. A few months earlier, Laughlin had come across a brief article in the *New York Herald Tribune*. He clipped it and placed it in his folder of articles that traced eugenic news around the world.

> *BERLIN, May 4—The first "race bureau" in Germany was established today at Dortmund, in the Ruhr district, by order of the city commissioner. Its object is to build up a "pure" Germanic race, free from Semitic influence. In pursuance of this aim, laws will be passed providing for "segregation and improvement of races."*
>
> *The population will be divided by families into two classes—families whose representatives are acceptable to the state and those whose members are "undesirable as being a burden upon the nation."*
>
> *Dr. Brauss, the Dortmund physician who has been placed in charge of the race bureau, announces that he has already on file data concerning 80,000 school children of the city.*

In the margin of the clipping, someone had written: "Hitler should be made honorary member of the E.R.A.!!" The handwriting wasn't Laughlin's—perhaps it was written by his wife, Pansy, or a secretary or another researcher at the office—but Laughlin had long admired the unsurpassed racial traits of the Teutonic German people, and was beginning to admire the German Reichschancellor. This "race bureau" idea was very similar to proposals he had outlined in his own eugenic writings for twenty years, and with such developments in Germany, he would have gladly given Hitler an honorary membership in the Eugenics Research Association.[9]

Laughlin had received the full text of the German sterilization law from the German consul Otto Kiep in July 1933, just days after the Nazis had passed it. (This did not include Gütt's commentary, however.) As he read the text, he saw startling similarities to his own "Model Sterilization Law," and felt there must have been a reason for this. In the next issue of *Eugenical News, Current Record of Genetic News and Racial Hygiene,* the monthly newsletter he had helped edit and publish since 1916, he described the new Nazi program with calm, measured enthusiasm:

> Germany is the first of the world's major nations to enact a modern eugenical sterilization law for the nation as a unit. . . . The law recently promulgated by the Nazi Government marks several advances. Doubtless the legislative and court history of the experimental sterilization laws in 27 states of the American union provide the experience, which Germany used in writing her new national sterilization statute. To one versed in the history of eugenical sterilization in America, the text of the German statute reads almost like the "American model sterilization law."[10]

Indeed, the German law included almost the exact list of unfit categories that Laughlin had cited: congenital idiocy, schizophrenia, manic depressive insanity, congenital epilepsy, Huntington's chorea, hereditary blindness, hereditary deafness, and "severe hereditary malformation." In a separate section, the German law also added that "anyone with severe alcoholism may be sterilized." Davenport and Laughlin had first studied each of these as specifically hereditary maladies in the labs at Cold Spring Harbor.

The Nazi law, however, was actually much more limited in its scope than Laughlin's Model Law. The Nazis, at this point, focused only on physical problems, while Laughlin had included the feebleminded, the criminally prone, and the "economically dependent." But Laughlin was most interested in the Nazi program for its "incurables census," a nationwide search that would comb cities and the German countryside for each type of social undesirable. According to the Nazi decree, all local physicians, heads of sanitariums, and private nurses were required to compile lists of every hereditary defective they knew. Any physician or nurse who failed to turn in a list of defectives would be fined 150 deutschmarks.[11] The Reich's minister of justice, Franz Gürtner, ordered all criminal justices, public prosecutors, and prison wardens "to sift all persons brought before them for candidates for sterilization."[12] Free citizens on the lists would be served with warrants and told to appear before special "Hereditary Health Courts." After being examined again by a district doctor, or special eugenics officer, they would then either be released or slated for sterilization. A State Sterilization Supreme Court would hear any appeals.

This had been Laughlin's idea all along. He was one of the first to suggest a statewide survey to identify hereditary defectives. He was the first to suggest a special state court to adjudicate the cases of citizens to be sterilized against their will. And he was the first to make "due process" the cornerstone of any sterilization law, so the constitutional rights of the citizen could be preserved. Now the German law followed his ideas almost to the letter. Since it included such careful legal protections, Laughlin defended the law from criticisms he knew quite well:

> To one acquainted with English and American law, it is difficult to see how the new German sterilization law could, as some have suggested, be deflected from its purely eugenical purpose, and be made "an instrument of tyranny," for the sterilization of non-Nordic races. One may condemn the Nazi policy generally, but specifically it remained for Germany in 1933 to lead the great nations of the world in the recognition of the biological foundations of national character.[13]

Across the United States, supporters and opponents of eugenic sterilization watched the German program take shape. In the first weeks of 1934, the Nazi Ministry of the Interior, headed by Wilhelm Frick and his assistant Arthur Gütt, quickly instituted the first local *Erbgesundheitsgerichte,* or Hereditary Health Courts, in Thuringia, a small state in central Germany known for its vast forests and farmland. Over the next few months, as they expanded this new legal bureaucracy throughout the country, fieldworkers began to identify Germany's "social undesirables" and doctors began to sterilize them in a systematic and efficient way. Children as young as ten, women up to the age of fifty, and men of any age fell under the auspices of the law. The program sterilized hundreds in the first few months, but by the end of the year, tens of thousands were filling operating rooms around the country. With all power consolidated, Hitler could begin the first stage in his quest. The country's surgeons were kept very busy.

As the procedure became more common, however, problems arose. In the state of Baden, where over three thousand warrants for sterilization had been issued by August, Nazi authorities had to issue a warning to anyone who ridiculed subjects of sterilization. Taunting had become such a widespread problem throughout the state that the efficiency of the program itself might be put in danger. Anyone who made fun of a sterilized person could be fined and imprisoned.[14]

In Berlin, the Hereditary Health Court ruled that even foreigners, whether resident aliens or simply transients passing through, were subject to the law. Foreigners must abide by German law, the court reasoned, and this was no exception. There was, of course, the question of how these foreign defectives were to be discovered. At this point the court simply ruled that obvious physical defects such as deafness or blindness would be one telltale sign of hereditary disease.[15]

The Nazi government soon felt the scope of its program was too limited. By November 1934, it expanded the law to cover what had long been American eugenicists' primary target: the feebleminded. The Ministry of the Interior soon maintained that even the "slightly feebleminded" should fall under the sterilization law. At first, doctors would diagnose this class of defectives using the Stern-Münsterberg intelligence test, the German version of the old Binet-Simon test. This diagnostic tool, as all German eugenicists knew, had been pioneered in the United States by Henry Goddard, Lewis Terman, and Robert M. Yerkes, the head of the U.S. Army intelligence testing program. But Nazi eugenicists soon felt these intelligence tests could not decipher the "moral element" of hereditary illness. In Danzig, home to one of the most magnificent synagogues in Europe, the eugenics officer Dr. Ernst Von Holst proclaimed that anyone who did not meet "German moral standards" could be sterilized, even if they tested at a normal or even a high level.[16]

By June 1935, the law was amended yet again. If the Hereditary Health Court found any pregnant woman unfit for procreation, it could provide for the compulsory termination of her pregnancy through the first six months. This policy of forced abortion did not take the genetic status of the father into account, however, since it was too difficult to prove the paternity of a child before its birth. According to the amendment, an unfit woman could still "voluntarily consent" to an abortion after six months, though after she gave birth, she could then be forced to undergo a sterilization operation.[17]

This was traditional eugenic social engineering, focusing on the genetically unfit. But later that year, German racial hygiene began to expand, taking on the old racial obsessions of the Reichschancellor, who had assumed the official title of German *Führer*. By September 1935, the Nazis began to pass a number of new racial hygiene laws. First came the *Gesetz zum Schutze des deutschen Blutes und der deutschen Ehre,* or the "Law for the Protection of the German Blood and German Honor." This law specifically banned any marriage or sexual relationship between a genetic German and a Jew. In October, again following the example of laws in many American states, the Nazis passed the *Gesetz zum Schutze der Erbgesundheit des deutschen Volkes,* the "Law for the Protection of the Genetic Health of the German People." This law required a certificate of racial purity to obtain a marriage license. In November, the Nazis passed the *Reichsbürgergesetz,* or "Reich's Citizenship Law," which stripped all non-Aryans of their German citizenship. Jews would now only be considered "inhabitants"; they had no rights as full German citizens.[18] Evolved from eugenic theory, these became known as the Nuremberg Laws.

As American newspapers and academic journals covered these developments, a growing revulsion toward this Nazi race policy, especially its unabashed anti-Semitism, began to take hold in the United States. Senator Millard Tydings of Maryland, for example, introduced a resolution urging President Franklin D. Roosevelt to send the Nazi regime "an unequivocal statement of the profound feelings of surprise and pain experienced by the American people" when seeing the brutal treatment of German Jews.[19] The front page of the *New York Evening Post* featured an excerpt from a newly published book, *Nazism: An Assault on Civilization,* which explained the history of the National Socialist Party and its expressed aim to eliminate the Jews from German life "on the ground that their racial heritage renders them inimical to the welfare of the state."[20] Yet, at this point, nothing in the eugenic Law for the Prevention of Hereditarily Diseased Offspring singled out the Jewish people. In fact, thousands of ethnic Germans were being sterilized against their will in an effort to purify the defects already within "Aryan" blood—a fact American eugenicists were quick to point out. But each month, dozens of articles were reporting ominous details of alleged atrocities against the Jewish "inhabitants" in Germany.

Amid the uproar, the *New York Times* decided to publish a sober-minded and balanced assessment of the sterilization law. A year before the Nazis passed the Nuremberg Laws, the newspaper had asked a renowned British geneticist, Dr. C. C. Hurst of Cambridge University, to write an article explaining the science behind the Nazi sterilization program. In his piece, "Germany's Sterilization Law: What It Might Accomplish," Hurst explained the elegant mathematical ratios of Mendelian theory, the crossbreeding of dominants and recessives, and concluded that, despite "some shortcomings," the Nazi law was based on sound genetic science. "Although the object of the German sterilization law will not be completely attained, there is no doubt that the German Government will be able to reduce considerably the frequency of all the scheduled diseases, provided that the law is carefully administered and rigidly enforced."[21] In other words, if done rigorously and efficiently, the German quest for racial purity might work.

If many of the public in the United States expressed outrage at Nazi racial policies, a significant number of doctors and scientists were evincing approval and even awe at Germany's rigid enforcement of their sterilization law. Some even tried to use the news from Germany as a call to pass new U.S. sterilization bills. In 1935, Dr. J. N. Baker, the state health officer in Alabama, addressed the state legislature as it considered a new eugenics bill. "With bated breath, the entire civilized world is watching the bold experiment in mass sterilization recently launched in Germany," Dr. Baker stated. "It is estimated that some 400,000 of the population will come within the scope of this law, the larger portion of whom fall into that group classed as inborn feeble-mindedness. . . . It is estimated that, after several decades, hundreds of millions of [deutschmarks] will be saved each year as a result of the diminution of expenditures for patients with hereditary diseases."[22]

Later that year, at the World Population Congress held in Berlin, Dr. Clarence Gordon Campbell, president of the Eugenics Research Association and a Manhattan socialite, gushed in admiration of Hitler. As *Time* magazine reported, German leaders had been "sore from the slings and arrows of foreign criticism," so they were grateful when Dr. Campbell proclaimed:

> *It is from a synthesis of the work of all such men that the leader of the German nation, Adolf Hitler, ably supported by the Minister of Interior, Dr. Frick, and*

guided by the nation's anthropologists, its eugenicists, and its social philosophers, has been able to construct a comprehensive racial policy of population development and improvement that promises to be epochal in racial history! It sets the pattern which other nations and other racial groups must follow, if they do not wish to fall behind in their racial quality, in their racial accomplishment, and in their prospect of survival.

After chiding the "sentimental and religious" objections to eugenic engineering, Dr. Campbell went on to say that "a decided tendency is now to be observed in enlightened minds no longer to place implicit faith in rhetorical principles which have no foundation in fact, and to explore the realities of nature." He even commented on the Nazi drive to purify the nation's racial stock of Jews. "The difference between the Jew and the Aryan is as unsurmountable as that between a black and white . . . Germany has set a pattern which other nations must follow!" And at the closing dinner of the conference, it was Dr. Campbell who raised his glass and toasted: "To that great leader Adolf Hitler!"[23]

Other Americans were less effusive and even somewhat frustrated. "The Germans are beating us at our own game!" exclaimed Dr. Joseph DeJarnette, the Virginia leader who had provided expert eugenic testimony at Carrie Buck's trial in 1924.[24] It was in Virginia, of course, that eugenic reformers had first won the battle for forced sterilization in the United States, and doctors here—especially those involved in *Buck v. Bell* almost ten years earlier—now watched the German program carefully. Dr. J. H. Bell, the man who had sterilized Carrie Buck, had already commented on the German law in the 1933 Annual Report of the Virginia Colony for Epileptics and Feebleminded. "The fact that a great state like the German Republic, which for many centuries has helped furnish the best that science has bred, has in its wisdom seen fit to enact a national eugenic legislative act providing for the sterilization of hereditarily defective persons seems to point the way for an eventual worldwide adoption of the idea."[25]

Scholars traveled to Germany to observe the administration of this sweeping sterilization program. In the summer of 1935, an American doctoral fellow named Marie E. Kopp received official permission to interview doctors and observe the *Erbgesundheitsgerichte* in person. When she returned to the United States, she wrote a glowing article for the *American Sociological Review,* offering a detailed description of what she had observed. "The leaders in the German sterilization movement state repeatedly that their legislation was formulated only after careful study of the California experiment as reported by Mr. Gosney and Dr. Popenoe. It would have been impossible, they say, to undertake such a venture involving some one million people without drawing heavily upon previous experience elsewhere."[26]

Indeed, a fire had been kindled, a purifying conflagration had been sparked by American eugenics, as Charles Davenport had predicted decades earlier. Under the leadership of Adolf Hitler, German race hygiene could finally begin to institute the types of systematic programs first outlined by thinkers such as Davenport, Laughlin, and Madison Grant. Now, from shore to shore, many of these American eugenicists toasted the great German leader.

But was sterilization the best solution to the problem? To be sure, many American eugenicists had often been ambivalent about what appeared to them to be the most efficient and final solution. The Committee to Study and Report on the Best Practical Means of Cutting Off the Defective Germ-Plasm in the American Population had mentioned it as a possibility. Henry Goddard, in his study of the Kallikaks, had mentioned it as a possibility. Davenport, Laughlin, and Grant had each alluded to it as a possibility. As Paul Popenoe wrote in his *Applied Eugenics,* "From an historical point of view, the first method which presents itself is execution. . . . Its value in keeping up the standard of the race should not be underestimated."[27] But alas, it seemed against the mores to revert to burning witches.

Still, by 1935, as the Nazis passed one new racial hygiene law after another, these Americans could feel a sense of pride at their influence on the German program. That year, another member of the Eugenics Research Association, Charles M. Goethe, a California banker and philanthropist, traveled to Germany to learn about the sterilization program. A resident of Sacramento, Goethe had helped organize the Human Betterment Foundation, and had helped—along with Dr. George Dock—to fund Gosney and Popenoe's study of six thousand sterilized men and women. When he returned, Goethe echoed Dock's enthusiasm, and sent a note to Gosney.

"You will be interested to know that your work has played a powerful part in shaping the opinions of the group of intellectuals who are behind Hitler in this epoch-making program," he wrote. "Everywhere I sensed that their opinions have been tremendously stimulated by American thought, and particularly by the work of the Human Betterment Foundation. I want you, my dear friend, to carry this thought with you for the rest of your life, that you have really jolted into action a great government of 60 million people."[28]

XIV

Harry's Secret

B y the spring of 1934, when Charles Davenport was celebrating his retirement after thirty years as director of the Carnegie research stations at Cold Spring Harbor, the American eugenics movement had begun a slow, although to many of its leaders imperceptible, decline.

Scientific research and statistical analysis, especially at the Eugenics Record Office, had begun to slacken, and Harry Laughlin and other scientists were devoting more time to developing eugenics as an "applied science," lobbying for government programs on better breeding rather than accumulating and analyzing human pedigrees. The field of "genetics," virtually indistinguishable from eugenics earlier, was distancing itself from the social engineering of Galton's science. Hard at work in quiet labs across the country, many geneticists were developing a new understanding of the complex chemical dynamics within the chromosomes, rendering Mendelian statistical ratios an incomplete explanation of hereditary forces at best. Eugenic scientists, participating little in this careful observation of living cells, had taken up the cause of racial purity, and were trying to implement national policies of forced sterilization, immigration restriction, and marriage counseling—each meant to ensure that only "citizens of the right type" would bear children.

Even as the fundamental premises of eugenics began to lose their credibility in important but little read scientific journals, Germany was embarking upon its highly publicized program of eugenic engineering. During the 1930s, after the decision in *Buck v. Bell,* other nations, including Canada, Denmark, Finland, Sweden, Norway, and Japan, had also passed eugenic sterilization laws. And the American public still recognized, for the most part, outspoken eugenic scientists as the keepers of the secrets of heredity.

Hope and despair, long the impetus for eugenic reforms, were again mingling in a highly restless era. The Great War had changed the very nature of human conflict, bringing the triumphs of technology and industry to the battlefield—mustard gas, engine-powered tanks, dirigibles and airplanes raining explosives from the sky—and leaving unprecedented millions dead. The Great Depression had devastated the Western world's economies, making even the basic necessities of food, clothing, and shelter precious commodities for many people. But at the same time, an inspiring new president, Franklin D. Roosevelt, a member of the American family long celebrated as the eugenic ideal, had promoted a memorable message of optimism at his first inauguration: "This great nation will endure as it has endured, will revive and will prosper. So, first of all, let me assert my firm belief that the only thing we have to fear is fear itself—nameless, unreasoning, unjustified terror which paralyzes needed efforts to convert retreat into advance." Invoking the "American spirit of the pioneer" and the "clean satisfaction that comes from the stern performance of duty by old and young alike," Roosevelt proposed a controversial "new deal" to redistribute the nation's wealth through a host of government-run projects and progressive social programs. For many, the plan smacked more of Old World socialism than the American pioneer, but this new charismatic president was invoking the long-held myth of a nation ever pressing forward to conquer the wilderness and become a shining city upon a hill.

But there were ominous signs in Europe again. Financial chaos and ethnic tensions were sweeping across the Continent. Germany, flouting the draconian Treaty of Versailles, a treaty which had left the nation humiliated and crippled with debt, was building a massive and technologically advanced military. Its domestic policies of racial hygiene were driving hundreds of thousands of citizens from the country. Most people in the United States preferred to remain isolated from the problems of the Old World, focusing on American social struggles instead. Yet a significant number, especially in New York, were demanding something be done about the growing refugee crisis in Europe. It was amidst such global anxiety that Harry Laughlin's career seemed to flourish as never before.

"There is a movement now to make special legislative provisions for the Jews persecuted in Germany," Laughlin wrote in 1934, introducing a scientific report commissioned by the Special Immigration Commission for the New York State Chamber of Commerce. As Jewish refugees fled the systematic brutality of Adolf Hitler and the new regime in Germany, a host of New York civic groups and private citizens demanded that the Chamber of Commerce, responsible for processing most immigrants after they had come to the country through Ellis Island, offer asylum to these foreigners driven from their homes. When the chamber's chairman, John B. Trevor, set up this special commission to study whether its members should grant any emergency exceptions to the Immigration Act of 1924, he asked Laughlin to prepare a scientific report on these questions. "Dr. Laughlin is beyond doubt the foremost authority in the United States today on the subject," Trevor said, "and his previous researches along similar lines for the Committee on Immigration and Naturalization of the House of Representatives peculiarly fitting him for the task."[1]

Laughlin presented his report in the first week in May 1934 and his recommendations were unambiguous: political oppression and forced emigration were not valid reasons to ease the nation's immigration curbs. "If, as a result of persecution or expulsion by any foreign country, men of real hereditary capacity, sound in physical stamina and of outstanding personal qualities, honesty, decency, common sense, altruism, patriotism and initiative, can be found, they should, because of such qualities, and not because of persecution, win individual preference within our quotas and be welcomed as desirable human seed-stock of future American citizens." These "personal qualities," which eugenicists believed they could measure with their special "Family Records" questionnaires, were vague and ill-defined. And scientists such as his mentor, Charles Davenport, had already claimed to discover the racial qualities of the Jews—which did not conform, as it turned out, to this list. So, Laughlin reasoned, the United States could not afford to offer political asylum just willy-nilly. "The Jews are no exception. . . ."

According to the strict quotas imposed by the Immigration Act, no more than 25,957 German citizens could emigrate to the United States each year. Even so, fewer than 1,000 German citizens were coming into the country each year—especially after President Hoover issued an executive order in 1930, barring anyone who could not prove beyond a doubt they would not become dependent on government aid. This order made immigration virtually impossible for everyone except the wealthiest immigrants. Still, despite such numbers, Laughlin insisted that any German Jew who did not measure up to American standards—defined in the immigration law in which he had played so essential a part—should not be allowed into the country. He did say "superior Jews" could, in theory, be allowed to immigrate, but his notion of racial purity was quite clear when he warned the committee: "If they who control immigration would look upon the incoming immigrants, not essentially as in offering asylum nor in securing cheap labor, but primarily as 'sons-in-law to marry their own daughters,' they would be looking at it in the light of the longtime truth: immigrants are essentially breeding stock."[2]

In not so subtle terms, using the inflammatory image evoking "the purity of our women," Laughlin was asking the members of the committee, Would you want your daughter to marry a Jew?

Based on Laughlin's report, the Committee on Immigration made three terse recommendations: "No exceptional admission for Jews who are refugees from persecution in Germany. No admission for any immigrant unless he has a definite country to which he may be deported, if occasion demands. No immigrant to be admitted whose ancestors were not all members of the white or Caucasian race."

This debate would rage on for years, in fact, but the United States continued to refuse to alter its 1924 immigration quotas. Even in 1938, when over 100,000 Austrian Jews were forced to leave their homeland after the German *Anschluss,* countries around the world still refused to take in many Jewish refugees. Responding to this growing crisis in Europe, President Roosevelt took the unprecedented step of calling for an international conference in Évian-les-Bains in France where leaders could discuss the mounting problem of "involuntary emigration" from Germany. The Évian Conference included thirty-two nations from around the world—though Nazi Germany was not invited to attend. As the American secretary of state Cordell Hull remarked, "You do not negotiate with the felon about his misdeeds." Still, little was done after the nine-day meeting. Participating nations simply agreed to urge the Nazis "to provide for orderly emigration" and ensure the transfer of emigrants' assets. Some nations also agreed to consider easing their immigration laws, without offering any concrete resolutions to do so. And in one popular proposal, nations floated an idea to establish a colony for Jewish refugees in Kenya. (The idea was later abandoned.) They did not, however, agree to grant immediate asylum to persecuted Jews. As Dr. Chaim Weizmann, the Jewish leader who would later become Israel's first president in 1948, quipped after the conference, "The world seemed to be divided into two parts: those places where the Jews could not live and those where they could not enter."[3]

Months later, after the murderous *Kristallnacht* pogrom of November 1938, the U.S. Congress still refused to ease the nation's immigration restrictions. The following February, Senator Robert F. Wagner, the New Deal Democrat from New York, introduced a bill to allow 20,000 German refugee children into the country. The bill was later introduced in the House of Representatives by Edith Nourse Rogers, a Republican from Massachusetts. A host of charity organizations and civic groups across the country were publicizing the plight of German refugee children (the Wagner-Rogers Bill did not mention that most of these children were Jewish) and urging passage of the bill. But other nativist, isolationist, and eugenic leaders—including the immigration expert Harry Laughlin—proved more powerful. They lobbied vociferously against the bill, some arguing, disingenuously, that these refugee children would deprive American children of needed aid. President Roosevelt's cousin, Laura Delano Houghteling, the wife of U.S. Commissioner of Immigration James L. Houghteling, even warned that "20,000 charming children would all too soon grow into 20,000 ugly adults." Congressmen debated the bill for several months, but they eventually killed it in committee.[4]

As the Nazi government expelled and killed its Jewish citizens and continued to build the most modern, technologically advanced military in history, the Allied nations did little but offer weak-willed condemnations. The numbing effects of the "World War" only two decades earlier had made another devastating global conflict seem inconceivable. Besides, the overwhelming consensus in established international relations was that a nation's sovereignty was always sacrosanct. Other nations could protest, but what Germany did within its own borders was ultimately its own affair. Every nation had the right to implement its own domestic policies as it saw fit, without outside interference.

Months after helping kill the Wagner-Rogers Bill, Harry Laughlin published his latest book, *Conquest by Immigration,* an expanded version of his report to the New York Chamber of Commerce. It was a study, he said, of "immigration as one of the three or four equally powerful forces which build up or destroy the inherent racial and family-stock qualities—physical, mental, and spiritual—of the people of the immigrant-receiving country."[5] His argument, clothed in eugenic science, called for the supremacy of the white race, and the duty to work to keep its Nordic blood pure.

Laughlin, of course, had long admired the Germans, and even now he felt the Nazis were doing something remarkable. Ten years earlier, in a display of solidarity with his German colleagues, he had revised the title of the Eugenics Record Office's main publication, *Eugenical News*, adding the subtitle *Current Record of Race Hygiene*—the German phrase for eugenics. And while he was lobbying against asylum for Jewish children before immigration committees, he was at the same time championing the Nazi sterilization law. Like many American eugenicists, he saw Germany as the new leader in applied eugenics.

In 1936, Laughlin was very excited about a new method in Nazi propaganda. He had purchased a copy of the Nazi film *Erbkrank* (*The Hereditary Defective*)—one of the first film documentaries on social problems. In the tradition of Leni Riefenstahl's documentary *Triumph des Willens* (*Triumph of the Will*, 1934), *Erbkrank* begins with a shot from the air. The camera pans over august and majestic buildings—then, ironically, reveals them as mental hospitals in Germany. A male voice-over dramatically intones: "A people that builds palaces for the descendants of drunks, criminals, and idiots, and which at the same time houses its workers and farmers in miserable huts, is on the way to rapid self-destruction." In the tradition of American eugenic storytelling, which had long contrasted pastoral, yeoman farmers with corrupt, urban riffraff, the film chronicles four feebleminded siblings who have cost the German state over 153,000 deutschmarks during their combined eighty years of institutionalization. These siblings, it turns out, are also Jewish. As the narrator explains, "The Jewish folk provides a particularly high percentage of feeble-minded people."

Near the end of the forty-minute documentary—which had been praised by the German Führer himself—the voice-over solemnly proclaims, "The prevention of hereditary sick offspring is a moral duty," which indicates "the highest respect for the God-given natural laws." In the closing scene, two farmworkers, a man and a woman, are seen tilling the soil: "The farmer, who prevents the overgrowth of the weed, promotes the valuable."[6]

After seeing the film, Laughlin quickly wrote to the Carnegie Institution in Washington, explaining how this innovative 16mm moving picture "leads the way for illustrating case-histories in pedigree analysis" and "expounds on the economic, moral, and biological costs of human handicap and inadequacy."[7] He was so enthusiastic about the film's power that he proposed showing it to every high school in the country. "If we advertised this film by a letter . . . sent to the teacher of biology in the high schools of each of the 3,000 county-seat towns of the country, certainly there would be more than 1,000 immediate applications for the loan of the film. With a little advertising and explanation doubtless practically all of this group of high schools would want the film." Having experienced the tremendous power of this new medium himself, Laughlin anticipated that moving pictures would become one of the most effective cultural forces for generations of Americans to come.

The members of the Carnegie Institution were rapt as well, but as a rule, the organization funded only research, not propaganda. So Laughlin began to raise money for his plan from the Eugenics Research Association and the new Pioneer Fund, an organization he had recently helped establish, which was devoted to funding projects for American racial purity. With the help of his assistant, Alice Hellmer, Laughlin re-edited the film, titled it *Eugenics in Germany,* and began distributing it to a number of local high schools. He even sent word to the SS Office for Race and Settlement in Berlin, explaining his plan. As a result of Laughlin's letter, a Nazi newspaper later blazoned the headline: "*Rassenpolitische Aufklärung nach deutschem Vorbild: Grosse Beachtung durch die amerikanische Wissenschaft*—Racial Political Enlightenment on the German Model: Great Reaction from American Science."[8]

But it was the Nazis' debt to Laughlin's Model Law that solidified his international reputation. During the celebrations of the 550th anniversary of Heidelberg University in 1936, the faculty conferred on Laughlin an honorary doctorate of medicine for his important contributions to "race hygiene." Laughlin was deeply moved by this award—he was one of only two Americans so honored—and he felt it was one of the greatest honors he had received. From New York, he sent Dr. Carl Schneider, the dean of Heidelberg University's faculty, two separate notes of gratitude:

I stand ready to accept this very high honor. Its bestowal will give me particular gratification, coming as it will from a university deep rooted in the life history of

the German people, and a university which has been both a reservoir and a foun-
tain of learning for more than half a millennium. To me this honor will be dou-
bly valued because it will come from a nation which for many centuries nurtured
the human seed-stock which later founded my own country and thus gave basic
character to our present lives and institutions. . . .

I consider the conferring of this high degree upon me not only as a personal
honor, but also as evidence of a common understanding of German and American
scientists of the nature of eugenics as research and the practical application of
those fundamental biological and social principles which determine the racial
endowments and the racial health—physical, mental and spiritual—of future
generations.[9]

This honorary degree, encased in the university's medieval coat of
arms featuring a lion brandishing a book with the words *Semper Apertus*
("The book of learning is always open"), was perhaps the highest point
in Harry Laughlin's long career. At fifty-six, he was a world-recognized
scientist, and the United States's most prominent expert on immigration and eugenics. Almost thirty years earlier, as a young man embarking on a career in science and teaching, he had told his pious mother, "If
I can't be great I certainly can
do much good. And I intend to
do it."

Still, there were ominous
signs for his future in 1936—as
for all the world.

An invitation from the University of
Heidelberg for its 550th Jubilee to confer
on Harry Laughlin an honorary Doctor
of Medicine for his contributions to
"race hygiene," 1936.

WHEN HARRY LAUGHLIN
heard the Carnegie Institution
was forming a special committee of scientists to evaluate the
work of the Eugenics Record
Office, he felt a bit confused. "I
am quite at sea concerning the
organization," he wrote to Dr.
A. V. Kidder, a member of the
institution's board of directors
and appointed chairman of the
committee. With Charles Davenport about to retire, Laughlin

Die Universität Heidelberg
die älteste Hochschule des Deutschen Reiches, begeht in den
Tagen vom 27. bis 30. Juni 1936 die Feier ihres
550jährigen Bestehens.
Ich würde es mir zur Ehre anrechnen
Herrn Prof. Harry H. Laughlin
in diesen Tagen als Gast der Universität begrüßen zu dür-
fen. Ihre Antwort erbitte ich möglichst bis 31. Mai 1936,
damit Ihnen rechtzeitig die näheren Mitteilungen zugehen
können.
Heidelberg, im Mai 1936
Der Rektor der Ruprecht-Karls-Universität

had expected to take the reins of the research lab himself. Now it seemed the Carnegie Division of Biology was going to swallow up the Record Office, stripping him of the authority and autonomy he thought he had earned. "I had felt that the advisory committee, of which you are chairman, would exercise a very substantial authority in your relation to the organization of the Eugenics Record Office and its coordination with other sections of the institution. I still hope that it may be possible to reestablish the Eugenics Record Office as a primary research section of the Carnegie Institution of Washington without subordinating it to any other department or division."[10]

As he tried to protect his place in the prestigious institution, Laughlin considered Dr. Kidder, a long supporter of eugenics, his ally. Still, he knew the president of the Carnegie Institution, the paleontologist John C. Merriam, who had also been an ardent supporter of eugenics throughout his tenure, had long disliked him. They had clashed numerous times, and Laughlin felt President Merriam did not fully understand the essential practical and "applied" aspects of eugenics.

Merriam was also receiving a number of critiques of Laughlin's work. Laughlin's research for the Committee on Immigration, in fact, was widely attacked by experts such as the Johns Hopkins geneticist Herbert Jennings. These critiques, however, had little effect on a U.S. Congress and president more than willing to believe theories of Nordic supremacy.[11] Yet by 1935, scientific criticism of eugenics was beginning to gain momentum, and Merriam had to respond.

A number of journalists and scientists had been attacking Galton's science for decades, of course. In addition to the fierce rhetoric of G. K. Chesterton in England, the American socialist Walter Lippmann, the son of German-Jewish parents in New York City, wrote a series of articles for the *New Republic* in 1922, attacking the eugenic mental tests of Stanford's Lewis Terman—the primary tool for eugenic reform. The fear that American intelligence was on the decline, as the massive Army testing program had seemed to indicate, had "no more scientific foundation than a hundred other fads, vitamins and glands and amateur psychoanalysis and correspondence courses in will power," Lippmann maintained, "and it will pass with them into that limbo where phrenology and palmistry and characterology and the other Babu sciences are to be found." Attacking a notion that went back to Francis Galton, he added, "Intelligence is not an abstraction like length and weight; it is an exceedingly complicated notion which nobody has as yet succeeded in defining." By 1930, many scientists were beginning to agree. Even the Princeton psychologist Carl Campbell Brigham, the scientist who had been one of the leading proponents of the differential fecundity of smart and stupid families, now switched his position in an article for *Psychological Review*, arguing that intelligence could not, in fact, be so easily measured.[12]

Some scientists began to attack the eugenicists' obsession with fecund, feebleminded females, as well as their ceaseless calls for forced sterilization. What were their real motives? "It is a well known psychological mechanism that hatred, which is repressed under normal circumstances, may become manifest in the presence of an object which is already discredited in some way," wrote Lionel S. Penrose, a British physician concerned with mental illness. A friend of Sigmund Freud, Penrose believed men such as Laughlin and Davenport might have unconscious reasons for being so obsessed with racial purity and human reproduction. "An excuse for viewing mentally defective individuals with abhorrence is the idea that those at large enjoy themselves sexually in ways which are forbidden or difficult to accomplish in the higher strata of society. The association between the idea of the supposed fecundity of the feebleminded and the need for their sterilization is apparently rational, but it may be emphasized by an unconscious desire to forbid these supposed sexual excesses. It has been pointed out that the advocates of sterilization never desire it to be applied to their own class, but always to someone else."[13] Since Galton, many eugenicists had been obsessed with women and statistics.

Still, the reification of abstract human traits—especially intelligence—lay at the heart of eugenic scientific theory, and it had become more than an unconscious desire to forbid freewheeling sex. The "unit character," initially proposed by Galton's disciple William Bateson, the first champion of Mendelian theory and the man who coined the term "genetics," had long been the bedrock assumption of eugenic statistical analysis. But in addition to funding the Eugenics Record Office, President Merriam had also been giving Carnegie funds to support the work of Thomas Hunt Morgan at Columbia University. (In an era before government grants, this type of private funding was crucial to university research.) During the 1920s, Morgan and a team of brilliant students, including the future Nobel Laureate Hermann Joseph Muller, began to revolutionize the study of chromosomes and the entire field of genetics. In 1907, one of Charles Davenport's researchers at the Station for

Experimental Evolution at Cold Spring Harbor had introduced Morgan to the fruit fly—*Drosophila*—a fecund, easy-to-breed, and easy-to-study organism. Morgan was dissatisfied with the statistical inferences geneticists made from simple Mendelian ratios, and he was more interested in the physical aberrations, or "mutations," in the fruit fly's offspring, wanting to study the chromosome under the microscope. What caused these unpredictable changes in heredity? What physical or chemical mechanisms determined these traits, and could they be altered? Beginning with his work on the white-eyed mutation of *Drosophila,* Morgan started to map the cytological structure of the chromosome, demonstrating how patterns in the cell were linked to inherited traits. These structures must contain some type of chemical code, he concluded. In 1933, Thomas Hunt Morgan was awarded the Nobel Prize in Physiology or Medicine for this discovery.[14]

Was Laughlin's Eugenics Record Office doing such important work? Not long after Morgan had won the Nobel, President Merriam asked a team of scientists to evaluate the scientific contributions of the twenty-five-year-old Carnegie-funded laboratory. In addition to Dr. A. V. Kidder, Merriam asked Johns Hopkins geneticist Adolph Schulz, University of Chicago biologist Hobart Redfield, and Harvard anthropologist Earnest Hooton to serve on the special committee, and each agreed. To Laughlin's dismay, Leslie C. Dunn, a geneticist at Columbia University and long a critic of eugenics, would also contribute a general appraisal of the laboratory's work, though he could not participate in the committee's evaluation of the Record Office since he was overseas.

Dunn wrote to Merriam from Oslo, Norway, giving a tempered assessment of the science of eugenics. At best, it should be a bridge, he felt, between medicine and genetics, showing how to use genetic discoveries to treat disease. "With genetics its relations have always been close," Dunn noted, "although there have been distinct signs of cleavage in recent years, chiefly due to the feeling on the part of many geneticists that eugenical research was not always activated by purely disinterested scientific motives, but was influenced by social and political considerations tending to bring about too rapid an application of incompletely proved theses."

Still, Dunn praised the groundbreaking work of the Eugenics Record Office, especially its collection of family pedigrees. "This method of collecting data on human heredity was begun before hospitals or vital statistics offices made any efforts to obtain or preserve family history data. It has been recognized as a pioneer work and has undoubtedly awakened many persons to the possibilities and need of a systematic collation of such records, and of the importance of a knowledge of human heredity."[15] The question seemed to be, How valuable are these archives now?

The other committee members visited the campus at Cold Spring Harbor in June 1935. Pansy Laughlin worked hard to be a gracious hostess, preparing lunch and dinner for the guests—a task she had always loathed. A worker named Dr. Steggerda presented the results of his comparative study on the development of Negro, White, Navajo, and Maya children. Laughlin provided numerous samples of the office's research over the years, and other relevant material. After two days, the committee members took the train to New York, meeting to discuss their findings at the Hotel Pennsylvania in Manhattan.

Their report was devastating. They had viewed samples from the Family Records questionnaires, the filed index cards, and the hundreds of thousands of pages of fieldworker reports—and they were flabbergasted by the astonishing eighteen tons of material.[16] "The records, the bulk of which consists of 'Record of Family Traits' questionnaires, supplemented by 'Family Tree Folders' and 'Individual Analysis Cards,' are unsatisfactory for the scientific study of human genetics," they concluded. Most of the questions asked were useless, since "so large a percentage of the questions concern traits of personality and character which, in the present state of our knowledge, cannot be measured in an impersonal and accurate manner." Traits such as "self-respect," "holding a grudge," "loyalty," or "sense of humor," for example, could hardly be honestly recorded, even if they were measurable. Even the data on "objective" traits, such as hair form, eye color, and tooth decay, were useless, since they were observed by little-trained fieldworkers who lacked the expertise and equipment to measure pigmentation or chemical constitution.

"The records, upon which so much effort and money has been expended," the committee concluded, "have to date been extremely little used, to judge by the number of publications based upon them. Thus the Office appears to be accumulating large amounts of material, and devoting a disproportionately great amount of time and money to a futile system for indexing it, without certainty, or even good probability, that it will ever be of value."[17] Even more devastating for Laughlin, however, the committee attacked the primary purpose of the lab: "The

Committee is of the opinion the Eugenics Record Office should devote its entire energies to pure research divorced from all forms of propaganda and the urging or sponsoring of programs for social reform or race betterment such as sterilization, birth control, inculcation of race or national consciousness, restriction of immigration, etc." Thus, the visiting scientists recommended the Record Office cut off its association with *Eugenical News*. Even more devastating, they recommended that "all current activities should be discontinued save for Dr. Laughlin's work in preparation of his final report upon the Race Horse investigation."

Merriam did not immediately heed the committee's recommendations. They had been funding the office for fifteen years, after all, and were still obligated by the endowment given by Mrs. Harriman in 1920, after she had funded the lab herself for ten years. So Laughlin was able to continue his work. But even as he was receiving accolades from American politicians and winning his honorary degree from Heidelberg, he was carrying a heavy and humiliating burden.

On New Year's Eve in 1938, as President Merriam was preparing to retire after eighteen years as president of the Carnegie Institution, he wrote Laughlin one final letter. There had long been a "welter of argument" in Washington, D.C., he explained, and "there are those who have not considered your work and your attitude, and perhaps your abilities, as representing the level of effectiveness which might be looked upon as the standard to be obtained in the Carnegie Institution." But there was a more serious issue.

"Last year I spoke with you especially concerning the matter of your health, with a view to obtaining data which would indicate one aspect of the problem of capability or efficiency, and you were kind enough to send me a statement from your physician," Merriam went on. "Examination of the statement from your physician has brought out a number of serious questions, one of which involves the competence of the physician."[18]

Harry's health, and his attempts to hide it: for years, the nation's most outspoken proponent of eugenic sterilization had been suffering seizures. Even in the laboratory at the Eugenics Record Office, he had fallen into embarrassing *grand mal* convulsions in front of fellow workers. One night, while driving home alone on a shore road near Cold Spring Harbor, Laughlin had a seizure and lost control of the car. It plunged down a hill and crashed into the corner of a house at the base of a seawall. If the house had not been there, Harry's car would have gone into the sea and he would have drowned.

Merriam waited until his last day as president to write this letter—a cowardly act. But his message to Laughlin was unambiguous.

You are incompetent, and you are an epileptic.

The new president of the Carnegie Institution, Vannevar Bush, did not dismiss Laughlin immediately. At first, he assured the longtime superintendent that his salary would be guaranteed for fiscal 1939. But Bush changed the name of the lab to the Genetics Record Office, cut its funding by 80 percent, and relieved most of the staff of their duties. By the end of the year, he decided to cease all operations at the Record Office, forcing Laughlin to retire. Bush agreed to pay to send Laughlin's papers—a full train boxcar—to his new home.

In January 1940, Harry and Pansy Laughlin moved back to Kirksville, Missouri, and settled in a small house not too far from where Harry had grown up. As the second world conflagration began, he lived quietly, visiting his brothers and sisters and their many children.

On January 26, 1943, the same day the Nazis began to clear out their "model ghetto" in Theresienstadt and transport its Jews to Auschwitz, Harry Laughlin had an unexpected stroke and died. He was buried near his beloved mother in Highland Park Cemetery in Kirksville. He was sixty-two.

Shortly after returning to Kirksville, however, Harry had received a letter forwarded to him by the Carnegie Institution. A professor at the University of Detroit, Dr. J. E. Coogan, had recently read Laughlin's book *The Legal Status of Eugenical Sterilization* (republished in 1930), and he had a few questions about Laughlin's testimony on behalf of the Virginia Colony for Epileptics and Feeble-minded:

> The booklet is concerned with the case of Carrie Buck, whose sterilization Justice Holmes declared constitutional, remarking, "Three generations of imbeciles are enough." . . . I find no reference whatever to imbeciles, properly so called; the three were morons at the worst. Too, the mother of Carrie Buck is dubbed a "mental defective" because of her grade in a much criticized intelligence test and for reason of "social and economic inadequacy." Carrie Buck herself is called a sterilizable mental defective for a like reason, although "She attended school five years and attained the 6th grade . . . no mental trouble attended her early years."
>
> . . . Can you tell me why counsel for Carrie Buck did not oppose her sterilization, at the several stages of her case through the courts, on the grounds that none of the three generations were proven to suffer from "mental defect," not to say

"hereditary"? Could you kindly say, too, what Carrie Buck's subsequent history has been? Also, and more particularly, in the thirteen years since the disposal of this case, has Carrie Buck's child proven to be a mental defective, and this despite reasonably good development opportunities?[19]

Laughlin, most likely, did not respond.

RETIREMENT DIDN'T SLOW Charles Benedict Davenport's ceaseless energy. Almost seventy, he walked briskly every day for miles, still collecting leaves and bugs, just as he had with Gertrude during their Harvard days in Cambridge. The aging eugenicist was named a Carnegie Institution associate and given a room in the Victorian house he had found to act as the Eugenics Record Office some twenty-five years earlier. He was still serving as a mentor to Harry Laughlin after 1934, and probably helped save the office after the special committee's devastating recommendations. Davenport also continued working on various research projects, publishing forty seven papers and one new book, and revising a new edition of his influential first book, *Statistical Methods,* initially published in 1899.

In the twilight of a long career, however, Charles was most interested in babies and young children. After leaving the arduous administrative tasks of running the research stations at Cold Spring Harbor, he devoted most of his time to studying the physical development of one-year-old infants. He published a number of papers on the postnatal development of the head and nose, as well as a more in-depth study of "bodily growth of babies during the first postnatal year." He spent hundreds of hours with these subjects, playing and laughing with the cooing babies and their mothers.[20] His last book, published in 1940, combined early child development with his life-long field. Titled *Medical Genetics and Eugenics,* it was published by the Women's Medical College of Pennsylvania.

Charles devoted much of his time to children even when he was not engaged in his research. When a new district school was built across the street from his home, he spent hours volunteering. He visited classes, talked to the children about the joys of science, and took them on field trips to gather frogs' eggs each spring. The children adored him, and when he wasn't helping teach a class, he would walk across the street to have lunch with the children by a nearby pond.[21]

But Charles had no grandchildren, no family to carry on his name or his genes. His younger daughter, Jane, had married a promising Cold Spring Harbor biologist named Reginald Harris. Ambitious like his father-in-law, Harris began a Symposium on Quantitative Biology in 1933 at the campus—a symposium of scientists that would last for decades to come.[22] In 1936, however, Harris became ill with a fever, and died. According to Cold Spring Harbor legend, he had worked so hard on the symposium that it eventually killed him. He and Jane never had a child.

Charles's daughter Billie, married to the Jewish businessman Walter Fleisher, had gone from the bohemian lifestyle of a Village artist to the posh lifestyle of a millionaire's wife. "Father darling, I have a fur coat, the first I ever had in my life!" she once exclaimed. "Nutria with a big dark fox collar. It's lovely, riding in the open car. And I have a dress from Bergdorf Goodman—not that that would mean anything to a man. But it's a nice one anyhow. . . . Wallie and I went to the Yale-Penn game, and I'm going over to Philadelphia to the Penn-Illinois game to see Red Grange play, and that's more than I can manage!" This lavish life didn't last, however, and she and Wallie divorced without having children.[23]

Davenport became more active in the local community—something he had never had time for before. He was a very conservative Republican in his later years, campaigning fiercely against the new progressive taxes of the New Deal. He organized taxpayers' associations and helped fight against the expanding role of the federal government—though he had once worked so closely with the Department of Agriculture to spread eugenics. "The expenses of government have thus multiplied due to the ever-increasing demands of the people, instigated by the welfare workers, the city planners, the engineers, and the technologists, who, through skillful advertising, create a public demand for their services and product . . . and in times of reduced personal income without reduction in the cost of government we are, many of us, on the verge of starvation,"[24] he wrote in one association pamphlet. During the war, Davenport served in the civilian defense as an air raid warden for the Nassau County Defense Council, and spent hours as a plane spotter for the Ground Observer Corps of the Army Air Forces.

Among these many activities, Davenport, approaching eighty, also acted as curator and director of the Whaling Museum on Long Island.

At the end of January 1944, he heard that a killer whale had been beached on a nearby shore. Hoping to secure the skull for the museum, he went to the beach, and in the cold of winter hacked off its massive head. Instead of soaking it for a few weeks, allowing the flesh to fall from the bones, he was impatient and insisted upon boiling it in his shed as soon as possible. He worked for nights over an open cauldron in the open shed, and became so infused with the smell of smoke and rancid blubber that people could barely stand to be near him.

Even when he caught a cold, Charles refused to stop working. Soon, the cold developed into pneumonia, and Gertrude had to bring him to the hospital as his condition worsened. On February 18, 1944, Charles Davenport, the nation's first geneticist, died.

Eugenics, the science of good genes, the yearning for the perfect family and for future generations living in a more perfect union, had long been a Davenport dream. As direct descendants of the original Puritan settlers in New England, they considered themselves a peculiar people, an authentic clan of Americans.

Despite their many clashes, Millia Davenport had loved her father dearly, and after his death, she began to change her life. When she was forty-three, she married for the third time—an Anglo-Protestant doctor named Edward Harkavy. Too old to have children, she devoted her time to her old passion, designing costumes for the theater, and again became one of the most successful designers in the United States. But she was also becoming more like her father.

In 1947, Millia declined an invitation from Orson Welles, a recognized genius in cinematography who had just made the revolutionary *Citizen Kane,* to design the costumes for his film adaptation of Shakespeare's *Macbeth.* She was working on a massive academic history of costume, she told him, and she didn't have the time. She was now fiercely committed to her research, just as her father had always been, and even an opportunity to work with Welles in Hollywood would not distract her.[25]

Her work, *The Book of Costume,* published in 1948, was a two-volume history of fashion throughout the ages, complete with over 2,800 illustrations. Orville Prescott of the *New York Times* raved that "the research which went into this work is monumental. The information it contains on a hundred subjects other than fashion must be seen to be believed."[26] The book was a smashing success. The first printing sold

out in a matter of months. Millia Davenport was now not only the foremost costume designer of her generation, she was also a recognized historian and intellect.

In 1949, an advertisement in the *Times* announced that *The Book of Costume* was "back in stock!" Just beneath the ad, in bold, dramatic letters, was another notice for a recently published book. Millia might have recognized the topic. It was a study of eugenics—her father's life's work—and the trials at Nuremberg.

A *New York Times* advertisement for Millia Davenport's groundbreaking two-volume study of the history of costume. Beneath it, coincidentally, was an ad for a book on Nazi medical crimes, which included eugenic practices pioneered by her father, Charles Davenport. (*New York Times*, March 27, 1949, p. BR18)

BOOK FOUR

GENERATIONS LOST

*It is common to think of our own time as standing at the apex of civilization,
from which the deficiencies of preceding ages may patronizingly be viewed in
the light of what is assumed to be "progress." The reality is that in the long
perspective of history the present century will not hold an admirable position,
unless its second half is to redeem its first. . . . If we cannot eliminate the
causes and prevent the repetition of these barbaric events, it is not an
irresponsible prophecy to say that this twentieth century may yet
succeed in bringing the doom of civilization.*

—U.S. SUPREME COURT JUSTICE ROBERT H. JACKSON,
IN HIS OPENING STATEMENT AS PROSECUTOR
AT THE NUREMBERG TRIALS

XV.

The Palace of Justice

In the late afternoon of December 13, 1945, nineteen days after the Nuremberg Trials had begun, an appalled hush hung over a small, crowded courtroom in the city's Palace of Justice. For most of the day, an assistant prosecutor for the Allied nations, the American Thomas J. Dodd, had been droning on, reading into the record a series of documents that described the horrendous conditions at Nazi concentration camps. Over these first few weeks, in fact, Allied prosecutors had been focusing their case on such written records rather than the emotionally charged testimony of eyewitnesses, a strategy that prompted a "savage impatience" in the courtroom and led one observer to label the palace "a citadel of boredom." But near the end of his presentation, Dodd unveiled one of the few pieces of material evidence offered during the entire eleven-month trial: a relic from the concentration camp at Buchenwald. In the center of the quiet courtroom, sitting on a wooden table, was a shrunken human head.[1]

The proceedings had already received a jolt of emotion moments earlier, when Dodd, sporting a snappy polka-dotted bow tie, presented a ninety-second amateur film showing Nazi soldiers beating naked Jewish women on the street. He had also shown a piece of flayed human flesh—a section of stretched skin tattooed with a naked female sprite with wings. Apparently, this macabre memento was an ornament for Ilse Koch, the wife of Camp Commandant Karl Otto Koch, and the notorious "bitch of Buchenwald." It was the shrunken human head, however, that left observers in the courtroom most appalled and stunned.[2]

The presentations lasted only a few minutes. Dodd did not explain the explicit purpose of this "evidence"—neither Ilse Koch nor anyone from the concentration camp was on trial here. Dodd did not, as he told the jury, dwell on this "pathological phase of Nazi culture," and he moved quickly to other matters the next day. But the "Shrunken Head of Buchenwald" would become one of the most well-known images at the Nuremberg Trials. One photo, featuring Dodd holding the head in his upraised hand, pondering it like a bow-tied professor at a museum, astonished the public. With its wavy blond hair and neatly trimmed beard, the head seemed to be a museum piece: a European explorer who must have encountered a tribe of head-hunting savages from a bygone era.

In fact, as the horrifying facts of Hitler's regime began to be revealed, this shrunken head could help explain the incomprehensible. How could such unspeakable crimes, on such a massive scale, have occurred in Europe? How could Germany, a nation and a culture which had, arguably, achieved more in the realms of art and science than any other, have committed such atrocious acts? Could this really be the

Allied counsel Thomas J. Dodd looks at the Shrunken Head of Buchenwald at the Nuremberg Trials. The image, used to illustrate the barbaric "pathological phase" in Nazi culture, belied the Holocaust's careful scientific planning.

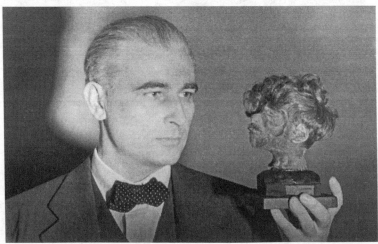

work of the nation that produced Wagner's operas, advanced physics, and enlightened eugenics?

As a symbol, the Shrunken Head of Buchenwald embodied one emotionally powerful answer to the genesis of Nazi "crimes against humanity"—a concept prosecutors would invent for the Nuremberg Trials. It was far easier to comprehend a shrunken head than a modern industry devoted to mass murder. As an emblem of Nazi atrocity, the Shrunken Head of Buchenwald could reveal a clue to the incomprehensible strangeness, the essence of Nazi crime and evil.

The Nazis, it appeared, had become not simply uncivilized, but *pre-civilized*. They had degenerated, as it were, and descended into a state of primitive lawlessness. In effect, the Nazis had been afflicted by a kind of cultural atavism. They were not modern Europeans anymore; they had evolved into the Other. Even for those who had not seen the shrunken head, this basic, simplistic understanding of Hitler and the Nazis would become the norm.

This understanding was wrong. Far more unbearable, far more incomprehensible were the true features of the Nazi crimes against humanity—a phrase which itself made Nazis seem "inhuman" or Other. To listen to the droning facts of their sophisticated plans, their modern transportation infrastructures, their intricate architectural designs for lethal chambers and crematoriums, their piles of shoes, bales of human hair, and boxes of gold-filled teeth waiting to be processed, their meticulous bureaucracies and endless paper records, their scientific medical experiments, including a search for a more efficient means of mass sterilization—indeed, to comprehend the sheer reason and order of it all—the Nazis could seem so recognizable, so banal, so boring. Using the tool of Method, impersonal, efficient, and bureaucratic, the Nazis had constructed a production line of mass murder.

These facts, somehow, did not fit into the millennia-long history of human cruelty, the sporadic massacres and human, all-too-human orgies of rape and murder that had always occurred during times of war. These lethal chambers and crematoriums were not the usual wartime aberrations or the familiar vicious atrocities unleashed in battle. They were manifestations of an entire society organized under the apparently rational goal of purifying the Nordic race and exterminating the unfit. Here, amid the drone of modern industry and science, was an unprecedented type of horror. If William Blake had once labeled the banal and dehumanizing landscape of modern British industry "those dark, treaties. "Humanity," the prosecution tried to prove, carried a meaning beyond the reductionist implications of the law of the survival of the fittest.

In his opening statement for the prosecution, the U.S. Supreme Court Justice Robert H. Jackson, in one of the great courtroom orations of the twentieth century, intimated how these trials must explore the meaning of human nature and society. He recognized that the tribunal was "novel and experimental," admitting that the prosecution could not simply base its case on certain positive "legalistic theories." Normally, a crime a government official committed against his fellow citizens would have to be adjudicated according to that nation's own laws. In this case, however, the German assembly lines of mass murder *were* the law of the land. War criminals were simply following the law. "Unfortunately, the nature of these crimes is such that both prosecution and judgment must be by victor nations over vanquished foes," Jackson explained. "Either the victors must judge the vanquished or we must leave the defeated to judge themselves."

But what was "the nature of these crimes"? In his opening statement Jackson frequently worried about the "uncivilized" danger of the strong merely dominating the weak. In fact, the unchecked domination of the strong was one aspect of the "nature" of these Nazi crimes. Mindful of the judgment of history, Jackson told the judges—and the world—that the Allies, these "four great nations, flushed with victory and stung with injury, [would] stay the hand of vengeance and voluntarily submit their captive enemies to the judgment of the law." But what law? These trials could not be seen as acts of passion, shams to justify revenge. So the nature of the crimes required something more than a typical legal proceeding in which the prosecution would prove its case and the court would mete out judgment. The sheer magnitude of these crimes, Jackson said, transcended the accused individuals and their victims and represented "crimes against the peace of the world" and "crimes against humanity." Indeed, as Jackson maintained, "The real complaining party at your bar is Civilization."[4]

The prosecution, therefore, had to provide a definition for "Civilization," and then portray the Nazis as having broken the self-evident laws that held this civilization together. To do this, Jackson asserted the apparently normative ideals of Western liberalism. He constantly compared the Nazis to non-European civilizations, saying at one point that

"National Socialist despotism [was] equaled only by the dynasties of the ancient East." Affirming the Enlightenment view of a civilized society, he argued that the Nazis

> took from the German people all those dignities and freedoms that we hold natural and inalienable rights in every human being. The people were compensated by inflaming and gratifying hatreds towards those who were marked as "scapegoats." Against their opponents, including Jews, Catholics, and free labor, the Nazis directed such a campaign of arrogance, brutality, and annihilation as the world has not witnessed since the pre-Christian ages. They excited the German ambition to be a "master race," which of course implies serfdom for others. They led their people on a mad gamble for domination.[5]

Indeed, Jackson portrayed the German Nazis as Other, as a strange, alien power that had somehow reverted to a pre-Christian barbarism, disrupting enlightened Civilization.

But this would also be the great irony of the Nuremberg Trials. To maintain the values of traditional European liberalism and to stand in judgment of the Nazi horrors, Jackson was forced to contend not with a clash of "civilized" and "savage" worldviews, really, but with the amoral implications of modern natural science and the pre-Darwinian liberal values of the Enlightenment. The prosecution at Nuremberg had to fight the notion that "the weaklings of the flock must perish." The Allies, "flushed with victory and stung with injury," could not simply banish their vanquished foes. They had to recognize their human dignity first.

Even so, dignity had its limits, as the prosecution's case would show. In order to sustain the legal traditions of Western Europe, the prosecution could not make the mechanism of the dark, satanic mills the primary focus of the trials. "Crimes against humanity," the phrase invented for the horrors of Auschwitz and Treblinka, were subordinated to "crimes against the peace of the world." To preserve the sacrosanct idea of state sovereignty—which few Allied nations in the world would ever abandon—the prosecution made the foremost Nazi crime waging "aggressive war." It would only prosecute those crimes against humanity that had occurred *after* Hitler first invaded Poland in 1938. Indeed, the crime of "genocide"—a term recently invented by the Jewish legal scholar Raphael Lemkin—was not, strictly speaking, prosecutable in and of itself at Nuremberg. Only as it occurred in the context of aggressive warfare or crimes committed against the peace of the world could the Nazi's systematic genocide be judged.[6]

As the Allied case at Nuremberg distanced Hitler and the Nazis from Western civilization, and as images such as the Shrunken Head of Buchenwald and the piles of ghostlike corpses made Nazis seem like atavistic savages, the recognizably European aspects of the National Socialist regime became distorted, transformed into the workings of an evil Other. The Nazi Nuremberg Code—the domestic laws that had stripped German Jews of citizenship and forbade racial intermarriage— became more sinister now. Even though the United States had similar (and much older) laws, these were barely mentioned. How could they be? At Nuremberg, the prosecution was describing the Nazi eugenics programs, including their practices of forced sterilization and marriage restriction, as "crimes against humanity."

In December 1946, in the section of the trials that became known as "The Medical Case," the Allied prosecution described how the Nazis had conducted various medical experiments on human beings. Placing these experiments within the context of waging aggressive war, these "crimes against humanity" included research for a more efficient means for mass sterilization. Some of the doctors on trial had experimented with the drug caladium seguinum—*Schweigrohr* in German—hoping it could prove a simple alternative to vasectomy and salpingectomy. At Nuremberg, Allied prosecutors considered this a crime against humanity.[7]

In July 1947, the prosecution put the Nazi quest for racial purity itself on trial. In the "RuSHA Case," held at the Palace of Justice, members of the SS Race and Settlement Main Office were accused of crimes against humanity for their systematic program of genocide. Using the full, expansive definition of Lemkin's new word, the prosecution defined the crime of genocide as much more than mass extermination. In the indictment, they listed the means of genocide as:

a. Kidnapping the children of foreign nationals in order to select for Germanization those who were considered of "racial value";
b. Encouraging and compelling abortions on Eastern workers for the purposes of preserving their working capacity as slave labor and of weakening Eastern nations;
c. Taking away, for the purpose of extermination or Germanization, infants born to Eastern workers in Germany; . . .
e. Preventing marriages and hampering reproduction of enemy nationals;

f. Evacuating enemy populations from their native lands by force and reset-
tling so-called "ethnic Germans" [*Volksdeutsche*] on such lands; . . .
i. Participating in the persecution and extermination of Jews.[8]

But the SS officers on trial were also accused of working to strengthen
and purify the Nordic race through "positive eugenic" efforts, including the office's *Lebensborn* program, which encouraged the breeding of
"pure Nordics." In describing the systematic programs of negative
eugenics, the prosecution also described the office's extensive genealog-
ical files, which attempted to trace the bloodline of non-Nordics in
Germany and the occupied territories. Many of these activities were the
exact same methods used at Cold Spring Harbor.

In one section of the "RuSHA Case," the prosecution again outlined
the Nazi program of forced sterilization. As lawyers cited SS memos
and described the Nazi program, they in fact outlined the classic
eugenics rationale in the United States. First, the prosecution showed
how the Nazis had been worried about illegitimacy and the fecundity
of the feebleminded in rural areas—the exact starting point of the classic American eugenics studies, Dugdale's *The Jukes,* Goddard's *The
Kallikaks,* and Davenport's *The Hill Folk.* Second, the prosecution
showed how the Nazis had set up a bureaucracy of fieldworkers to go
out into these rural areas and compile a register of all non-Nordics and
"unfit" people. Third, the prosecution, citing Nazi memos, revealed
that "comprehensive *sterilization* of such men and women of alien blood
in German agriculture who, on the basis of our race laws—to be
applied even more strictly in these cases—have been declared inferior
with regard to their physical, spiritual, and character traits."[9]

Physical, spiritual, and character traits. At the Palace of Justice in
Nuremberg, the Allied prosecution had just described the essence of
the American eugenics program, but in the context of the Nazi genocide.

The defense, to a certain extent, recognized this. To counter the
charge that the Nazis' quest for racial purity constituted a crime of
genocide, defense lawyers cited the numerous anti-miscegenation
statutes in some thirty American states:

> Alabama: *Prohibition of marriage between a Negro or a person of Negro origin
> and a white person. A marriage contracted in spite of this is regarded as a crime.*
> Arizona: *Marriage between a white person on one side and a Negro, Mongolian,
> or Indian on the other side is considered null and void.* Arkansas: *Marriage*

between a white person on one side and a Negro or Mulatto on the other side is considered illegal and void. California: *Marriage between a white person on one side and Negroes, Mongolians, or Mulattoes on the other side is considered illegal and void.* Florida: *Marriage between a white person and a person who has one-eighth or more of Negro blood is considered null and void.* Louisiana: *Marriage between a colored person and a white, as well as marriage between Indians and blacks, is forbidden.* Maryland: *Marriage between a white person and a Negro or a descendant of Negroes back to and including the third generation is void and considered a crime. This far from complete enumeration should meet the wishes of our collaborators for some examples.*

As for forced sterilization, the defense again pointed out that it was a common procedure in the United States of America. And they quoted a very famous American jurist to justify this eugenic practice:

Since 1907 sterilization laws have been passed in 29 States of the United States of America. Those affected by the law were primarily criminals, feeble-minded, insane, epileptics, alcoholic and narcotic addicts, as well as prostitutes. Although almost all states try to carry out sterilization on a voluntary basis the courts have more than once ordered compulsory sterilizations. In a judgment of the Supreme Court of October 1926 it says, among other things: "It is better for everybody if society, instead of waiting until it has to execute degenerate offspring or leave them to starve because of feeble-mindedness can prevent obviously inferior individuals from propagating their kind. The principle justifying compulsory vaccination is broad enough to cover the severing of the Fallopian tubes."[10]

The Nazi defense counsel got the date wrong, and left out the resounding closing: "Three generations of imbeciles are enough." And the quote's translation from English to German and back to English again made the language inexact. The argument did not persuade the Allied judges, and the Nazi doctors on trial were convicted of crimes against humanity for forcibly sterilizing their fellow German citizens.

In the United States, however, the precedent set by *Buck v. Bell* would protect American doctors and superintendents. Sterilization in the first country to pass eugenic legislation was not, after all, administered within the context of waging aggressive war. Still, as the Nazis cited the reasoning of Justice Oliver Wendell Holmes, Jr., and the forgotten case of Carrie Buck, they knew they were not alone in their belief that some people were "manifestly unfit from continuing their own kind."

Preventing them from bearing children, the Nazis too believed, would be better for all the world.

AS REVELATIONS OF the Holocaust stunned the world, it became impossible for eugenic reformers to carry on their message of racial purity in the United States. Their foremost leaders, Charles Davenport and Harry Laughlin, had passed away, and many of their organizations soon folded or simply removed the term "eugenics" from their names. After Ezra Gosney, the founder and president of California's Human Betterment Foundation, died in 1942, the sterilization program in the most eugenically active state began a precipitous decline. By 1972, the American Eugenics Society would change its name to the Society for the Study of Social Biology—a new euphemism for eugenics. It no longer sponsored Eugenics Sermon contests or Fitter Family competitions. Other eugenic organizations became devoted to a new cause: world popluation control.[11]

Yet many U.S. states would continue their policies of forced sterilization in their mental institutions. The culture of the mental health profession had been enveloped by eugenic teaching, and many superintendents, who did not have the time or inclination to update their professional skills, still operated according to the scientific beliefs of the 1920s and 1930s. In 1946, a person such as Frederick Thorne, the superintendent of Brandon Training School in Vermont, could argue that the Nazi genocide resulted from the German failure to provide a "compassionate and scientifically enlightened means" to prevent the reproduction of these "hereditary defectives." Defending the sterilization procedures he performed in his Vermont institution, he wrote:

> There are others who would exterminate the unfit or support them on minimal levels of existence, oblivious to the inevitable human degradation which results from allowing any underprivileged group to grow like a cancer threatening the health and welfare of the whole society. The only alternative to the moral collapse which inevitably overwhelmed Europe following Fascistic methods of handling these problems is the development in America of truly enlightened methods. . . .
> We in Vermont must not fall behind in facing our responsibilities to the large mentally defective group while, at the same time, protecting society against the vicious circle of poverty, degeneracy, disease, and delinquency, which this group becomes involved in unless properly supervised.[12]

In some Southern states, officials would actually expand their programs. By 1958, Georgia, North Carolina, and Virginia became the most active, sterilizing a reported 574 of their "unfit" citizens.[13] At the height of the Civil Rights era, these states also began to target black women, especially mothers dependent on state welfare. In South Carolina, which operated a relatively small program, almost all of the people sterilized were black females. These programs continued quietly for decades, with little attention and little controversy.

There were a few exceptions, however, and a number of citizens again tried to sue the states that had legalized forced sterilization.

On a summer afternoon in 1979, Carrie Buck's sister, Doris, was sitting in her living room weeping with her husband, Matthew Figgins. The superintendent of the Lynchburg Training School and Hospital—formerly the Virginia Colony for Epileptics and Feeble-minded—had come to talk to her about her time in the institution some fifty years earlier. Dr. K. Ray Nelson had been doing research on her sister, Carrie, and the nine thousand people who had been sterilized in Virginia over the last six decades. As he reviewed some of the significant dates from Carrie's file—including her birth date, which Doris had never known—he briefly mentioned the day she had been sterilized. At first Doris and Matthew were stunned. Then they both began to sob. No one had ever told them Doris had been sterilized.[14]

In 1928, one year after Carrie had undergone a salpingectomy, doctors prepared Doris for a simple surgery, telling her she only needed to have her appendix removed. Not too long after this surgery, Doris escaped from the Colony and soon married. For years, she and Matthew had hoped to have a child, and had been heartbroken when they were unable to conceive as they grew old. It wasn't until this moment that they understood the reason for their infertility.

Dr. Nelson's research helped prompt a rediscovery of Virginia's forgotten eugenic history. After locating Doris, he found Carrie, too, living in a state-run nursing home. Soon, historians and local newspapers began delving into the history of forced sterilization, and dozens of former inmates at Virginia hospitals began to tell their stories. After *Buck v. Bell,* it was clear, Virginia authorities had become more aggressive and more proactive in their quest for racial purity. In the 1930s, for example, the Montgomery County sheriff conducted raids on communities of country folk living on Brush Mountain—the kind of people labeled "poor white trash" by many Virginians. He rounded up groups of men and women and brought them to the Western State Hospital in Staunton to be sterilized. In one report, Howard Hale, a former candy store owner and Montgomery County supervisor, recalled how these mountain misfits, most supported by the state's relief services, were simply arrested and taken to the mental hospital. "Everybody who was drawing welfare then was scared they were going to have it done on them. They were hiding all through these mountains, and the sheriff and his men had to go up after them. . . . They really got them up on Brush Mountain. The sheriff went up there and loaded all of them in a couple of cars and ran them down to Staunton so they could sterilize them." Hale added: "People as a whole were very much in favor of what was going on. They couldn't see more people coming into the world to get on welfare."[15]

With the storm of media attention, a number of citizens sterilized in Virginia institutions, including Doris Figgins, filed suit against the state. Helped by attorneys from the American Civil Liberties Union and the National Law Center, they sought only modest compensation: First, a full accounting of the circumstances of their sterilizations. Second, any medical, surgical, or psychological assistance required to help them cope with the harm done to them, or any means to reverse the procedure, if possible. Finally, compensation for attorneys' fees and other appropriate costs. They were not seeking to become rich.

The case was brought on behalf of a man dubbed "James Poe" and others who wished to remain anonymous, naming the Lynchburg Training School and Hospital and other Virginia state officials as defendants.[16] According to one anecdote in the complaint, doctors at the old Virginia Colony had sterilized "Judith Doe" against her will in 1949, when she was only fourteen. She had been admitted to the Colony shortly after giving birth to a son, who was conceived when her stepfather raped her. As with Doris, doctors simply told Judith that she had to have her appendix removed, and that she couldn't leave the hospital unless she submitted to the surgery. She had not been given any medical or intelligence tests beforehand, according to her Lynchburg records, and doctors even noted that she was not thought to be "defective." Still, doctors sterilized her anyway since she had had a child out of wedlock.

Poe v. Lynchburg revisited the constitutionality of Virginia's sterilization law—which had been repealed by the State Assembly in 1974. Again, plaintiffs argued that the procedure violated the Fourteenth

XVI.

"What They Did to Me Was Sexual Murder"

On a warm Sunday morning in June 2002, a woman named Lucille was leaning on an aluminum walker, shuffling slowly toward the doors of the Church of the Master. Two tattered tennis balls cushioned the back legs of the bars that supported her, and as she looked down, concentrating on the step-and-drag ritual of moving herself forward, a man in a blue suit walked down the steps to offer his arm. She shook her head, still gazing down, polite but aloof. She doesn't like when people touch her.

Lucille was wearing the same black cotton sweatpants and red sweatshirt she had worn for the past two days, and her feet, cracked and callused, were wrapped in brown plastic sandals. She's diabetic, so she's had problems with her feet since she was young, and when she isn't leaning on her walker, she uses a motorized scooter to get around. She has bright blue eyes, the kind that glow, and though her long blond hair has long turned gray, hanging tangled and matted down to the middle of her back like moss from a magnolia, its disheveled length, rarely seen on a woman her age, hints at lost youth and beauty. But her face is ruddy, still flushed with rage and bitterness and memories of a past that continues to ravage her.[1]

This is a family neighborhood in Denver, Colorado—quiet, almost like a suburb. The church is tucked into a neat row of manicured lawns and two-story houses, with a view of the snow-capped Rockies rising to the west. Most of the houses on the street display American flags, patriotic sentiments after the terrorist attacks on September 11, 2001—an event Lucille gives only a passing thought. Strewn in a few front yards, tricycles, Big Wheels, and other toys sit abandoned from the day before.

Lucille lives thirty-one blocks away in a private church-run nursing home, and though it's always a bit of an effort to get here, she comes whenever she feels well enough. She usually avoids being around people and tends to shoot hard, suspicious stares at those she doesn't know, yet something has always drawn her here. A church member usually comes to pick her up on Sunday mornings; sometimes, when the weather is nice, she'll drive her electric scooter down the parkway sidewalk to worship with this small Baptist congregation.

Entering the sanctuary this morning, struggling to fold her walker, Lucille backs her way into the last pew. There is a large space of empty pews between Lucille and the rest of the elderly congregation, and she sits conspicuously alone. Now seventy-eight, Lucille refers to the thirty-one others in front of her as "the geriatric crowd." When one young family with a little boy finally walks in, Lucille watches them, following the boy as he walks down the aisle holding his mother's hand.

Despite her gruff demeanor to those who come to greet her, she seems to feel a measure of comfort here, and she always hopes they'll sing her favorite hymn. Even at the weekday chapel services at the nursing home, Lucille always raises her hand and asks to hear the one song that moves her more than any other:

> *I'm so glad I'm a part of the family of God,*
> *I've been washed in the fountain, cleansed by His Blood!*
> *Joint heirs with Jesus as we travel this sod,*
> *For I'm part of the family, the family of God.*

> *You will notice we say "brother and sister" 'round here,*
> *It's because we're a family and these folks are so near;*
> *When one has a heartache, we all share the tears,*
> *And rejoice in each victory, in this family so dear.*

> *From the door of an orphanage to the house of the King*
> *No longer an outcast, a new song I sing;*
> *From rags unto riches, from the weak to the strong,*
> *I'm not worthy to be here, but, praise God, I belong!*

Lucille was sterilized against her will by the state of Colorado over sixty years ago, and she still considers herself an outcast. She has not seen any of her own family for over thirty years, and though she thinks she has nephews and nieces living somewhere in Wyoming, she says she wouldn't want to see them. "I'm insignificant," she says. "I ain't nobody worth carin' about." She has spent decades trying to forget her ordeal at the Colorado State Hospital, where doctors sterilized her in 1941, when she was seventeen. "I've kept it a dark, dark secret, and if the people here knew—my life, it would be a hell—a living hell, you see?" Yet, though she's reluctant to recall it now, she remembers writing a diary about her experience years ago, documenting everything so she would never forget. Later, after showing her scrawled notes to a friend, she was overcome with shame, and she asked the person, a woman she trusted, to destroy it for her. Lucille also decided to rip up and throw away every picture she had of herself, and she refuses even now, in rage, to have a photo taken of her. When people ask her why, she simply says, "Fools want their faces always shown in public places." She often responds to questions with such ready-made aphorisms.

Almost forty years ago, Lucille sued the superintendent of the Colorado State Hospital, a man named Frank Zimmerman, and the three resident doctors who helped sterilize her. Her case became notorious, followed for weeks by local papers. "I didn't get a dime, not a penny from them!" she says. "If I saw them now, I'd spit in their faces. What they did to me was sexual *murder.*" She hisses these words, almost shouting, but then becomes quiet and sullen. "I'm just like a female spayed animal. They made me half a woman. They took my heart and left a stone, you hear me?" Her face is flushed a ruddy red, and her bright blue eyes begin to well with tears.

On this morning, after the hymns and scripture readings, the minister preaches a sermon on "A Question: To Whom Shall We Go?" Taking off his jacket and loosening his tie, the young pastor steps down from the pulpit, Bible in hand, and paces back and forth in front of the congregation. He is trying to tell his congregation how they can minister to those in pain, and how they can justify the goodness of God in the face of human suffering. It's the classic "problem of evil," a problem theologians have been wrestling with for millennia. Can a loving, all-powerful God really exist when people suffer so much?

The minister feigns his voice with the pitiful wonder of a child, but he isn't trying to comfort his flock. He doesn't assume his own congregation might be grappling with these kinds of questions themselves. "How do we show compassion? Are we willing to allow God to use us in these hard-hitting questions of faith? How do we make ourselves ready for this? I don't expect us to have easy answers for these questions, but I want to challenge us to be prepared."

Lucille isn't really listening—she's hard of hearing anyway. But, perhaps like most who come to church here, and perhaps like the subtle assumption underlying the pastor's sermon, she doesn't come to church to wrestle with her own hard-hitting questions of anger and pain and bitterness. This is a sanctuary. This is a place to feel protected from the wilderness, to escape history and recover lost innocence, and to be a beacon of faith to the world outside. As the pastor continues to preach, Lucille is gazing at the young family near the front, watching the boy as he fidgets next to his mother.

After the sermon, Lucille peers into the bulletin, trying to read the words of the closing chorus of worship:

> *As the deer pants for the water, so my soul longs after you,*
> *You alone are my heart's desire, and I long to worship you.*
> *You alone are my strength, my shield, to you alone may my spirit yield.*
> *I want you more than gold or silver, only you can satisfy;*
> *You alone are the real joy-giver and the apple of my eye.*

It's a uniquely American spirituality expressed in the chorus—an individual longing for solitude with God—and Lucille has embraced its comforting, transcendent isolation. The heart's desire to yield, to find real joy in God alone, is a source of comfort for many people like Lucille. For others, this solitary longing, quenched by an intense personal relationship with Jesus, can be a shining example of faith to the world, a transforming purity in a world of betrayals and broken promises.

Americans usually look to the future and strive for self-improvement. But very few citizens of Colorado know the secret history of eugenics and forced sterilization in their state, and like many Americans, few, really, would want to know.

THE COLORADO STATE HOSPITAL, a sprawling mental institution in the old Gold Rush town of Pueblo, was founded in 1896 by Dr. Hubert Work, a man who would become one of Colorado's most eminent and nationally renowned public servants. Dr. Work had first come west around 1885, after studying medicine at the University of Pennsylvania, in his home state. It was still a time of Western expansion, and after the young doctor settled here with his wife and son, he became fiercely committed to the social well-being of this new "Centennial State," which he saw as part of the manifest destiny of the United States to "extend the boundaries of freedom." He worked for years in humble private practices, caring for farmers and townsfolk in the small Colorado settlements of Greeley and Fort Lupton; but he had always had much grander ambitions.

After ten years as a rural physician, Dr. Work came to believe there was a pressing threat to the country's destined greatness: the "social curse" of feeblemindedness. While others saw peril in foreign entanglements, problems with public education, or even a decline in Christian virtue, he believed he was one of the few enlightened enough to discern the real menace lurking in the human protoplasm.[2] Committed to the future of American freedom in the midst of such danger, Dr. Work founded the Woodcroft Sanatorium for Mental and Nervous Disorders. Securing funds for a magnificent Tudor Revival building, he formed a new hospital, instituting the most up-to-date methods of modern mental health. This sanatorium would become the Colorado State Hospital in 1923.

Like many institutions being built around the country, the purpose of the sanatorium was not simply to care for the mentally ill. In fact, Dr. Work believed such care was actually tearing away at the fabric of the country. "Indoor relief" for the poor—including mental hospitals and insane asylums like Woodcroft—was an "inefficient" use of taxpayers' money, he always explained. The ultimate goal of such institutions should be to eliminate, and not simply treat, the problem of mental weakness. The primary purpose should be to purify the American bloodline and allow the nation to use its wealth on more important matters.

By 1912, Dr. Work's colleagues elected him president of the American Medico-Psychological Association, one of the nation's largest and most prestigious guilds for mental health professionals. (The association would change its name to the American Psychiatric Association in 1921.) At the sixty-eighth annual meeting, in his first President's Address, Dr. Work presented a sweeping vision of national eugenics and proclaimed the essential role doctors must play in the future well-being of America.

> *Our national prosperity, largely due to its climate, religious freedom and virgin soil, has enabled us to carry the financial burden and social curse of feeble-mindedness amazingly well but:* Ill fares the land to hastening ill a prey, Where wealth accumulates and men decay. *Our dangers from unfriendly nations are minimized by the greater danger that we may impoverish ourselves by taxation for the support of our mental derelicts, by direct enfeeblement of individuals with indirect national efficiency, against which our attitude as superintendents must be that of physicians, alienists, political economists, sociologists and through it all humanitarians.*[3]

It was a familiar American refrain, now proclaimed as the findings of science. The social curse of feeblemindedness was threatening to consume this good land and make America a story and a byword throughout the world, and Dr. Work and other social reformers felt they alone could meet this epic task. Only they could understand the mystery of this root cause of social ills, and only they could find the social policies and medical solutions needed to protect the great promise of America.

In his address, entitled "The Sociologic Aspect of Insanity and Allied Defects," Dr. Work explained how the menace of the feebleminded arose from a strange irony in evolution. "Our high intellectual development has devised means of keeping alive a defective type of humanity which has already impressed itself upon the race, as four percent of our children to-day are feeble-minded, with a larger unknown proportion defective to a lesser degree."

Dr. Work then offered what he believed was the only reasonable solution to the growing population of the "unfit." He explained first how states had enacted marriage restriction laws, and permitted marriage licenses only to those deemed fit enough to procreate. But he also noted how these laws hardly prevented the propagation of the feeble-minded, who mostly bred out of wedlock anyway.

"Apparently there are but two remedies for this social pollution: The segregation of the abnormal at puberty by holding them in institutions during their procreative period, or rendering them sterile," he said. But these two remedies actually came down to only one. "Segregation has been attempted for a hundred years with the practical result that insanity has increased 25 percent, while the whole population has increased

11 percent, to say nothing of an annual, direct, increasing tax on the people of millions of dollars. This alone is proof of its failure, not to mention a greater public expense for its congeners, the criminal and pauper class. We are then perforce compelled to invoke the alternative of sterilization."

For the next thirty minutes, Dr. Work summarized the work of Harry Sharp in Indiana, and listed all the states that had already passed sterilization laws. The essence of what he called "prophylactic psychiatry," the future of their profession, he believed, must be founded upon a national program of eugenic sterilization, which only they were equipped to undertake. "The country owes it to itself as a matter of self-preservation that every imbecile of productive age should be held in such restraint that reproduction is out of the question. This having proven impracticable through institutional seclusion, because of expense and the interference of relatives, then sterilization is necessary. *Where the life of the state is threatened, extreme measures may and must be taken up*" (his italics).[4]

As president of the American Medico-Psychological Association, Dr. Work had also become one of the most politically influential citizens in Colorado. That same year, the Colorado Republican Party appointed him state chairman, and one of his first priorities was to spearhead a campaign for this "extreme measure" to be made law in his state. In 1913, the Colorado state legislature considered a sterilization bill for the first time. It did not arouse much interest, however, and legislators killed the bill.[5]

Still, since Dr. Work was the state's unquestioned expert in mental health care, he may have instituted a quiet policy of sterilization anyway. He was the legal guardian of the patients at the Woodcroft Sanatorium, and he had the authority to provide whatever medical care he deemed necessary for their well-being—as did the heads of similar institutions in other states. According to the testimony of a later superintendent at the Colorado State Hospital, sterilization had long been a "routine operation" for many patients, a procedure that could be administered at the superintendent's discretion. Indeed, since they believed so much was at stake for the welfare of their states and nation, doctors in Colorado, Indiana, Kansas, Virginia, and elsewhere had already begun quietly to sterilize the patients they deemed unfit for procreation, even without explicit legal authority.

After his tenure as president of the American Medico-Psychological Association, Dr. Hubert Work became more involved in national politics, developing close relationships with powerful Republican leaders. In 1921, President Warren G. Harding appointed him Postmaster General, and after the Teapot Dome Scandal, named him Secretary of the Interior—the federal department that had long promoted eugenics in the United States with the help of Charles Davenport and the American Breeders' Association. After he reformed the Department of the Interior's well-publicized corruptions, Dr. Work stayed in this post through the administration of President Coolidge. In 1928, when his distinguished service had earned him a reputation as a man of absolute integrity, the Grand Old Party, then in political disarray, named him chairman of the Republican National Committee. He continued to urge eugenic reforms after the decision in *Buck v. Bell,* calling for the sterilization of those who were "manifestly unfit from continuing their own kind."

After a lifetime of public service, Dr. Work retired to his home in Pueblo, near the Colorado State Hospital. But on August 22, 1934, tragedy struck this successful American pioneer. Dr. Work's son Robert, now a thirty-seven-year-old vice president for a moving and storage concern in Hollywood, California, and the father of the third generation, the doctor's namesake and grandson, Hubert Work II, checked into a Hollywood hotel room and shot himself in the head. Robert Work had been suffering from a mental malady psychiatrists had recently described as an organic disorder of the brain, probably passed down in the genes. They called it "depression."

A FEW MONTHS AFTER Robert Work's suicide, on the night of November 19, 1934, ten-year-old Lucille was wandering in the dark, clutching a two-foot baby doll. The lifelike toy made a squeaky crying noise when turned upside down, and Lucille had longed to have the miraculous plastic baby for herself the first time she saw it at the Golden Eagle Department Store. Hours earlier, no one noticed when the blond-haired little girl walked out of the store with the toy in her arms. Now, on a street called Golden Road, in front of the towering antennae of Denver's oldest radio station, KFEL, a police officer finally found her shivering in the cold and took her home.

Lucille was so small the officer thought she might be undernourished. But he quickly discovered how sassy she could be. She didn't want to go home, she said, because her family didn't like her. She started to kick and yell when he brought her to the house. Once inside, the officer told her mother she should bring Lucille to the station the next day. This was a case that required the expertise of Denver's social services—it was the second time in a week the girl had run away and stolen something. When the officer left, he took the doll.

Lucille had long sparked with a nervous energy unusual for a girl her age. She loved to run, even when alone. At Maria Mitchell Elementary School, she sometimes played tag with the boys or pulled at girls' pigtails. But her teachers also reported they saw her running in the field alone. She would run until flushed with exhaustion, they said. Other students didn't like her much, and teachers could barely control her as she ran down the halls. Her nervous energy had become a serious problem. For the past two years the girl was constantly running away from school—as well as her home—and becoming a serious troublemaker.

The most serious incident had occurred a year earlier, in the fall of 1933. Nine-year-old Lucille ran away from school, only to return clutching a one-year-old baby. The child had been left unattended as it slept on a front porch ten blocks from the school, and Lucille had simply picked it up and carried it proudly back to school. The principal, Miss Keller, immediately called the police, and the baby was returned to its frantic mother.

After this incident, workers at the Children's Aid Society in Denver told Lucille's mother, Annabelle, she should take her daughter to the Colorado Psychopathic Hospital for evaluation. The doctors gave Lucille a mental test, and found she had an IQ of 89—within the range of a feebleminded child. They also discovered a startling fact: the nervous nine-year-old "practiced abuses against herself," masturbating frequently. After a few days of tests, however, Annabelle said she could not afford to take Lucille to the hospital, so she stopped the evaluation before the doctors could make any recommendations.

A year later, when she took her daughter to the police station, Annabelle felt hopeless. Lucille had wandered away from the house earlier in the week, taking her four-year-old brother Paul with her. When they came across a car parked in an alley, Lucille simply opened the unlocked door, climbed in, and found a purse with two dollars. She took the purse, and after meeting a thirteen-year-old boy in front of the corner drugstore, took him to a moving picture show, bringing her brother along. Later that evening, after Annabelle had called the police, Lucille finally came home with her brother, but refused to say where they had been. Little Paul told their mother everything. Lucille remained defiant and refused to name the boy, whom she said she had just met anyway.

Annabelle told the social worker she hoped they would place Lucille in the State Industrial School for Girls, an institution in Morrison for troubled youth. Her oldest daughter, Louise, now a married twenty-five-year-old woman, had spent time at this detention home, and Annabelle said it had done her some good. Annabelle and her husband, Louis, had six children: Louise; another twenty-year-old daughter, Anna; two teenage sons, Frank and Carl; and Lucille and little Paul. In tears, Annabelle told the social worker how, in the midst of the Great Depression, their large family was struggling to get by. Her husband, Louis, was a professional photographer, and she herself was working full-time at a bakery. But these constant problems with her youngest daughter were interfering with her job. Her oldest son, Fred, had also been getting into trouble, and he too was facing a delinquency charge. But he was doing much better now, Annabelle insisted. What was she supposed to do with this uncontrollable ten-year-old?

The case worker with the Children's Aid Society could see this was a poor, unruly family. There had been complaints about Annabelle and Louis always fighting, and Lucille was not the only troubled child in this family. Still, the worker thought Lucille might best be served by the Home of the Good Shepherd, an institution housing orphans and runaways, run by Catholic Charities. Ten years old was much too young for the State Industrial School—which had a reputation as a rough and dangerous place. After the case worker called Good Shepherd, however, the nuns told her Lucille was too young for them as well.

The only other option was foster care, so the case worker scheduled a hearing before a local magistrate. But when Lucille's father heard of this plan, he refused to let his daughter be cared for by a family of strangers.

For the next few weeks, as they waited for the hearing, Lucille kept running away. First, she stole a bicycle. Then she took a rare wire-haired puppy from a pet store. Police found the bicycle, but the puppy was never found. By this time, Annabelle just wanted Lucille taken away. Each time at the station, she begged police to keep her daughter in custody. Couldn't the State Industrial School at Morrison just admit her?

When the date of the hearing finally came on December 18, 1934, the judge decided Lucille should be given one more chance to stay with her family. She was too young, he reasoned, and if the Home of the Good Shepherd would not take her in, there was really no place more appropriate than her own home. Foster homes were difficult to find, and the Industrial School was indeed a home for troubled teens, not a little girl. So Annabelle took her daughter back, feeling weary, frustrated, and hopeless.

Three days later, Lucille ran away again. This time, after the police found her, the case worker convinced the Catholic home to accept the troubled girl. The judge, too, now approved the decision.

On December 21, Annabelle took her youngest daughter out for an ice cream cone at Shorties, which always made Lucille excited and happy. Afterwards, she drove her to the Home of the Good Shepherd.

Lucille, who had just turned eleven, would never live with her mother again.

BACK IN THE SUMMER of 1898, the State Industrial School for Girls at Morrison was in a shambles. It had been founded nearly two decades earlier to segregate young "incorrigibles"—runaways, petty thieves, moral deviants—and train them to be useful citizens in the future, if possible. But the superintendent, Captain John Smithers, a veteran of the cavalry, had found himself unable to maintain discipline among the many teens housed here. There had been a number of escapes, and girls told stories of beatings. When newspapers reported these problems, Governor William H. Adams fired the aging superintendent and turned to one of the school's famous founders.[6]

It was a shrewd political move, for Dr. Minnie C. T. Love was a Colorado legend. A member of the wealthy Roosevelt family, she had been one of the first women in the United States to earn a degree in medicine. After she came to Colorado with her husband and three sons in the late 1870s, she quickly became one of the most visible civic leaders in the region. Dr. Love had already founded the Florence Crittenton Home for troubled girls, serving for years as its chief physician, and then helping organize the State Industrial School for Girls. She was also an active social reformer, a leader in a number of Denver community groups. Like most progressive women of the time, Dr. Love wore the white ribbon of the Women's Christian Temperance Union, and led the crusade against drinking. She had also helped found the Denver Women's Club, becoming a leader in the suffrage movement. Governor

Adams turned to Dr. Love because she had been so successful in bringing a woman's right to vote to Colorado, nearly thirty years before the passage of the Nineteenth Amendment.

The Denver Women's Club had been deeply involved in charity work—like many women's organizations around the country. But under the leadership of Dr. Love, its members soon became active proponents for a sweeping overhaul of the state's system of charity and corrections. Meeting at Dr. Love's home, women discussed the ideas of eugenics and better breeding, and began to see themselves as guardians of the "germ-plasm," the microscopic elements that determined good and bad qualities in human beings. As one newspaper described the silver-haired leader, Dr. Love was "the stormy petrel for fearless stands in the vanguard of modern and untried theories."[7]

After agreeing to help Governor Adams reorganize the State Industrial School, Dr. Love spoke to the newspapers, and said the problem did indeed lie with the superintendent. But she also thought the person in charge had long been hampered by outside interference. To run an institution effectively, she said, the superintendent must be given "supreme authority" to manage the girls as he or she deemed fit.

"Above all things I believe that the home should be conducted on scientific principles, in which science and practical knowledge are combined," she told the *Denver Times*. "I want the board to produce an experienced person, to whom to give the active control of the institution, and one who has been successful in the care of girls such as are to be found in an incorrigible girl's home. Then, let the scientific principles apply, and the result will be satisfactory, no doubt."[8]

Though committed to scientific principles and eugenics, Dr. Love's ideas had been evolving since she was a young idealist living on the East Coast. She had grown up in Washington, D.C., the daughter of William Henry Tucker, who, though from a prominent Virginia family, fought for the Union during the Civil War, and of Lizzie Leticia Roosevelt, a noted writer. Committed to the anti-slavery movement and active in the social issues of the day, the couple instilled in their daughter a belief in the great promise of America and its unique destiny in the world. They also provided Minnie with the best education in the city, even hiring private tutors to teach her science. When Minnie was eighteen, she enrolled in Howard University, a school for black "freemen" founded in 1866 by a group of Congregationalists. Dedicated to progressive principles of equality and freedom, Howard University became one of the few colleges to open its doors to women at the time. When Minnie received her degree from the School of Medicine, hers was the only white face in the class.[9]

The bright and energetic young doctor set up a private practice in Washington and developed a reputation as a physician who sought out the poor. Though trained as a surgeon, she also developed an interest in osteopathy, the nonsurgical techniques of Andrew Taylor Still. Eager to embrace new and modern ideas, she was committed too to her eminent American heritage, becoming an active member in such patriotic organizations as the Daughters of the American Revolution, the Daughters of 1812, and Eastern Star.

On August 16, 1876, just after the centennial celebrations in Washington, D.C., Dr. Minnie Tucker married Charles Guerley Love, a clerk in the War Department. They moved to San Francisco, but soon left for London, where Minnie would study obstetrics and children's diseases. There, she began to learn about the new theory of evolution and the ideas of Francis Galton. By the end of 1879, after she gave birth to three sons in three years, the family moved back to Denver, where Charles took a position with the Colorado Supply Company.

Nearly thirty years later, after Dr. Love had become one of the most famous progressive doctors in the Rocky Mountain region, Governor Jesse McDonald appointed her head of the Colorado Board of Health. But just as she began this new phase in her career, Minnie lost her husband, Charles, to tuberculosis. Her three sons were now grown men, and she found herself alone for the first time in her life.

Over the next few years, hundreds of influential citizens began to discuss the ideas of eugenics—especially after Dr. Love's colleague, Dr. Hubert Work, began promoting them around the country. In September 1910, a group of prominent Denver citizens formed the Eugenics Club, inviting professors from the Midwest to come speak at meetings, held on Sundays at local churches in lieu of evening services.[10] Another of Dr. Love's colleagues, Dr. Mary E. Bates, began organizing Better Baby Contests, which had been popular in states such as Kansas and Missouri. These competitions, which were meant to educate the public on the importance of heredity and better breeding, became popular all through Colorado as well.[11]

By 1913, as a host of Colorado leaders were becoming excited about the promise of eugenics, a number of them decided to form a new national organization they planned to call either the "American Society

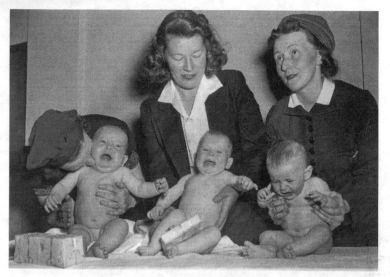

Mothers display their "perfect" babies in a contest at Methodist Hospital, Brooklyn, New York, in 1949. (Arthur Fellig Weegee, courtesy of the International Center of Photography and Getty Images)

of Eugenics" or the "Society for Conservation of Human Life." Denver was booming, and many of its leaders hoped to make it a city on the vanguard, a leader in modern ideas. Dr. Mary Bates, in fact, was to become vice president of the new society. Their plan was to form the organization, develop a network of like-minded societies throughout the United States, then turn it over to David Starr Jordan, the president of Stanford, and Charles Davenport, the head of the Eugenics Record Office in New York State.[12]

This was part of a heady, ambitious plan: many progressives hoped to make Denver a modern, cosmopolitan metropolis in the American heartland, rivaling the great cities on the East Coast. David Starr Jordan, a prominent figure in the Western states, was intrigued by the idea, but when he wrote in February 1913 to Davenport to ask for his opinion, the nation's top eugenicist replied: "I, for one, fear that if a national society were to start out with the burden of primary association and promotion of baby shows that it would be unduly handicapped. . . . I should bid them God Speed in their undertaking, but for myself would not enter into it."[13]

Without the backing of these prominent eugenic leaders, the Denver organization never materialized. Despite the wild successes of the Better Baby Contests, more serious efforts for eugenic reform in Colorado often failed. The eugenic sterilization bill of 1913 stalled in committee, and another bill, introduced in 1918, was handily defeated.

In 1921, as the nation began to debate the perils of immigration and the dangers posed by unfit foreign races, Dr. Love was elected to the Colorado state legislature. By this time, she had become a fierce proponent of the concept of "racial purity." She had also become ashamed of her degree from Howard University—when people asked, she told them she had attended Georgetown.[14] Devoted to eugenic reform, in 1925 Dr. Love sponsored yet another bill to sterilize Colorado's unfit citizens. Her bill was easily defeated.[15]

That same year, the "stormy petrel" joined a nationwide organization dedicated to the racial purity of the American people. The group had only recently established a chapter in Denver, and it was drawing thousands of followers, including a number of prominent women, like the legendary doctor. After decades of public service and civic involvement, Dr. Minnie C. T. Love added another title to her leadership roles in Colorado: "Excellent Commander" of the Ku Klux Klan.[16]

SEPARATED FROM HER FAMILY, the teenage Lucille's nervous energy alternated between rage and depression. There was something wrong with her, she thought. Everyone seemed to hate her.

She wasn't doing poorly at Good Shepherd. She was passing her classes, doing adequate work. She kept to herself most of the time, but she had made a few friends in two years. Still, whenever the nuns tried to tell her what to do, or whenever they tried to discipline her, Lucille flew into a rage. By her third year, she began to scream at teachers and classmates. Then she stopped doing her schoolwork. To hide from the nuns, she began crawling into laundry machines. In the summer of 1937, administrators at Good Shepherd told the municipal court they could no longer keep Lucille. She was becoming a danger to herself and others, they said, and beyond their control.

Now nearly fourteen, Lucille was sent back home. Her parents had divorced a year earlier, and her mother was living in Cheyenne, Wyoming, with her oldest daughter, Louise. Lucille's father had moved to Kansas. So when Lucille came home, her older siblings, Anna and

Fred, were living at the house, looking after Paul. Now entrusted with the care of Lucille, they enrolled her in the public school.

The arrangement lasted less than six months. A few weeks after New Year's Day, 1938, Anna and Fred filed a formal complaint against their little sister. She was still unmanageable, always staying out late. After the police picked her up on the streets at 1:00 a.m. and brought her to the Denver Juvenile Detention Home, her siblings decided they could not look after Lucille. Given her history, the court committed her to the State Industrial School.

Established for incorrigibles, this detention home was a nightmare. The girls were angry and violent in a way Lucille had never seen. Some beat her, since she was small and slight. The workers, too, were angry and violent, and if Lucille refused to do what she was told, they would simply yank her by the arm or pull her hair. Lucille's nervous energy, so long a source of her troubles, finally seemed to break. She became sullen, depressed, and apathetic, less defiant and sassy, crying more easily instead. Finally, in August 1940, after a year and a half, Lucille developed what doctors diagnosed as a "hysterical cough." Then she stopped eating.

After she refused food for over a week, administrators sent her to the Denver General Hospital. Physicians there quickly transferred her to the Colorado Psychopathic Hospital—the place she had been sent as a child.

The little blond-haired girl who loved to run, who loved to cling to dolls and babies, had changed. The doctor in charge of her physical examination jotted a few cursory notes: "She is a 16-year-old female of good skeletal and muscular development. She is rather stupid in appearance and quite untidy. . . . Neurological examination is negative." In other words, though she looked "stupid," he found no apparent physical problems. The doctor in charge of her mental examination jotted similar observations: "She is extremely frowzy in her appearance. Her hair is stringy and matted. She pays no attention to her clothes and spends most of her time stretched out on a bed, with a stupid, open-mouthed expression, demonstrating pain and grief. During interviews she slouches in her chair with her hair hanging down in front of her eyes, half crying, half whining, and complaining about her back and exhaustion. . . . Subjectively, she says her spirits are O.K. She does admit she becomes angry, but admits no true depressive feelings. Shows no hallucinations or illusions. The patient has a host of hypochondriacal ideas." After giving her a cursory mental test, he concluded: "She reads the donkey story easily and repeats it adequately. Judgment on abstractions shows no impairment." A standardized mental test revealed an IQ of 99—average intelligence, and an improvement on the 89 scored when she was nine.

In fact, Lucille seemed quite normal, though a depressed, tearful hypochondriac who cared little about her appearance. What could be wrong with the girl? What sort of demons afflicted her, making her so unhappy? Doctors did not believe her problems were necessarily organic or hereditary but thought they were the result of emotional reactions to her environment, combined with what they assumed to be a "deficient" intellect. Her medical file summed up: "The past history of the patient shows that she has been a problem since early childhood. She was seen and followed by the psychiatric clinic from the age of 8. Much of her behavior could be explained upon a reactionary basis. She was in competition in the home with girls considerably her intellectual superiors. She received no intelligent personal training, nor was there any attempt for adequate habit formation. Her response to the whole thing was runaways, delinquencies, disobedience, nervousness, and masturbation." Yet despite her test results and doctors' medical observations, the final diagnosis was a "hysterical reaction in a mentally deficient individual."

The doctors decided they could do nothing to help Lucille in Denver. At this point, they could only send her to an asylum; no other place would take her. On August 13, 1940, they sent Lucille to the Commission on Lunacy, as required by law. The Commission declared her legally insane—a bureaucratic step necessary to commit her to the largest asylum in the state, the Colorado State Hospital in Pueblo.

ON AUGUST 27, 1928, a year after the landmark Supreme Court case *Buck v. Bell,* Dr. Frank Zimmerman, the stocky, chain-smoking, bald-headed and bespectacled superintendent of the Colorado State Hospital, wrote a letter to the state's attorney general, William L. Boatwright. A year earlier, the Colorado legislature had finally passed a sterilization bill—a bill similar to the Virginia law vindicated in the case of Carrie Buck. Governor William Adams, however, under enormous public pressure from the Knights of Columbus and the Catholic

Archdiocese, vetoed this historic law, infuriating the state's eugenic reformers, including Dr. Minnie Love.[17]

Still, Superintendent Zimmerman believed that if the hospital could obtain the written consent of the patient or the patient's relatives, it should have the legal authority to sterilize inmates. But he wanted to be sure, so he wrote to the attorney general.

Dear Sir:

The question of sterilization of mental patients comes up quite frequently, and as we are not quite sure as to what our responsibility is in these cases, we are submitting to you a number of different conditions. We would like to know just what our responsibility would be under the following circumstances:

A. A patient makes a written request to be sterilized.
   (1) Patient has relatives who agree to the operation.
   (2) Patient has relatives who refuse to permit the operation.
   (3) Patient has relatives who have neither communicated with the patient since in the hospital, nor with the hospital authorities, and who ignore all communications from the hospital to them.
   (4) Patient has no relative insofar as is known to patient or hospital authorities.

B. Relatives of a patient make a request that said patient be sterilized.
   (1) Patient agrees to operation.
   (2) Patient refuses to permit operation.
   (3) Patient is unable to realize, because of mental condition, the nature of the operation.

Very truly yours,
Dr. F. H. Zimmerman.

The response was not what he had hoped, however. The attorney general's letter came quickly, dated August 30, 1928:

Dear Sir:

I have your favor of the 27th instant requesting an opinion of this office as to your legal authority and responsibility concerning the sterilization of mental patients in the Colorado State Hospital.

I find no law in Colorado governing this matter. You will recall that an attempt was made to enact a statute regulating this practice at the last session of the Legislature.

It is my opinion that there is no authority at the present time for performing such operations. I suggest that if you are sufficiently interested in legislation governing the matter that you sponsor a movement similar to that of two years ago. This office will be glad to cooperate with you in any way that it can.

> Very truly yours,
> William L. Boatwright
> Attorney General[18]

This was a problem. Dr. Zimmerman had just been appointed superintendent, but he knew that the hospital had been sterilizing patients for years. Like many institutions around the country, the Colorado State Hospital had instituted a quiet policy of sterilizing individuals after obtaining a "written request." The hospital was teeming with thousands of patients, including hallucinating psychotics, chronic alcoholics, and "child morons," and Zimmerman had been begging the legislature for better funding. Sterilization was one way to discharge certain patients, especially feebleminded girls. When the attorney general's statement put this practice in jeopardy, Zimmerman launched another crusade to legalize sterilization.

The state had rejected eugenic sterilization on four separate occasions. In 1913, 1918, and 1925, legislators had refused to pass proposed legislation. In 1927, though the governor had vetoed the measure, the House and Senate had finally embraced the idea, so perhaps it was time to try again. In March 1929, four House members introduced a bill to sterilize feebleminded and epileptic individuals. The legislation passed the House, but this time it languished in the Medical Affairs Committee in the Senate, despite active lobbying from Zimmerman.[19] For the fifth time in sixteen years, Colorado rejected eugenic sterilization.

Yet sterilizations at the Colorado State Hospital continued. Dr. Zimmerman believed he had a legal loophole: There was no law granting him authority to sterilize patients, but then again, there was no law forbidding it, either. He was the legal guardian of the patients committed to his institution, and he had the discretion to give his patients whatever medical care he deemed fit. Besides, he was obtaining either his patients' or their relatives' consent, and he believed this would shield him from litigation. And the medical records of these individuals remained strictly confidential, far from public scrutiny. This time, Dr. Zimmerman did not risk asking the attorney general for his opinion.

Two decades after Zimmerman had first written to the attorney general, however, other Colorado doctors were still wondering whether they had the legal authority to sterilize fecund, feebleminded females. In May 1948, Dr. Ward Darley, director of the University of Colorado Medical Center, wrote to Attorney General H. Lawrence Hinkley asking if he could legally sterilize a nineteen-year-old female patient. Her parents had consented to the procedure, he explained. Dr. Darley was not a part of the established culture of mental health, so he was not aware of the surreptitious practice of sterilization. On May 10, Attorney General Hinkley responded:

> Dear Sir:
> Colorado does not presently have such a sterilization law and, therefore, sterilization such as the one herein contemplated would not have any statutory authorization.
> While the parents are the natural guardians of a minor child, it has been generally held that an insane or mentally incompetent person is the ward of the court and it is its duty to protect this person.
> It is therefore my conclusion that:
> 1. In Colorado, there is no statutory authorization for the sterilization of mental incompetents.
> 2. Since the mental incompetent is a ward of the court, the consent of the natural guardians would not afford any protection to those performing the operation.
> 3. The consent of the incompetent would likewise afford no protection.[20]

Zimmerman's loophole, according to two Colorado attorneys general, was invalid. Since the state was the legal guardian of these patients, neither the patient nor her parents could legally "consent" to a sterilization operation—and thus get around the fact that Colorado had explicitly rejected the procedure and its eugenic rationale. Nevertheless, despite these clear conclusions from the state's top legal officers, Colorado institutions, like so many around the United States, continued to sterilize their patients quietly, secure in the hope that they were working to better the welfare of their country, and, indeed, the world.

ON THE MORNING OF May 14, 1942, a nurse wheeled seventeen-year-old Lucille into the operating room at the Colorado State Hospital.

Two weeks earlier, her parents, though divorced, had both signed a consent form authorizing her sterilization. The hospital's resident surgeon, Dr. Irving Schatz, gave Lucille a dose of anesthesia, prepped her for surgery, and when he saw his patient was ready, made a small incision in her lower abdomen. He had done this many times.

After the operation, he jotted the surgical report for her medical files: "Lower right rectus incision made and appendix located. Meso appendix cut and ligated with transfixed ligatures. Stump crushed, ligated, cut and treated with phenol and alcohol and buried with purse string. Uterus seized, each tube cut at uterine end. Uterus sutured over and free end of tubes turned into meso salphinx. Abdominal wall closed in layers."

There had been no indication of appendicitis, and no reason to perform this appendectomy along with the salpingectomy. Lucille had not complained of pain, nor had any previous problems been noted in her medical file.

It was a curious fact that throughout the United States, institutions like the Colorado State Hospital were performing an extraordinary number of appendectomies—far more than would be statistically probable. In later years, some would suspect that these procedures (which happened to be recorded in Lucille's case) were actually clandestine sterilizations, dual procedures often done without the knowledge of the patient.[21] Though in Lucille's case this two-in-one procedure was recorded without fear, it seems to have been a way in which many American hospitals hid their practice of eugenic sterilization.

Less than two months later, after Lucille had been at the asylum for over a year and a half, she was discharged. No psychological improvement had been noted, and she was not declared "restored to reason." Once they had sterilized her, however, doctors no longer saw a reason to keep her confined.

As war was raging in the Pacific and European theaters, Lucille's life became a series of short-term stays. She lived with her sister Anna for a while, but left after an argument with her brother-in-law. She found a boardinghouse for young women, but when she couldn't afford the rent, she moved in with her brother Carl. She waitressed at a local diner, then clerked for Walgreen Drug Company on Broadway. When her brother found her a better job, clerking at St. Anthony's Hospital in Denver, she started right away. It paid more than she had ever earned.

After only two days at her new job, however, Lucille quit. When she saw the families with small children, she broke down.

The despair and shame was unbearable. She became deeply depressed again and soon was sent back to the Colorado State Hospital in Pueblo. But her former guardians did not want to keep her there, and they immediately transferred her to the Home for Mental Defectives in Grand Junction, an institution for more severely retarded adults. This was hardly a place for her, so after a month, Lucille stole nine dollars, escaped, and hitchhiked to Florida. She had heard people talk about what a paradise it was down there.

Again, she waitressed. Again, she quit. Again, she left for somewhere new. She decided to hitchhike up to Missouri, and try to find a former nurse she had known at the Colorado State Hospital, a woman who had always been kind to her and who had written to her after her discharge. Lucille made her way to the town of Stockton, and knocked on the woman's door. The nurse took her in, helped her find a job, and even worked to encourage her and make her feel better about herself. Lucille responded, and started going to school with the nurse's younger sister. But the woman's husband was an officer in the Navy, fighting in the Pacific. When he was wounded and sent to San Diego, the nurse left to join him there. Lucille hitchhiked back to Denver.

During her travels, Lucille had found a solitary pleasure. She had become a voracious reader, and she bought nearly every issue of *True Detective Mysteries,* a series of splashy magazines put out by Bernarr Macfadden, one of the most successful publishers in American history.[22] She read stories about the shooting of John Dillinger in Chicago, a daring jewel robbery in New York City, and a crazy man who kept a house full of captives in Georgia. She also devoured the novels of Erle Stanley Gardner, who wrote mystery stories featuring "that fantastically unconventional criminal lawyer," Perry Mason. Her favorite book was *The Case of the Sulky Girl.*

She stayed a few months in Denver before deciding to hitchhike to San Diego, hoping to be near this nurse, the only person she felt had ever been kind to her. When Lucille arrived at her front door, the nurse again tried to help her. But this was a difficult time for the woman and her husband, and they were expecting their first child. San Diego was teeming with soldiers coming in and shipping out, and it was the primary transfer point for hundreds of tons of military supplies. Their living space was very small. Still, she helped Lucille find a boarding room, as well as a job serving food. But the eighteen-year-old found it hard to live alone, and she hated the work. So, the nurse telephoned Lucille's father, Louis, one day, without her knowledge, and explained the situation. After they spoke, she sent him a letter.

"It is good to know that you are interested in Lucille," she wrote.

I felt that you would be, even if her story does give the other impression. She definitely needs some help and understanding, along with a lot of love and patience from someone. She has a lot of good qualities when they are brought forward; but it is going to take a lot of reasoning to get her straightened out completely. She has that "What's the use of it all?" attitude toward life in general now. Do not misunderstand the way that phrase sounds, please. Lucille is no more anxious to die than either you or I even if she does talk that way. . . .

As you may have heard San Diego is very crowded. If we had large enough a place for all of us to live comfortably I should be very glad to try and get Lucille on the right track, but under the circumstances I feel that my first duty is to my own little family. I am interested in Lucille and would like for her to have as useful and happy a life as other people do. . . . Right now she needs someone to talk with her and reason things out; she needs real love and companionship with some discipline perhaps.[23]

Moved by her concern, Louis immediately sent money to bring his daughter home, and Lucille boarded a bus for Denver. At first, things went well; it seemed a new beginning. Louis paid his daughter twelve dollars a week to be an assistant at his new photography store on Welton Street. For the first time, Lucille felt someone in her family wanted to be with her.

And Lucille, too, wanted to change. Perhaps because of the influence of this nurse, she wanted to live a happier life, and she tried hard to be "normal," less depressed and angry. She even wanted to look pretty, to make a good impression on people, and wear nice clothes and cosmetics.

But she became frustrated. She couldn't afford the nice clothes she wanted. She had to pay four dollars every week to rent a room at a boardinghouse, and the remaining eight dollars was barely enough to cover her meals and other necessities—including her *True Crime* magazines and mystery novels. So, in the summer of 1944, Lucille decided to steal again.

At first, she simply snuck into an unlocked house, quickly grabbing some dresses and bedsheets. Then she burglarized two other homes, taking fancy dresses and some bottles of perfume. She longed to look pretty, to be like the pretty girls she always saw on the street, and she only stole items that she thought would help her make this happen. She gained a bit more courage, but when she broke into a fourth house, the family came home as she was rifling through their drawers, looking for jewelry. Ashamed, she didn't try to run away when they called the police.

Lucille was humble and cooperative with the officers when they arrived—something she had never done before. She told them she "wanted some clothes that were nice in order to make a good impression on other people." In the police report, a clerk noted: "The arresting officers state this applicant was truthful in her statements, was cooperative, and had a good general attitude regarding their investigation. . . . [She was] helpful in recoveries and full recovery made in all jobs." Lucille explained that she never stole from anyone who could not afford to lose the articles she had taken.

Later, at the police station, when she was being interviewed by a probation officer, Lucille burst into tears, telling the startled cop how she had lived in mental institutions for years, and how doctors at the Pueblo hospital had made her "less than a woman," sterilizing her so she couldn't have children. The officer noted this in his report.

Lucille's family, however, couldn't see the changes in her. Despite their own stormy histories, they saw Lucille's arrest as just another incident in a long, frustrating battle. Her father blamed the burglaries on the *True Detective* magazines, telling the probation officer that his daughter had been reading them incessantly. He also told the officer she had become so embittered toward the family that it would be best if she not work for him anymore. His hope for a new beginning gone, he also said it would be best for both Lucille and the rest of the family if she did not associate with them again.

When Lucille went before the judge, the state attorney's office recommended she not be sent to jail, explaining that "the applicant is willing to try hard to abide by the rules and provisions of probation if same is granted. She states she has learned her lesson. This girl also states she realized at the time she committed these burglaries she was doing wrong."

The changes in Lucille had been profound. This was the first time she had accepted responsibility for her actions, the first time she did not remain defiant when someone in authority told her what to do, and the first time she did not respond with indifference or despair. The judge granted her probation.

Years later, Lucille cannot recall what happened after her probation. There are no memories, no records describing this crucial moment in her life. But she does remember that most of her family stopped speaking to her after this arrest. She was a young woman completely alone.

On May 5, 1945, less than a year after being granted probation, Lucille returned to the Colorado State Hospital. She would remain there, confined, for seven years.

WHEN JANE WOODHOUSE, the Colorado House member who represented Denver, first heard that the attractive, blond-haired woman in her office had been sterilized against her will, she was flabbergasted.

Lucille had come to see her in November 1954, trying to obtain a formal order declaring her "restored to reason"—thus freeing her from the custody of the state. Miss Woodhouse had been part of a long tradition of socially active women involved in the state's social welfare system, and she had worked with the House's committees on mental health and other hospital-related issues. She was especially interested in the ways the state's institutions cared for troubled women, so when Lucille began to ask her doctors how she could obtain her formal restoration and independence, one of them told her to go see the well-known representative in Denver.

Lucille was just about to turn thirty. She had been paroled from the Colorado State Hospital exactly two years earlier and since then had been trying to piece together a normal life. She had lived with an elderly couple in Pueblo at first, and then thought she would try to join the Women's Army Corps and travel overseas. To enlist, officers told her, she would need to have her high school diploma—something she had never attained. So she enrolled in Pueblo Junior College, earning a General Education Degree.

Lucille liked school, and her teachers encouraged her to continue. Abandoning her plans to join the Army, she enrolled in Midwest Business College in Pueblo, a two-year secretarial school. She did well, taking courses in shorthand, accounting comptometry, English, and spelling. After she graduated, however, it was difficult to get a job, since she was still a ward of the state—and legally insane.

Lucille told Representative Woodhouse her story. She told her of her years at Good Shepherd, the State Industrial School, and the Colorado State Hospital. In tears, she described how she had been sterilized there against her will. Woodhouse was stunned. She had never heard of such a thing, even though she had been involved in mental health issues for years. This smacked of something from Nazi Germany, not an institution in Colorado.

With Woodhouse's help, Lucille was "restored to reason" on January 5, 1955, after being examined by two psychiatrists in Denver. But Woodhouse was determined to discover what had happened. She made a number of inquiries to the Colorado State Hospital and to Superintendent Frank Zimmerman. When he refused to give her any information and seemed rude and evasive, she convinced Lucille to file suit.

Woodhouse asked two of the state's best attorneys to represent Lucille. She first spoke to Molly Edison, a five-foot firebrand, a woman who played cards, smoked, and drank with many of Colorado's most powerful politicians, and one of the few female lawyers in the state. Woodhouse also spoke to one of the state's finest trial lawyers, Norman Berman, who would later become president of the Colorado Trial Lawyers Association and a judge on the Colorado Court of Appeals. Both agreed to take the case.

In December 1955, Berman and Edison filed suit in the Denver District Court. They claimed that Superintendent Zimmerman and three other doctors at the Colorado State Hospital had instituted a quiet policy of eugenic sterilization, in violation of state law and the U.S. Constitution, which had disfigured their client and deprived her of the right to have children. They were seeking $250,000 in damages.

They had worked on the case for months, but Berman and Edison hurried to file because they believed the statute of limitations would expire one year from the day Lucille was declared "restored to reason." Colorado law stipulated that any civil suit must be filed within a year of the time when the "cause of action" occurred, and since Lucille had been sterilized in 1942, they knew this could be a legal technicality.

In fact, it would be three years before the case went to trial. During this time, Lucille lived in a number of different places—sometimes staying with elderly couples, helping them with daily routines, sometimes in boardinghouses, working again as a waitress. In 1956, however, she got a job as a clerk with the Motor Vehicles Department of Denver and made a decent wage: $196 a month, more than she had ever made before. By the time the trial began in the fall of 1958, she had become a good typist and secretary.

It was clear from the beginning that the hospital's attorneys, Vasco G. Seavy and his son Jack, would adopt a strategy that focused on the technicalities of the law. First, they succeeded in getting a change of venue, arguing the case should be tried in Pueblo, the location of both the State Hospital and Lucille's place of residence at the time she filed suit, rather than Denver. The 6,000-inmate hospital had long been a significant part of the Pueblo economy, employing many locals— potential jury members. The Seavys also worked to have the case dismissed, arguing Lucille had been competent for years and had long lost her right to sue years before being declared sane. The judge in the case, the Hon. S. Philip Cabibi, however, ruled Lucille's competence would be an issue for the jury to decide.

During the pleadings before the trial, Norman Berman outlined the sweeping issues of their case. Lucille's sterilization was far more than a tort action. Crucial constitutional issues were at stake, issues concerning the fundamental rights of a human being.

> *I need hardly tell Your Honor that this case involves what we consider to be one of the most substantial principles of law that will be enunciated by all courts in many and many a long day. It involves a basic, inherent right in all human beings: the right to have children. . . .*
>
> *Your Honor, of course, is aware of the fact that for many years long past, statutes have attempted to be enacted in certain states regarding the sterilization of certain types of individuals—some mental defectives, some insane persons, some with sexual proclivities which are not normal, and some just habitual criminals. Your Honor, of course, is also aware, I am sure, of the case of* Buck v. Bell, *which first enunciated the principle by the Supreme Court of the United States that gives the proper procedure and proper protection for the unfortunates who were to be sterilized. They could not say that there was an absolute denial of due process, providing of course the statute was such as to come within the province of the due state process.*
>
> *We are dealing here with legislation which involves one of the basic civil rights of man, life and procreation fundamental to the very existence and survival of the race. That is the question which, eventually, this Court and this State is going to have to decide in this matter. . . .*[24]

Since Colorado didn't have a sterilization law in the tradition of *Buck v. Bell,* and since Lucille had never received any type of due process other than a consent form signed by her parents, their case seemed strong.

When the trial began before the twelve Pueblo jurors—seven men and five women—distinct strategies unfolded as each side examined its witnesses. The Seavys wanted to portray Lucille as a normal, competent individual who had been working and living alone for years, thus having lost her standing to sue under the statute of limitations. Berman and Edison were out to prove the Colorado State Hospital had been sterilizing inmates for decades, as a matter of policy, despite the fact the state had explicitly rejected the rationale for the procedure. The evidence they had, including the devastating letters from the Colorado attorneys general, was powerful.

When Berman examined Superintendent Zimmerman, the first witness, he asked about the hospital's policy with regard to sterilization. Zimmerman testified he did not have to approve every "minor" surgery, such as sterilization, so Berman asked:

"Do I understand, then, the doctors in the hospital had *carte blanc* to perform sterilization operations without your permission?"

"I wouldn't put it that way."

"How would you put it?"

"I would put it this way. If consent has been obtained from the next of kin, the operation would be performed."

"Now, that method of procedure, was that enunciated by you as Superintendent of the hospital?"

"No."

"How did that method of procedure with regard to operations, where there is a consent, come about?"

"I think it was in force when I took charge of the Hospital, and we just followed along."

Having established that sterilization was a *policy* at the hospital, Berman continued, asking how often doctors were sterilizing patients, and for what purpose.

"Would you consider a sterilization operation then as being routine in the hospital?"

"I would say it isn't dangerous and I would say in years gone by, we consider it a minor operation."

"Was such an operation routine in your hospital?"

"You mean by that every day, every week, every month, or what?"

"Well, how often do you do them, Doctor?"

"I wouldn't know."

"Were they done on many patients?"

"I think we have probably had a number of different patients sterilized there, because I definitely believe in sterilization in certain cases. . . ."

"And in what cases do you believe in sterilization?"

"In these cases where you suspect mental deficiency, or we feel that the patient is going to be involved in giving birth to illegitimate children."

It was clear from Zimmerman's testimony that the rationale behind the sterilizations at the Colorado State Hospital was based on eugenic theory, not the health of the individual patient. As Berman continued to focus on this rationale, Zimmerman did his best to evade the question.

"Doctor, would you tell me the basis, the medical basis for recommending sterilization operations on mental deficients?"

"I think it is pretty well known throughout the United States and there is [sic] a good many states that have sterilization laws, making it mandatory. . . ."

"What are your medical reasons for recommending sterilizations?"

"I am telling you, I am just going along with accepted opinion in the United States."

"It was discussed and you knew, did you not, Doctor, that this state had no sterilization law?"

"I don't think that had anything to do with it at all."

"Just answer my question. You knew that this state had no sterilization law; that this state had no sterilization law?"

"No law authorizing it?"

"Yes."

"And no law forbidding it."

"But you knew we had no law regarding sterilization?"

"Yes, because that bill was killed in the legislature in 1929."

"Actually, Doctor, you were instrumental, were you not, in trying to get the bill passed for sterilization?"

Zimmerman denied this repeatedly, though his testimony before the Senate was a matter of public record. When this was pointed out, he claimed he could not remember. Berman continued:

"Now, Doctor, you still haven't answered the original question asked you, what was your medical reason for advocating sterilization of mental deficients?"

"So they will not produce more mental deficients."

"Is it true that mental deficients will always produce mental deficients?"

"Not necessarily, but we have three or four generations out at the State Hospital."

With this perhaps unintended allusion to Oliver Wendell Holmes's final words in the decision in *Buck v. Bell,* Zimmerman made it clear that Lucille was not sterilized for her own health: she was sterilized for social reasons, ostensibly to protect society either from the birth of mental deficients or from "illegitimate" pregnancy.

Berman grilled Zimmerman and another doctor from the Colorado State Hospital for almost two days. Eugenics had been a discredited theory for nearly twenty years, but Zimmerman's testimony indicated that the superintendent still assumed sterilization was "accepted opinion" in the United States. And it was clear Lucille was not sterilized for "therapeutic purposes," when doctors believed cutting the Fallopian tubes would calm a patient's psyche or keep her from masturbating. Zimmerman and the doctors at the Colorado State Hospital had one aim: to keep Lucille, a supposed mental defective in danger of "getting into difficulty," from having children. On the third day of the trial, the plaintiff finally took the stand.

Lucille was nervous. She did not like the attention—she never had—and the story she had to tell was humiliating. She had to explain the most intimate matters of her life to the people in this court— the judge, the jury, the opposing counsel, the people in the courtroom, and even newspaper reporters. Molly Edison tried to calm and reassure her, leading her through the details of her story.

Lucille again explained how she had grown up, how she had spent many years in private and state facilities. She went through the details of her humiliating ordeal, including the morning she was sterilized, answering Edison's questions in as straightforward a manner as she could. When Edison asked about her job at St. Anthony's Hospital after she was paroled, Lucille fell apart.

"Why did you leave St. Anthony's?"

"Because there were children," Lucille said, and then burst into sobs.

Judge Cabibi called a short recess, but when the trial resumed, Edison continued to ask Lucille the intimate details of her life.

"Are you married?"

"No."

"Would you like to get married?"

"Yes. But I'm afraid to go out with a man because I might fall in love. I can't have children, and no man would want me."

When Vasco Seavy cross-examined Lucille, he took her step by step through the places she had lived and the jobs she had worked. His strategy was to show that this woman had lived not only a functional life for the last six years, but a perfectly sane and normal life. If he could convince the jury she could have sued years before she did, he could ostensibly prove that the statute of limitations applied to this case. He was subtle, too, and he found ways to show how intelligent Lucille really was. In a line of questions about an eye operation Lucille had undergone, he asked:

"After you were in the State Hospital in 1945, was an operation performed upon you then?"

"Yes, sir."

"What was that?"

"Strabismus—S-T-R-A-B-I-S-M-U-S."

"Now, what is that?"

"I had my right eye straightened."

"And when did you learn that the correct term for that operation is the word you used?"

"They had discussed it and I remembered it."

"Who had discussed it?"

"The nurses and the doctor that did the operation."

"And when did you learn to spell that term?"

"I looked it up in the dictionary."

The irony of Lucille's lawsuit, in fact, was that her intelligence, which said to be deficient, was demonstrably normal, and it hurt her suit for damages. She had taken yet another intelligence test in 1952, just before she was paroled for the second time, and it revealed an IQ of 105, considered just above the national average. This was the third time she had taken an intelligence test, and she had climbed from an IQ of 89 when she was nine years old to an IQ of 99 when she was sixteen and now an IQ of 105 as an adult. (IQ, according to theory, is an innate constant; it is not supposed to change.) The low test scores when she was young contributed to the reasons she was sterilized. Now her high scores were being used against her as she sued for damages.

"How do you arrive at your claim for damages in this case?"

"I beg your pardon, sir?"

"How do you arrive at your claim for damages in this case?"

"Well, all the money in the world would not compensate me for the terrible loss I am suffering."

"What loss have you suffered?"

"The loss of the right and the joy and the privilege of bearing children and being a mother. . . ."

"How much, dollar wise, are you asking for pain and suffering?"

"Sir, there isn't enough money in this world to make up for the pain and suffering and humiliation which I have gone through and am going through now and will for the rest of my life."

"What pain are you going through?"

"The humiliation of knowing I am only half a woman."

Testimony continued two more days. The Seavys brought workers from the hospital to testify that they had never heard of a policy to sterilize girls before they were to be paroled. Berman and Edison brought Lucille's mother, Annabelle, to testify the hospital had told her and Louis the sterilization operation would not be permanent. After four days, the case went to the jury.

Judge Cabibi's instructions to the jurors were crucial. He gave them a list of thirty-one statements to consider in their deliberations—most of which indicated that Berman and Edison had won key points in the trial's legal maneuverings. Cabibi agreed with the plaintiffs that, as a matter of law, "for a person under the care of the state, relatives cannot supplant the jurisdiction and authority reposed in the courts over such persons," and that "parents cannot legally consent to the sterilization of the plaintiff." Zimmerman's loophole was moot. Judge Cabibi went on, telling jurors that "when a doctor performs a surgery without the consent of the person or someone legally authorized to consent for her, the acts of both doctors in so doing are an assault and battery, and both will be liable in damages."

Given the facts presented, Lucille's sterilization was indeed illegal on this basis alone. But in what turned out to be the most crucial instruction in the trial, Judge Cabibi also told the jurors: "You are instructed that the defendant Zimmerman is liable for damages if you find that he authorized and approved of the operation upon the plaintiff; and if you further find that this action has been brought within the period of time as provided by the Statute of limitations as defined by these instructions." Sympathy for the plaintiff could play no part, he explained, and they should consider the "cause of action" to be the date Lucille was sterilized, in 1942. However, jurors had to determine whether Lucille was sane, whether she was competent enough to have filed her lawsuit within the statute of limitations.

The question could have been settled by the simple fact that Lucille was legally insane before being restored to reason in January 1955. But despite her unambiguous legal status, the judge told jurors *they* must decide if she was sane during this time, despite the legal technicality. "In determining whether or not she was sane before that date . . . if you find by a preponderance of the evidence that she was sane and not falling in any of the classifications of [mental deficiency], as hereinbefore defined, prior to December 28, 1954, then, and in that event, your verdict should be for the defendants."

On Tuesday, November 4, 1958, the day of the midterm elections, the jury decided for Superintendent Zimmerman. Lucille was obviously sane, they thought. She was bright, lucid, sassy, and had maintained employment for years. So Lucille, institutionalized and sterilized for being "mentally deficient," lost her case because she was mentally bright and intelligent. The doctors who sterilized her were acquitted of a civil crime—a crime proven at the trial—because she was able to work without the burden of rearing children.

"They had made it very plain that she had gotten out of the hospital, she had gotten jobs, supported herself," said one juror, years after the trial. "It was quite obvious that she could have filed sooner. And that was the whole thing in a nutshell. Right or wrong, we really didn't have a case to judge."[25]

LUCILLE SITS UPRIGHT IN HER BED, squinting at the decades-old Sylvania television on the other side of her room. The picture flickers full of grainy electric snow, but she doesn't want to miss the Perry Mason rerun on a local Denver station. "My sister gave me his books when I was young, and I just fell in love with him!" she says, smiling slightly. Her TV is large, and it sits on a wood chest of drawers almost fifteen feet from her bed. There is no other place to put it, she explains. Besides, she really can't see that well, anyway. She's seen each episode so many times that she only has to imagine the picture, and she turns the volume loud to hear the dialogue. It's a soothing drone for her, even with the hiss of static.

Lucille has been in the nursing home for three years now. She usually has a roommate, but the bed next to her is empty at the moment.

Lucille never asks what happened when her roommates don't return, and she never talks to them much when they're around. She lived alone in the adjacent assisted-living complex for the previous twenty-three years, and still hasn't gotten used to sharing her space. She does have a few friends she sits with in the dining room—residents have assigned seats so orderlies can bring their specific medications and food—but she doesn't tolerate outsiders. Once, when a confused old woman, probably suffering from Alzheimer's, wandered to the table where she was sitting, Lucille became angry. "You don't belong here," she told the woman. "Go on now, git. This ain't no place for you."

Though she's suspicious of strangers, sometimes cruel, trying to keep outsiders away, Lucille can also be interested in people she learns to trust. When a visitor she had never met told her he had never seen the Rocky Mountains before coming to Denver, she said she would be happy to take him to see the grave of William "Buffalo Bill" Cody, who is buried on Lookout Mountain in the nearby town of Golden. The grave, she may have known, was not too far from the former State Industrial School for Girls, where she had been confined sixty-five years earlier. But she beamed with pride, feeling this was a significant place in her home state. They drove together to the mountains with another nursing home resident, but when they arrived, Lucille said she'd rather wait alone in the van.

Almost eighty now, her life has become a predictable routine, just as it had when she was confined to state-run institutions. The daughter of a couple she lived with in the 1950s comes to visit her sometimes, and the pastor of her church pays a visit every week or so. But few others. She doesn't know if any of her brothers or sisters are still alive—her younger brother, Paul, would be seventy-three now—and she hasn't seen any of them for over thirty years. She may have nieces or nephews living somewhere, perhaps even in Denver, but she says she would not want to see them.

"I loved my mother. I thought a lot of her," she says. "She's gone now, but just a few years before she passed away, she apologized and said she wished she hadn't signed those papers. My mother told me that my dad said sign it, or it would be too bad for her. He would leave her. I did forgive her, but not fully."[26] Lucille did not remember that her parents had already been divorced at this time, and had already remarried other people.

Lucille's faith reflects her ambivalence about her family. Bitterness

and shame mingle with a longing, a longing she knows she can never fully satisfy. She keeps aloof from other church members, including her pastor, but when she hears her favorite hymn, she closes her eyes, humming along as she listens to the words: *From the door of an orphanage to the house of the King, / No longer an outcast, a new song I sing; / From rags unto riches, from the weak to the strong, / I'm not worthy to be here, but, praise God, I belong!*

But it is not necessarily heaven Lucille envisions when she thinks of the family of God, but a past that can never be. When asked if she is looking forward to going to heaven, she replies, a bit manipulatively, "I don't know, 'cuz I ain't goin' there. They'll just say, 'Who the blazes are you?' " And when asked why she enjoys going to church so often, she says it's so God can "forgive my sins." Then, considering the question, she adds: "It's so people can connect to a higher power, really. So people can forget about this dang-blame life."

Lucille refuses to talk about children now. In many ways, the men who sterilized her, the men who believed so strongly in the theory of eugenics, achieved their primary objective in her case. Not only did she never have children, she never engaged in any type of promiscuous, wanton behavior. She has never had a boyfriend, and she has never been touched by a man. "Who would want me?" she says. "All those years I did not date. Why should I? Anytime a man got close, I froze up to them. I didn't have any feeling for any of them." When doctors took her fertility, they also took her intimacy.

"Not too long ago, a man who lives here took me shopping. He was a perfect gentleman. I respect that. And when they do a kindness for me, I do hug them. You know what I mean? To show my appreciation for their kindness."[27]

The Apex of Civilization

The United States was the first nation to experiment with a primitive form of genetic engineering. The social policy of forced sterilization was neither an aberration arising from a strange pseudo-science nor an obscure idea springing from the radical fringes of a racist American intelligentsia. The promise to breed a more healthy, wealthy, and wise citizenry touched a deep American longing, enthralling a host of thinkers, including the nation's best physicians, scientists, and politicians. As a means to better the human race, the eugenic sterilization of those deemed "a shiftless, ignorant, and worthless class of people" became a legal practice in at least thirty American states, and a surreptitious practice in many others.

But nature and the tide of human history, it seems, are not without a sense of irony.[1]

The smirking genius Francis Galton first founded a science based upon the concept of the Good, imagining an aptly named "Kantsaywhere" where *eugenes,* good origins or good genes, were passed on by fit and virile males and massive, buxom females—girls just short of being heavy, he said, and potential mothers of a "noble race." For a thinker like Aristotle, the Good was based not on biological fitness but on the human longing for happiness, a secular version of *eudaimon,* the good virtues that pleased the gods. But Galton proclaimed his notion of the Good when scientists like his cousin Charles Darwin were probing the mysteries of evolution, discovering how mankind was not "fallen" from a higher state, nor created perfect through the breath of God then cast

from Paradise because of sin, but was rising up from a lowly state, moving toward perfection with the crawl of time. With man no longer seen as little lower than the angels, nor sparked with the divine, but simply a biological creature more evolved than apes, Galton could begin to build a science of better breeding upon a simple intuition, an intuition so disarmingly uncomplicated and eminently rational that it still makes sense today. If pigs and peas and flowers can be engineered for better traits, why not human beings? "Could not the undesirables be got rid of and the desirables multiplied?"

Desire and longing. These are peculiar human traits, transcending basic instincts, native intelligence, and natural physical strength. They are much too imprecise to be quantified or measured, or brought before the great god Science. Ancient epics told stories of the complex interplay between the known and the unknown, the mysterious clash of human pride and weakness with the godlike virtues of courage, beauty, and strength. Better breeding, however, is a theme in the modern epic quest to build a more efficient civilization, to battle social ills and human suffering, and to make life less nasty, brutish, and short. Hubris plays no real part in Hobbes's view of civilization, which he understood as a bulwark against Nature's brutality, a collective use of technology to fend off her constant blows. Modern civilization has in many ways sprung from the epoch-changing successes of science and technology, and all hope for salvation is now placed in the tool of Method—its calm, impersonal detachment, its relentless pursuit of efficiency, its emotionless processes.

Galton's extraordinary and eccentric mind made lasting contributions to such tools of method. His tremendous physical prowess and sharp, biting wit made him both a legendary explorer and a famous Victorian socialite. Yet the founder of eugenics, the science which would later condemn people with mentally troubled lives and statistically low IQs, suffered his own debilitating mental breakdowns and deep depressions. The person who had urged the most eminent men and women of his day to be fruitful and multiply—for the health of the human race depended on this, he believed—would himself live a life without the joys of rearing children.

The life of Galton's disciple Charles Davenport, the direct descendant of an original American family that had helped shape the country' destiny, was also filled with similar desire and longing. Davenport

own eugenic vision of a nation purified, a nation with citizens living lives of simplicity, sobriety, and restraint, had been first expressed in the religious vision of his father, Amzi, a vision at once passionate and enthusiastic, Puritan and severe. But in the Davenport family saga, Amzi's scientifically learned children ridiculed the faith of their fathers in the end, and Amzi died a dejected, unhappy man. Charles's daughter Billie, too, later satirized her father's devotion to eugenics in her countercultural magazine. Like Galton, Davenport had urged the best and brightest to bear many offspring; but his two surviving daughters never had children of their own, and he died without the joys of grandchildren. As a scientist, Davenport had argued, for the first time, that genetic weaknesses, running in certain families, determined who would live and who would die in the struggle against deadly viruses, the ravages of cancer, and even the congenital disposition to drink to excess. When Davenport's only son, his beloved and precocious Charlie Junior, lost his struggle against the poliomyelitis virus and died, the implications of his eugenic teachings were clear enough. His weak and sickly son had inherited bad genes.

Harry Laughlin, too, longing to accomplish something great and lasting, to impress his mother with the modern tool of Method, crusaded for the eugenic sterilization of every feebleminded, epileptic, and congenitally weak citizen in the United States. He wanted to purify the white, Nordic race, and subjugate all others. Laughlin died with no children of his own, ravaged by the convulsing effects of epileptic seizures.

In part, the history of forced sterilization in the United States evolved from what could be called a genelike notion, an idea passed on for generations, recombining, mutating, and intermingling with foreign genelike notions, yet maintaining its own essential traits. This notion, a theological imperative implanted at the nation's birth, would help shape an American yearning for centuries. It would promise the settlers in this new Promised Land that if they would only be faithful, hardworking, and true, they would become a new Israel, a city upon a hill, a beacon to the world. Health, wealth, and national prosperity were contingent upon both sober-minded common sense and upright moral character. Unfaithfulness, laziness, and debauchery, however, would make the land but a story and a byword in the world. In many ways, this theological imperative helped shape what would be called

the distinctive "Protestant work ethic" and contribute to America's unsurpassed economic enterprise, technological ingenuity, and drive toward constant self-improvement.[2] Indeed, Americans continue to see themselves as a "peculiar people," a nation set apart by destiny and charged with a unique mission in the world. In the first half of the twentieth century, eugenic theory and forced sterilization merged with this larger quest for a national purity, having evolved in a secular sense, and prompted certain American reformers to use eugenic methods to ensure the country's destined, though morally contingent, greatness.

Their eugenic ideas have long been discredited, of course, exposed as simplistic and slipshod, shown to be a science infected by social prejudice, condemned as an ideology of racial supremacy. But the fundamental hypothesis may yet be proved correct. In the past, science often worked through incomplete inferences and unproven probabilities, finding the most likely explanation for statistical correlations and apparent causation. Isaac Newton's theory of gravity was once derided as an "occult science," an impossible and fictitious "force at a distance." His ideas have been refined for centuries, but the hypothetical graviton particle has still never been directly observed. In a similar way, Gregor Mendel merely discovered the predictable patterns of crossbreeding certain traits, a discovery which seemed to imply a biological force for heredity, but far from explaining how it worked, exactly. Indeed, before the discovery of the structure and behavior of DNA, eugenicists extended the Mendelian analogy far too confidently, proclaiming that complex human behaviors could be crossbred like the dominant and recessive traits of peas. Their simplistic assumptions seem foolish now, in light of further science.

And after the prosecution at the Nuremberg Trials deemed similar German eugenic practices as crimes against humanity, or aspects of the crime of genocide, most people have been content to dismiss eugenics as a "pseudo-science," an idea forever tied to Nazi barbarism. Eugenics, too, is now considered something like an "occult science."

A misguided or corrupted method, however, does not necessarily discredit an idea. An early inspiration, a simple intuition can evolve, becoming more refined over the decades. Its early incarnation may have been premature, but the basic premise of eugenics may prove valid still. Indeed, the methods of genetic engineering and better breeding are no longer as primitive as surgical sterilization or marriage restriction, and

America's drive for perfection and its ongoing quest for purity, effi-ciency, and constant self-improvement remain potent cultural impulses.

In the present epoch-changing age, an age of dizzying biological dis-covery coupled with the unprecedented power of the microchip, the simple eugenic intuition that human "undesirables" can be got rid of—or at least prevented from being born—and that human "desirables" can be multiplied is emerging once more as a seemingly rational idea. The methods of biotechnology, genetic engineering, and stem-cell research promise once again that humankind can take charge of its evo-lution and engineer a more perfect species.

It is possible, if not certain, that scientists will discover not only cor-relations between intelligence and family heredity—as the old eugeni-cists had—but actual chemical and molecular pathways between genes and such complex human traits as intelligence, criminal and sexual behavior, and the relative certainty of disease. It is possible, if not cer-tain, that doctors will be able to find revolutionary cures for diseases that have afflicted human beings for millennia. It is possible, if not certain, that biologists will be able to engineer our offsprings' bodies, creating significantly stronger muscles, sharper memories, and even happier moods.

The perfection ostensibly promised by the great god Science today is no different from the perfection promised by eugenics in the first half of the twentieth century. So the specter of better breeding is thrust upon us once again.

As scientists begin to unlock the secrets of the human genome and discover new methods to prevent disease and engineer human behavior, there remains a fundamental but unresolved question. This question deals not with origins or destiny, necessarily, nor with the existence of a god, but with the moral status of the human individual. The question sprang up in the late nineteenth century with the identity-shattering discoveries of Charles Darwin, but even now presents a challenge to the very heart of the theory of liberal democracy.

What is the foundation of human dignity in light of evolution? Or, more precisely, what is the scientific basis of individual rights in light of the malleable human genome?

It is a question that in many ways verges on theology, and secular sci-entists and political philosophers have been loath to confront it. Indeed, contemporary debates over biotechnology tend to feature the

calm, objective methods of science versus the deeply held moral commitments of religious communities. The issues surrounding the biological enhancement of a species and the dignity of a human individual are far more complex, of course, and are far from being divided into two diverging camps, one secular and the other religious. Even so, a certain understanding of human nature has long informed the foundations of liberal democracy and human rights, a foundation that can seem alien to the basic assumptions of scientific method.

Human dignity and individual rights have in many ways always been tied to theological notions. Even in their purely secular forms, ideas of *inalienable* rights were founded upon a "natural law," a universal principle most Enlightenment thinkers believed was woven into nature by a creating Deity, and which the human mind could come to comprehend. These ideas, too, were primitive at first, limited to white males who owned property and slaves. But the autonomy and dignity of the human individual, the pinnacle of Creation, whose mind was akin to the divine, formed the cornerstone of liberal democracy. Political equality, as it evolved, was based not upon the relative social or biological value of an individual, but upon his or her ontological status as a human being with inherent dignity.

Darwin's theory of evolution profoundly challenged the Enlightenment's confidence in those natural rights, seen as "self-evident" in the light of reason. Human nature became understood as changeable, and "the survival of the fittest" seemed to imply that inherent dignity was a quaint, outdated notion. Indeed, many elite thinkers in the United States began to question the ideals of democracy at the turn of the twentieth century, as "fitness" was reduced to the new biology.

As a legal matter, the concept of human rights in a liberal democracy has had to exist for over a century either as a useful fiction, a rhetorical device replacing natural law, or simply as a body of positive law, enshrined in authoritative documents. The problem with positive law, however, is that rights are only guaranteed by the relative strength of the powers that enforce them. If overthrown, if changed by a plurality of voters, or even if simply ignored, "human rights" can cease to exist. There is no "higher power," no natural law that makes individual rights *inalienable.* They are simply subject to the vicissitudes of power, evolving constitutional hermeneutics, and shifting cultural values. Indeed, in the sweeping course of evolution, if nature does not condemn the lion for slaughtering a weaker rival's offspring and usurping his harem

and his pride, why should it condemn a human creature for doing the same? Genocide can be a perfectly natural and even perfectly rational objective, in terms of the survival of the fittest. The only fundamental law, perhaps, as Thucydides said long ago, is that the strong do what they can, and the weak suffer what they must.

This fundamental problem of human dignity clashing with the danger of ultimate moral nihilism emerged clearly at the Nuremberg Trials. The prosecution, with no body of positive law upon which to base its judgments, unable to appeal to the scientifically discredited notion of natural law, simply asserted human dignity as a tradition of Western liberalism. The prosecuting Allies fretted out loud about history viewing their verdicts as simply victors vanquishing the defeated, the strong dominating the weak, without a solid foundation of human justice. In many ways, they were forced to contend with Darwinian reductionism and the myth of human progress, both of which challenge the ideals of liberalism to this day.

As he wrestled with this problem, Robert H. Jackson, acting as chief prosecutor at the Nuremberg Trials, proclaimed an Enlightenment view of humanity, contrasting it with the ostensible horror of Nazi crimes. "It is common to think of our own time as standing at the apex of civilization, from which the deficiencies of preceding ages may patronizingly be viewed in the light of what is assumed to be 'progress,' " he told the trial's judges—and those who would evaluate their actions in history. "The reality is that in the long perspective of history the present century will not hold an admirable position, unless its second half is to redeem its first. . . . No half-century ever witnessed slaughter on such a scale, such cruelties and inhumanities, such wholesale deportations of peoples into slavery, such annihilations of minorities. . . . If we cannot eliminate the causes and prevent the repetition of these barbaric events, it is not an irresponsible prophecy to say that this twentieth century may yet succeed in bringing the doom of civilization."[3]

Indeed, the irony of the epic modern quest of science is that, as it smashes the old authorities, discovering the secrets of nature and unleashing the unprecedented forces of new technologies, it could in fact bring about the doom of civilization as we know it. Better breeding necessarily assumes a better breed, and notions of fitness necessarily imply notions of unfitness. There is no guarantee that liberal democracy will survive humankind's power to direct its own evolution. The

inherent danger in engineered enhancement, the inherent danger in ideas of evolutionary fitness—for those considered "unfit," at least—is the threat of an accompanying idea that there are "undesirables" who, in the end, deserve to be got rid of. Eugenics naturally breeds contempt for "those manifestly unfit from continuing their kind."

A nation like the United States could reach a "tipping point," a movement in which a few eloquent prophets of better breeding can proclaim life-changing promises, promises that stick to a great majority of Americans living with a yearning for perfection, and again inaugurate a sweeping program to get rid of those they consider "unfit."[4] Biotechnology and genetic engineering present an epic, eschatological vision of paradise where every tear shall be wiped away and where weakness and imperfection shall be eliminated in purified bodies. This is a particular American theological vision, a progressive perfectionism driven by a long-held cultural yearning to build a more perfect union. It is not an irresponsible prophecy to say that ideas of better breeding could again lead to the horrors witnessed in the twentieth century.

The sense of power engendered by revolutionary technologies may betray their creators in the end. Like a defeated virus that mutates and returns more powerful than before, nature, with its sense of irony, mocks the hubris of human strength, sometimes turning epic desire and longing to tragedy.

Will genetic enhancement enhance human peace and repose? Or, will it ratchet up the law of competition, creating a social context in which the genetic "haves" develop a natural contempt for the genetic "have-nots," ushering in a tipping point in which genocide—cultural, ethnic, or genetic—can seem a rational and desirable goal? Will the goals of bioengineering and better breeding, armed with the rational premise that greater intelligence and strength can alleviate social ills and human suffering, again seek to eliminate the least of the human race?

The answer is not a religious Great Awakening, certainly, nor a movement to condemn the methods of science. Positive law, ensuring the rights and dignity of all human beings—including the right to bear and enjoy children—remains the most important guarantee of human freedom. The traditions and institutions of liberal democracy must be nurtured and protected, even as the inevitable technologies of bioengineering begin to change the course of human evolution.

As human beings enter this new era considering the stunning prom-

Fitter Families Contest medal awarded to families scoring B+ or better, ca. 1920. It reads: "Yea, I have a goodly heritage."

ises of science and technology, as they contemplate the possibilities of directing their evolution and moving toward a more perfect state of being, the history of forced sterilization and America's quest for racial purity is worth remembering. And as Americans consider again policies they believe will be better for all the world, they should remember, too, that the apex of civilization might actually spell its doom.

ACKNOWLEDGMENTS

A book that tells a story spanning 150 years and including dozens of complex characters must by necessity stand upon the work of others. The studies of scholars and other writers—in particular Allan Chase, Nancy Gallagher, Daniel Kevles, Wendy Kline, Mark Largent, Paul Lombardo, and David Smith—were in many ways my starting points, and my own research often followed the winding trail of sources they had already laid out in their footnotes.

This trail brought me to a number of archives around the country, and I remain astonished at the dedication, generosity, and enormous talent of the people I encountered. At the American Philosophical Society Library in Philadelphia, Rob Cox and Valerie-Anne Lutz not only helped me through weeks of research, they also offered encouragement and a number of helpful suggestions. At the Pickler Memorial Library at Truman State University in Missouri, Judy May-Sapko was wonderfully helpful, and her knowledge of the Harry Laughlin Papers made my time there both pleasant and efficient. At the Cold Spring Harbor Laboratory Archives in New York—a treasure trove of untapped sources on the family life of Charles Davenport—Clare Bunce assisted me greatly, and later made getting photos for this book a relatively easy matter. At the California Institute of Technology Archives in Pasadena, the staff was also very helpful and professional.

The story and legacy of Carrie Buck has been covered in a number of newspaper articles over the decades, as well as in a few books and even a made-for-TV movie. But every researcher or storyteller of the Buck saga begins—and often ends—with the meticulous scholarship of Paul Lombardo, especially with the unorganized bog of papers Aubrey Strode left with the University of Virginia. Dr. Lombardo first revealed the ironies of Strode's life and the role he played in Carrie's story, and the shape of my narrative owes an enormous debt to his work. When I began my research, Carrie and Emma Buck's confidential medical records were, by law, sealed from public view, but when the law was changed in early 2002, the director of the Central Virginia Training Center, Judy Dudley, and her staff were helpful in providing access to these records to me. In addition, the staffs at the Alderman Library Special Collections, the Circuit Court of Amherst County, and the State Records Center in Richmond each provided valuable assistance during my research in Virginia.

In Colorado, the scholar Rob Prince was exceedingly generous when I came to research the history of forced sterilization in a state that had never passed a eugenics law. He helped

ACKNOWLEDGMENTS

outline the broad contours of the story and then provided many of the obscure sources he had found. Mike Anton, at the time a reporter with the *Rocky Mountain News,* first broke the story of Lucille, which drew me to Colorado, and he was more than generous as he helped me understand her ordeal. The staff at the Colorado State Archives in Denver also provided valuable help during my time there.

But my deepest gratitude in the course of researching this book extends to Lucille, who provided the most meaningful experience of my professional career so far. She not only shared painful parts of her life from more than sixty years ago, she also allowed me to spend nearly a week with her and showed me places from her past and present. The story of forced sterilization and America's quest for racial purity is her story, and my hope, above all else, is that this book will be a record of the wrongs done to her and others like her.

In addition to those who helped my research, I owe an enormous debt to the mentors who have shaped and supported my work as a writer. Ron Scherer at *The Christian Science Monitor* gave me my first break as a journalist, taking me on as an intern when I hadn't a sliver of training, and taught me how to be a careful reporter and concise news writer. Carole Agus, too, took this academically trained rube and taught him how to hit the streets and talk to people, how to look beyond ideas and find the heart of a story. Michael Shapiro guided me through my first long-form narrative about eugenics in Vermont, and taught me how to ask the right questions and how to make sense of the complex motives that drive people. At the same time, Ron Rosenbaum, whose writing first inspired me to leave the halls of academia and become a journalist in New York, taught me how to write engaging stories about complex ideas. His encouragement when I was a fledgling writer paved the way for me to embark on this task to write a book of narrative nonfiction.

This book would never have been written, however, without the tireless support of Sam Freedman. This work is a product of his legendary book writing seminar at the Columbia Graduate School of Journalism, and during its early stages Sam observed and critiqued every jot and tittle of the initial proposal—or, as he would say, took a *schochet's* blade to my bad writing. Along with throngs of others, I owe much of my professional success to the commitment and dedication Sam continues to give his students.

I am fortunate, and forever grateful, for the dream team that put this book together. My agent, Tina Bennett, worked tirelessly to make my proposal presentable. Her keen intellect, eye for sloppy prose and weak argument, and prescient editorial suggestions made writing the book far easier, saving me from pitfalls I would not have seen without her. My editor at Knopf, Vicky Wilson, helped condense a sprawling story into a more coherent narrative, and helped put together a physical book that I find simply stunning. Zachary Wagman also helped with the hundreds of details necessary to put this book together, and I am grateful for his steady presence during its production.

The process of writing a book can be lonely and exasperating, and I am fortunate to have a number of friends who offered desperately needed support and distraction. I especially want to thank Mike Clancy, Steve Hudson, Raabiya Hussain, Mitra Kalita, and Aram and Dunia Sinnreich for their love and support through this project. I also want to express my gratitude to Mariam Dilakian, who read and edited every line of the manuscript as it came fresh off the computer screen, and offered the most profound encouragement.

Finally, I want to express my love and gratitude to my parents, Harry and Judy Bruinius. To my mother, especially, who dedicated her life to her children, I must say I could write this book only because of what you have taught me over the years, and for the love you continue to give.

NOTES

I. Prologue: A Simple and Painless Procedure

1. From the microfilmed medical records of Carrie Buck, kept at Central Virginia Training Center, formerly the State Colony for Epileptics and Feebleminded. Though Bell's tone was unusual in its attention to the authorization and details of the surgery, the reference to the Act of Assembly in 1926 was an oversight. The law was passed in 1924.
2. Ibid.
3. Letter from Paul Popenoe to Bell, May 23, 1933. Copy: Microfilm Record, Carrie Buck, Central Virginia Training Center Records.
4. Letter to Charles Davenport, January 3, 1913, Charles B. Davenport Papers, American Philosophical Society Library, "Theodore Roosevelt" file.
5. Quoted in Clive Ponting, *Churchill* (London: Sinclair Stevenson, 1994), p. 100.
6. "The Biological Relationship of Eugenics of [*sic*] the Development of the Human Race," *Virginia Medical Monthly* (February 1931), p. 728 (italics in the original). This is a copy of Bell's address to the sixty-first annual meeting of the Medical Society of Virginia, in Norfolk, October 21–23, 1930.
7. Ibid., p. 731.

II: An Epic Quest in the Modern World

1. See *Eugenic Sterilization,* ed. Jonas Robitscher (Springfield: Charles C. Thomas, 1973). A table listing state-by-state figures is in Appendix 1, pp. 118–19. See also Philip J. Reilly, *The Surgical Solution, A History of Involuntary Sterilization in the United States* (Baltimore and London: Johns Hopkins University Press, 1991), p. 97. Reilly draws on the data gleaned by the eugenicist Harry Laughlin, who made his own surveys of institutions around the country. See Laughlin's "The Legal, Legislative and Administrative Aspects of Sterilization," *Eugenics Record Office Bulletin No. 10B* (February 1914). See also his book, *Eugenical Sterilization in the United States* (Chicago: Psychopathic Laboratory of the Municipal Court of Chicago, 1922).
2. See Sacvan Bercovitch, *The American Jeremiad* (Madison: University of Wisconsin Press, 1978), p. 3.
3. See, for example, the speeches of Ronald Reagan. On January 25, 1974, he delivered a

speech entitled "'The Shining City Upon a Hill'" before the First Conservative Political Action Conference at the Mayflower Hotel, Washington, D.C., in which he stated: "We cannot escape our destiny, nor should we try to do so. The leadership of the free world was thrust upon us two centuries ago in that little hall of Philadelphia. In the days following World War II, when the economic strength and power of America was all that stood between the world and the return to the dark ages, Pope Pius XII said, 'The American people have a great genius for splendid and unselfish actions. Into the hands of America God has placed the destinies of an afflicted mankind.' We are indeed, and we are today, the last best hope of man on earth."

4. See the defense exhibits at the trial of Otto Hofmann in *Trials of War Criminals Before the Nuremberg Military Tribunal,* 15 vols. ("Green Series") (Nuremberg, 1949), Vol. 4, Case no. 8, p. 1159.

III: The Purity of Our Women

1. The scene that follows was reconstructed from that week's issues of *The Daily Progress,* Charlottesville's main newspaper. See, for example, "Town Buzzing with Excitement," July 3, 1906, p. 1. No quotes that appear here have been invented; they are taken from the paper's ads and editorials, though placed in a living context. For descriptions of Charlottesville at the time, see John Hammond Moore, *Albemarle, Jefferson's County, 1727–1976* (Charlottesville: University Press of Virginia, 1976), p. 288.

2. See the institution's Web site at http://carnegieinstitution.org/about.html.

3. Taken verbatim from an advertisement in *The Daily Progress,* July 3, 1906, p. 7.

4. Albemarle County Court Records, Will Book #27, Chancery Orders Book #6. See J. David Smith and K. Ray Nelson, *The Sterilization of Carrie Buck* (Far Hills, N.J.: New Horizon Press, 1989).

5. Information about this is sketchy. Carrie's medical file notes that her father was "accidentally killed."

6. See Arthur James, *Virginia's Social Awakening* (Richmond: Garrett & Massie, 1939), pp. 16–17.

7. See photos of Strode included with his papers at the University of Virginia Manuscripts Collections. Accounts of the Loving trial, including excerpts from Strode's closing argument, can be found in the *New York Times,* June 29, 1907, p. 2; also May 29, p. 1, and July 6, p. 4.

8. See Paul Lombardo, *Eugenic Sterilization in Virginia: Aubrey Strode and the Case of Buck v. Bell,* unpublished dissertation, University of Virginia, 1982, p. 60.

9. See the family genealogy in the Strode Papers, Box 1.

10. Strode to Voorheis, December 19, 1902, and Voorheis to Strode, December 30, 1902, Strode Papers, Box 81. This influence on Strode was first suggested by Lombardo.

11. See *Proceedings of the National Education Association* (1898), pp. 1048–49, quoted in Lombardo, *Eugenic Sterilization,* p. 86.

12. Herbert Spencer, *Social Statics* (New York: D. Appleton, 1865), pp. 353–56.

13. Ibid., pp. 454–55. For a discussion of Spencer's influence in America, see the standard work on the subject, Richard Hofstadter's *Social Darwinism in American Thought* (Boston: Beacon Press, 1992). See also Robert C. Bannister, *Social Darwinism, Science and Myth in Anglo-American Social Thought* (Philadelphia: Temple University Press, 1979).

14. See J. M. Boies, *Prisoners and Paupers* (New York: Putnam, 1893), p. 269. Also quoted

in Phillip R. Reilly, *The Surgical Solution* (Baltimore: Johns Hopkins University Press, 1991), p. 9.

15. Quoted in Henry H. Goddard, *Feeble-mindedness: Its Causes and Consequences* (New York: The Macmillan Company, 1914), p. 4.

16. See Reilly, *The Surgical Solution,* p. 15.

17. Quoted in ibid., p. 14.

18. James, *Virginia's Social Awakening,* p. 2.

19. Ibid., pp. 14, 43.

20. See Lombardo, *Eugenic Sterilization,* p. 65.

21. *Acts of Assembly,* 1912, chapter 196.

22. See *The Daily Progress,* April 6, 1920, p. 1.

23. See Emma Buck's Medical Records file, microfilm copy, Central Virginia Training Center Records.

24. See Lewis Terman, *The Measurement of Intelligence* (Boston: Houghton Mifflin, 1916), pp. 6–7.

25. *The Kallikak Family, A Study in the Heredity of Feeble-Mindedness* (New York: The Macmillan Company, 1912), pp. 104–5.

26. Copies of these annual reports are kept at the Alderman Library at the University of Virginia.

27. See Charles Carrington, "Sterilization of Habitual Criminals with Report of Cases," *Virginia Medical Semi-Monthly* 13 (1908–9), p. 389.

28. *Virginia Medical Semi-Monthly* 15 (1910–11), p. 4.

29. Quoted in Lombardo, *Eugenic Sterilization,* p. 117.

30. See Paul Lombardo, "Three Generations, No Imbeciles: New Light on *Buck v. Bell,*" *New York University Law Review* 60 (April 1985), pp. 30–62.

31. See Emma Buck's Medical Records file, microfilm copy, Central Virginia Training Center Records.

IV: A Forgotten Gravestone

1. See John Hammond Moore, *Albemarle, Jefferson's County, 1727–1976* (Charlottesville: University Press of Virginia, 1976).

2. See Democratic Primary Pamphlet, 1907, Strode Papers, University of Virginia Manuscripts Collections, Reference file. See also 1905 Pamphlet, Box 58.

3. See J. David Smith and K. Ray Nelson, *The Sterilization of Carrie Buck* (Far Hills, N.J.: New Horizon Press, 1989), p. 3. Information based on Smith's interviews with Carrie Buck in her later years.

4. Transcript of the Dobbses' testimony and their petition to commit Carrie are kept in the *Buck v. Priddy* file at the Circuit Court of Amherst County.

5. For the correspondence that follows, see Carrie Buck's Medical Records file, microfilm copy, Central Virginia Training Center Records.

6. See Paul Lombardo, *Eugenic Sterilization in Virginia: Aubrey Strode and the Case of Buck v. Bell,* unpublished dissertation, University of Virginia, 1982, p. 149.

7. Letter, November 17, 1919, Strode Papers, Box 80.

8. Letter, September 24, 1923, Strode Papers, Box 57.

9. See Strode's letter to Don Preston Peters, July 19, 1939, Strode Papers, Box 30.

10. Ibid.

11. Ibid.

12. See Smith and Nelson, *The Sterilization of Carrie Buck,* p. 40.

13. Carrie Buck Records file, microfilm copy, Central Virginia Training Center Records.

14. For this and the quotes that follow, see the *Buck v. Priddy* file in the Amherst County Courthouse records room.

15. For this letter and the quotes that follow, see the Carrie Buck Records file, microfilm copy, Central Virginia Training Center Records.

16. See John and Margaret Peters, *Virginia's Historic Courthouses* (Charlottesville: University Press of Virginia, 1995).

17. Based on personal observations of the Amherst County Courthouse and conversations with court clerks. Transcript of the trial taken from the *Buck v. Priddy* file in the court's records room.

18. See Mark De Wolfe Howe, ed., *The Holmes-Laski Letters: The Correspondence of Mr. Justice Holmes and Harold J. Laski*, Vol. II (Cambridge, Mass.: Harvard University Press, 1953), pp. 939–41, 964.

19. Francis Galton, *Memories of My Life* (London: Methuen, 1908). See chapter 21, "Race Improvement," p. 311.

20. Carrie Buck Records file, microfilm copy, Central Virginia Training Center Records.

21. Ibid.

22. See Paul Lombardo, "Three Generations, No Imbeciles: New Light on *Buck v. Bell*," *New York University Law Review* 60 (April 1985), p. 60.

23. See Stephen J. Gould, "Carrie Buck's Daughter," *Natural History* (July 1984).

V: Hottentots in Kantsaywhere

1. Francis Galton, *Memories of My Life* (London: Methuen, 1908). See chapter 21, "Race Improvement," p. 315.

2. Ibid., p. 25.

3. Karl Pearson, *The Life, Letters and Labours of Francis Galton*, 3 vols. (I, II, IIIA, IIIB) (Cambridge: The University Press, 1914), Vol. I, pp. 231–32.

4. The word first appears in Galton's work, *Inquiries into Human Faculty and Its Development* (London: The Macmillan Company, 1883).

5. Francis Galton, "Eugenics as a Factor in Religion," in *Essays in Eugenics* (London: Eugenics Education Society, 1909), p. 70.

6. Pearson, *The Life, Letters, and Labours of Francis Galton*, Vol. IIIA, p. 348.

7. Ibid. See also Daniel Kevles, *In the Name of Eugenics* (Cambridge, Mass.: Harvard University Press, 1995), p. 3.

8. Galton, *Memories*, p. 288.

9. Henry Chadwick, ed., *Lessing's Theological Writings* (Stanford: Stanford University Press, 1956), p. 55.

10. See S. T. Coleridge, *Confessions of an Inquiring Spirit* (Philadelphia: Fortress Press, 1988), "Letter IV."

11. See John van Wyhe, "The History of Phrenology," http://pages.britishlibrary.net/phrenology/constindex.html.

12. Quoted in Nicholas Wright Gillham, *A Life of Sir Francis Galton: From African Exploration to the Birth of Eugenics* (Oxford: Oxford University Press, 2001), p. 17.

13. Desmond King-Hele, *Erasmus Darwin* (New York: Charles Scribner's Sons, 1963), p. 73.

14. Pearson, *The Life*, Vol. IIIA, p. 207.

15. Galton, *Memories*, pp. 287–88.

16. Francis Galton, *Hereditary Genius* (Honolulu: University Press of the Pacific, 2001), p. 45.

17. Francis Galton, "Hereditary Talent and Character," *Macmillan's Magazine* 12 (1865), pp. 157–66, 318–27.

18. Galton, *Hereditary Genius*, p. 72.

19. Ibid.

20. Ibid., pp. 122–23.

21. Ibid., p. 399.

22. Galton, "Hereditary Talent and Character," p. 165.

23. Galton, *Hereditary Genius*, p. 415.

24. Ibid.

25. Galton, *Memories*, p. 290.

26. See Emel Aileen Gökyigit, "The Reception of Francis Galton's *Hereditary Genius* in the Victorian Periodical Press," *Journal of History of Biology* 27 (1994), pp. 215–40. See also Gillham, *A Life of Sir Francis Galton*, p. 171.

27. See Galton's description of his breakdown in *Memories*, pp. 154–55. For the description of his house, see Pearson, *The Life*, Vol. II, pp. 11–12.

28. See Galton's *English Men of Science: Their Nature and Nurture* (London: Frank Cass, 1970), p. 259.

29. Quoted in Daniel J. Kevles, *In the Name of Eugenics* (Cambridge, Mass.: Harvard University Press, 1995), p. 12.

30. Galton, *Memories*, p. 154.

31. See Gillham, *A Life of Sir Francis Galton*, p. 24.

32. See ibid., p. 22.

33. Galton, *Memories*, p. 2.

34. Pearson, *The Life*, Vol. II, p. 119.

35. Ibid., Vol. I, p. 137.

36. Ibid., p. 144.

37. Galton, *Memories*, p. 85.

38. Pearson, *The Life*, Vol. I, p. 200.

39. Quoted in Gillham, *A Life of Sir Francis Galton*, p. 55.

40. Galton, *Memories*, p. 152.

41. See Michel Foucault, *The History of Sexuality*, Vol. 1 (London: Penguin Books, 1976): "Toward the beginning of the eighteenth century, there emerged a political, economic, and technical incitement to talk about sex. . . . This need to take sex 'into account,' to pronounce a discourse on sex that would not derive from morality alone but from rationality as well, was sufficiently new that at first it wondered at itself and sought apologies for its own existence. How could a discourse based on reason speak like that?" (p. 25).

42. *Nation*, April 6, 1893. Also quoted in Gillham, *A Life of Sir Francis Galton*, p. 171.

43. See Galton, "A Theory of Heredity," *Contemporary Review* 27 (1875), pp. 80–95.

44. Pearson, *The Life*, Vol. IIIA, p. 422.

VI: A City Upon a Hill

1. See Diary, 1879, in the Charles B. Davenport Papers, American Philosophical Society Library. What follows is taken from Davenport's special entry at the back of his diary, in the "Memo" section. Some details are taken from other entries.

2. See Amzi Davenport, Composition Book, Cold Spring Harbor Laboratory Archives, Davenport/Harris Papers, Box 3. A picture of Amzi Davenport is included in *A Supplement to The History and Genealogy of the Davenport Family in England and America, From A.D. 1086 to 1850* (Stamford, Conn.: W. W. Gillespie & Co., 1876).

3. Quoted in Ric Burns, et al., *New York, An Illustrated History* (New York: Alfred A. Knopf, 1999), p. 174.

4. See Richard Hofstadter, *The Age of Reform* (New York: Vintage Books, 1955), p. 9.

5. Quoted in Justin Kaplan, *Mr. Clemens and Mark Twain: A Biography* (New York: Simon & Schuster, 1991), p. 96.

6. Amzi Davenport, *Supplement,* p. 324.

7. Quoted in Burns, et al., *New York,* p. 158.

8. See "Autobiographical #1," Davenport Papers, APS Library, "Autobiographies" file.

9. See E. Carlton MacDowell, "Charles Benedict Davenport, 1866–1944, A Study in Conflicting Influences," *Bios,* Vol. XVII, No. 1 (March 1946).

10. Here and on, see Amzi Davenport's letters to Jane, Cold Spring Harbor Archives, Davenport/Harris Papers, Box 3.

11. See Jane Davenport's diaries, Cold Spring Harbor Laboratory Archives, Davenport/Harris Papers, Box 3.

12. MacDowell, "Charles Benedict Davenport," p. 8.

13. Davenport Papers, APS Library, "A. B. Davenport" file.

14. Letter, July 6, 1890, Davenport Papers, APS Library, "Mrs. A. B. Davenport" file.

15. Letter, June 3, 1888, Davenport Papers, APS Library, "Mrs. A. B. Davenport" file.

16. Letters, July 7, 1884, and November 23, 1888, Cold Spring Harbor Archives, Davenport/Harris Papers, Box 3.

17. Letter, February 9, 1889, Cold Spring Harbor Archives, Davenport/Harris Papers, Box 3.

18. See Oscar Riddle, "Biographical Memoir of Charles Benedict Davenport," in *National Academy of Sciences Biological Memoirs* (Washington, D.C., 1949), Vol. XXV, p. 85.

19. See December 23, 1894, letter to his mother, Davenport Papers, APS Library, "Mrs. A. B. Davenport" file. Also, July 2, 1893, letter to Gertrude, Davenport Papers, APS Library, "Mrs. Gertrude Crotty Davenport" file.

20. Letter, September 13, 1893, Davenport Papers, APS Library, "Mrs. Gertrude Crotty Davenport" file.

21. Letter, May 13, 1894, Cold Spring Harbor Archives, Davenport/Harris Papers, Box 3.

22. Letter, April 6, 1897, Davenport Papers, APS Library, "Francis Galton" file.

23. Letter, June 26, 1895, Davenport Papers, APS Library, "Mrs. Gertrude Crotty Davenport" file.

24. Davenport Papers, APS Library, "Family Records" file. This contradicts the standard interpretation of Davenport's family "influences," first presented by E. Carlton MacDowell in his article "Charles Benedict Davenport, 1866–1944, A Study in Conflicting Influences." MacDowell, who was a colleague of Davenport and who had interviewed him, claimed his mother was the person who was "inclined toward skepticism" and who first encouraged his love of science. This standard interpretation of Davenport is simply repeated throughout the enormous academic bibliography of eugenics literature, including Daniel Kevles's *In the Name of Eugenics.* And it is completely wrong.

VII: *The Hideous Serpent of Hopelessly Vicious Protoplasm*

1. Karl Pearson, *The Life, Letters and Labours of Francis Galton.* 3 vols. (Cambridge: The University Press, 1930), Vol. IIIA, p. 323.

2. See *The Times* (London), June 1, 1906, pp. 3, 6; June 2, p. 9; June 6, p. 4.

3. Pearson, *Life,* Vol. II, p. 132.

4. Robert Rentoul, "Proposed Sterilization of Certain Mental Degenerates," *American Journal of Sociology* 12 (November 1906), p. 326.

5. See J. Arthur Thomson, *Heredity* (London: John Murray, 1912), pp. 528–29. See also Richard A. Soloway, *Demography and Degeneration* (Chapel Hill: University of North Carolina Press, 1995).

6. Quoted in Michael W. Perry's foreword to G. K. Chesterton, *Eugenics and Other Evils* (Seattle: Inkling Books, 2000), p. 7.

7. Herbert Spencer, *Social Statics* (New York: D. Appleton, 1865), pp. 353–56.

8. Havelock Ellis, *The Task of Social Hygiene* (Boston and New York: Houghton Mifflin, 1912), p. 401.

9. Quoted in Daniel J. Kevles, *In the Name of Eugenics* (Cambridge, Mass.: Harvard University Press, 1995), p. 85.

10. See *American National Biography* (New York: Oxford University Press, 1998). Dugdale's entry accessed at www.anb.org.

11. See the discussion of the Jukes in Elof Axel Carlson, *The Unfit, A History of a Bad Idea* (Cold Spring Harbor, N.Y.: Cold Spring Harbor Laboratory Press, 2001), pp. 162–72.

12. R. L. Dugdale, *The Jukes: A Study in Crime, Pauperism, Disease, and Heredity,* 4th ed. (New York: Putnam, 1910), p. 167.

13. See Isaac Kerlin, "Report to the Eleventh National Conference of Charities and Reforms," in *Proceedings of the Association of Medical Officers and American Institutions for Idiotic and Feeble-minded Persons* (Philadelphia: Lippincott, 1885), p. 404.

14. See Martin Barr, *Mental Defectives, Their History, Treatment and Training* (Philadelphia: P. Blakiston's Son, 1904), p. 189. See also Jesse Spaulding Smith, "Marriage, Sterilization and Commitment Laws Aimed at Decreasing the Mentally Deficient," *Journal of Criminal Law* 5 (September 1914), pp. 364–66.

15. See *New York Times,* April 1, 1909, p. 8.

16. Quoted in "Emasculation of Masturbators—Is It Justifiable?" *Texas Medical Journal* X (1894), p. 240.

17. Quoted in E. Stuver, "Would Asexualization of Chronic Criminals, Sexual Perverts and Hereditary Defectives Benefit Society and Elevate the Human Race?" *Texas Medical Journal* XII (1896), p. 226.

18. Excerpted in "Emasculation of Masturbators," p. 244.

19. A. J. Ochsner, "Surgical Treatment of Habitual Criminals," *Journal of the American Medical Association,* April 22, 1899, p. 867.

20. H. C. Sharp, "The Severing of the Vasa Deferentia and Its Relation to the Neuropsychopathic Constitution," *New York Medical Journal,* March 8, 1902, p. 412.

21. See Barr, *Mental Defectives,* p. 194.

22. Quoted in Harry H. Laughlin, "The Legal, Legislative and Administrative Aspects of Sterilization," *Eugenics Record Office Bulletin No. 10B* (February 1914), p. 32.

23. The actual text reads: "AN ACT to prevent procreation of confirmed criminals, idiots, imbeciles, and rapists; Providing that superintendents or boards of managers of institutions where such persons are confined shall have the authority and are empowered to appoint a committee of experts, consisting of two physicians, to examine into the

mental condition of such inmates. WHEREAS, heredity plays a most important part in the transmission of crime, idiocy, and imbecility; THEREFORE IT BE ENACTED BY THE GENERAL ASSEMBLY OF THE STATE OF INDIANA, that on and after the passage of this act it shall be compulsory for each and every institution in the state, entrusted with the care of confirmed criminals, idiots, rapists, and imbeciles, to appoint upon its staff two (2) skilled surgeons of recognized ability, whose duty it shall be, in conjunction with the chief physician of the institution, to examine the mental and physical condition of such inmates as are recommended by the institutional physician and board of managers: If, in the judgment of this committee of experts and the board of managers, procreation is inadvisable, and there is no probability of improvement of the mental and physical condition of an inmate, it shall be lawful for the surgeons to perform such operation for the prevention of procreation as shall be decided safest and most effective. But this operation shall not be performed except in cases that have been pronounced unimprovable: *Provided,* That in no case shall the consultation fee be more than three dollars to each expert, to be paid out of the funds appropriated for the maintenance of such institution." Quoted in Laughlin, "The Legal, Legislative and Administrative Aspects of Sterilization," p. 14.

24. See William J. Robinson, *Eugenics, Marriage, and Birth Control (A Practical Guide)* (New York: The Critic and Guide, 1917), pp. 74–76.
25. George Bernard Shaw, *Complete Plays with Prefaces,* 6 vols. (New York: Dodd, Mead & Co., 1963), Vol. III, pp. 503–4.
26. See "Report of the Committee on Eugenics," *Proceedings of the American Breeders' Association* 6 (1909), p. 94.

VIII: But, Oh, Alas for Youthful Pride

1. In "Charles Benedict Davenport, 1866–1944, A Study in Conflicting Influences," *Bios,* Vol. XVII, No. 1 (March 1946), E. Carlton MacDowell reports that "a clambake of legendary proportions" occurred at the end of each season at Cold Spring Harbor (p. 15). The Annual Reports for the Experimental Station are in the Charles B. Davenport Papers, American Philosophical Society (APS) Library, Cold Spring Harbor Series #1.
2. Ibid., p. 35.
3. A copy of this song exists in the Becker Medical Library Archives at Washington University (St. Louis). See the Mildred Trotter Papers, Series 2, Correspondence, 1922–1937, "Botany Department" file.
4. Davenport Papers, APS Library, Cold Spring Harbor Series #1, "Beginnings" file.
5. See August Weismann, "The Continuity of the Germ-plasm as the Foundation of a Theory of Heredity," in his *Essays Upon Heredity and Kindred Biological Problems* (Oxford: Clarendon Press, 1889).
6. See W. S. Sutton, "The Chromosomes in Heredity," *Biological Bulletin* 4 (1903), pp. 231–51.
7. *New York Times,* May 20, 1906, p. 8.
8. Davenport Papers, APS Library, "Family Records" file.
9. Quoted in MacDowell, "Charles Benedict Davenport," p. 26.
10. Charles Davenport, *Heredity in Relation to Eugenics* (New York: Henry Holt & Co., 1911), p. 1.
11. Ibid.
12. Ibid., p. 83.

13. Ibid., p. 216.
14. Ibid., p. 258.
15. Ibid., p. 263.
16. Quoted in Daniel J. Kevles, *In the Name of Eugenics* (Cambridge, Mass.: Harvard University Press, 1995), p. 52.
17. For a description of the Congress, see the accounts in *The Times,* July 25, 26, and 31, 1912. See also *Problems in Eugenics* (London: Eugenics Education Society, 1912).
18. Quoted in Nicholas Wright Gillham, *A Life of Sir Francis Galton* (Oxford: Oxford University Press, 2001), p. 351.
19. "Preliminary Report," *Problems in Eugenics,* p. 465.
20. Ibid., p. 479.
21. *The Times,* July 31, 1912, p. 4.
22. Ibid.

IX: Oh, the Bliss of Being a Mother!

1. Harry H. Laughlin Papers, Truman State University (Kirksville, Mo.), Box E-1-1, file #10.
2. See Frances Hassencahl, "Harry H. Laughlin, 'Expert Eugenics Agent' for the House Committee on Immigration and Naturalization, 1921 to 1931," unpublished doctoral thesis, Case Western Reserve University, 1971, p. 42.
3. For most of the information that follows, see Mark H. Laughlin, "Deborah Laughlin, A Life Story," a pamphlet in the Harry H. Laughlin Papers, Truman State University.
4. Ibid.
5. For a general discussion of women's clubs at this time, see Gail M. Beaton, "The Widening Sphere of Women's Lives: The Literary Study and Philanthropic Work of Six Women's Clubs in Denver, 1881–1945," *Essays in Colorado History* 13 (1992).
6. Charlotte Perkins Gilman, *Women and Economics* (Boston: Small, Maynard & Co., 1899), p. xxxix.
7. Laughlin, "Deborah Laughlin, A Life Story," p. 24.
8. Ibid., p. 8.
9. See Elizabeth N. Armstrong, "Hercules and the Muses: Public Art and the Fair," in *The Anthropology of World's Fairs: San Francisco's Panama Pacific International Exposition* (Berkeley: Scholar Press, 1983), pp. 122–23; and Robert W. Rydell, *World of Fairs: The Century of Progress Expositions* (Chicago: University of Chicago Press, 1993), pp. 40–42. See also Wendy Kline, *Building a Better Race: Gender, Sexuality, and Eugenics from the Turn of the Century to the Baby Boom* (Berkeley: University of California Press, 2001), pp. 7–8.
10. Laughlin Papers, Truman State University, Box E-1-1, file #10.
11. Ibid., file #15.
12. Hassencahl, "Harry H. Laughlin," p. 50.
13. Letter, March 30, 1908, Charles B. Davenport Papers, APS Library, "Harry Laughlin" file.

X: Citizens of the Wrong Type

1. Charles B. Davenport Papers, American Philosophical Society Library, "Theodore Roosevelt" file.
2. *New York Times,* January 12, 1913, E1.

3. See Letter to Jane, January 28, 1914, in Cold Spring Harbor Laboratory Archives, Davenport/Harris Papers, Box 1; Letter to Charles, Davenport Papers, APS Library, "Charles Jr." file.

4. Charles B. Davenport Papers, American Philosophical Library, Cold Spring Harbor Series #2, "Play" file.

5. "The First Quarterly Report of the Eugenics Record Office," Davenport Papers, Cold Spring Harbor Series #2.

6. See "Training Course for Field Workers," Harry H. Laughlin Papers, Truman State University (Kirksville, Mo.), Box C-2-6, file #17.

7. See "Directions for the Guidance of Field Workers," Eugenics Record Office Records, Manuscript Collection 77, Series VII, at the American Philosophical Society Library.

8. Letter, September 15, 1910, Davenport Papers, "Henry H. Goddard" file.

9. See Nicholas Lehman, *The Big Test, The Secret History of the American Meritocracy* (New York: Farrar, Straus & Giroux, 1999).

10. *New York Times,* February 8, 1913, p. 10.

11. H. H. Goddard, *The Kallikak Family* (New York: The Macmillan Company, 1912), pp. 102–3.

12. Quoted in John David Smith, *Minds Made Feeble: The Myth and Legacy of the Kallikaks* (Rockville, Colo.: Aspen Systems Corp., 1985), pp. 62–65.

13. See Frances Hassencahl, "Harry H. Laughlin, 'Expert Eugenics Agent' for the House Committee on Immigration and Naturalization, 1921 to 1931," unpublished doctoral thesis, Case Western Reserve University, 1971, pp. 100–101.

14. See *Proceedings of the First National Conference on Race Betterment* (Battle Creek, Mich.: Race Betterment Foundation, 1914), pp. 243–45. See also the discussion in Steven Selden, *Inheriting Shame, The Story of Eugenics and Racism in America* (New York: Teachers College Press, 1999), p. 8.

15. *Proceedings of the First National Conference on Race Betterment,* pp. 412, 470.

16. Ibid., p. 478.

17. Ibid., p. 480.

18. Ibid., p. 484.

19. Ibid., p. 490.

20. See Harry Laughlin, *Eugenical Sterilization in the United States* (Chicago: Psychopathic Laboratory of the Municipal Court of Chicago, 1922), pp. 438–40.

21. Letter, February 21, 1911, Davenport Papers, APS Library, "Mrs. E. H. Harriman" file.

XI: Catechisms Old and New

1. See *New York Times,* August 8, 1915, p. 2.

2. See Harry Laughlin's analysis in *Eugenical Sterilization: Historical, Legal, and Statistical Review of Eugenical Sterilization in the United States* (New Haven, Conn.: The American Eugenics Society, 1926). See also Philip R. Reilly, *The Surgical Solution, A History of Involuntary Sterilization in the United States* (Baltimore and London: Johns Hopkins University Press, 1991), pp. 50–55.

3. See "Preliminary Report of the Committee of the Eugenic Section of the American Breeders' Association to Study and Report on the Best Practical Means for Cutting Off the Defective Germ-Plasm in the Human Population," in *Problems in Eugenics* (London: Eugenics Education Society, 1912), p. 477.

4. See E. S. Gosney and Paul Popenoe, *Sterilization for Human Betterment; A Summary of Results of 6,000 Operations in California, 1909–1929* (New York: The Macmillan Company, 1929). See also charts in Reilly, *The Surgical Solution,* pp. 49, 97. Reilly draws on the data gleaned by Harry Laughlin's surveys.

5. A copy of the lecture can be found at the Charles B. Davenport Papers, American Philosophical Society Library, "Lectures" file.

6. See Letters, September 25, 1915, and September 30, 1915, in the Davenport Papers, Cold Spring Harbor Series, No. 2, APS Library, "Eugenics Record Office: Board of Scientific Directors" file.

7. See Letter, February 18, 1916, Davenport Papers, APS Library, "Irving Fischer" file.

8. See phone call record, April 8, 1916, Davenport Papers, Cold Spring Harbor Series, No. 2, APS Library, "Eugenics Record Office: Board of Scientific Directors" file.

9. Quoted in Daniel J. Kevles, *In the Name of Eugenics* (Cambridge, Mass.: Harvard University Press, 1995), p. 58.

10. See Frances Hassencahl, "Harry H. Laughlin, 'Expert Eugenics Agent' for the House Committee on Immigration and Naturalization, 1921 to 1931," unpublished doctoral thesis, Case Western Reserve University, 1971, pp. 62–63.

11. See Letters, January 4, 1914, July 22, 1914, Cold Spring Harbor Laboratory Archives, Davenport/Harris Papers, Box 1.

12. See Letter, September 9, 1916, Cold Spring Harbor Archives, Davenport/Harris Papers, Box 1.

13. See Charles Davenport, *Heredity in Relation to Eugenics* (New York: Henry Holt & Co., 1911), pp. 253–54.

14. See E. Carlton MacDowell, "Charles Benedict Davenport, 1866–1944, A Study in Conflicting Influences." *Bios,* Vol. XVII, No. 1 (March 1946), p. 36.

15. *The Life, Letters and Labours of Francis Galton,* 3 vols. (Cambridge: The University Press, 1930), Vol. IIIA, pp. 218–19.

16. See Billy Sunday, "Historical Fabric of Christ's Life Nothing Without Miracles," *Commercial Appeal,* February 7, 1925, p. 13.

17. See Richard Hofstadter, *Anti-Intellectualism in American Life* (New York: Alfred A. Knopf, 1963). See also Mark Noll, *The Scandal of the Evangelical Mind* (Grand Rapids, Mich.: W. B. Eerdmans, 1994), and George Marsden, *Fundamentalism and American Culture* (New York: Oxford University Press, 1980).

18. See *The Fundamentals: A Testimony to the Truth.* An online version of the tracts can be found at http://www.xmission.com/~fidelis/.

19. See *Denver Catholic Register,* March 24, 1927, p. 1; April 14, 1927, p. 1.

20. See *The Commonweal,* September 20, 1935. Roswell H. Johnson, "Legislation," *Eugenics,* Vol. 2, No. 4 (1927), p. 64.

21. See *Five Great Encyclicals* (New York: Paulist Press, 1939), p. 96.

22. G. K. Chesterton, *Eugenics and Other Evils: An Argument Against the Scientifically Organized State* (Seattle: Inkling Books, 2000), pp. 46, 102.

23. See George Hunter, *A Civic Biology* (New York: American Book Co., 1914), p. 261.

24. See Davenport's syndicated article "Evidences for Evolution," in *Nashville Banner,* June 1, 1925, p. 6. Also quoted in Edward J. Larson, *Summer for the Gods: The Scopes Trial and America's Continuing Debate Over Science and Religion* (Cambridge, Mass.: Harvard University Press, 1997), p. 115.

25. Albert Edward Wiggam, *The New Decalogue of Science* (Indianapolis: Bobbs-Merrill Company, 1923), p. 109.

26. See W. S. Evans, *Organized Eugenics* (New Haven, Conn.: American Eugenics Society, 1931), p. x. See also Steven Selden, *Inheriting Shame: The Story of Eugenics and Racism in the United States* (New York: Teachers College Press, 1999), pp. 22–23.

27. See the American Eugenics Society Papers, Boxes 11 and 14, at the American Philosophical Society Library. Also quoted in Kevles, *In the Name of Eugenics*, p. 61.

28. See Kevles, *In the Name of Eugenics*, pp. 61–62. Also, Selden, *Inheriting Shame*, pp. 22–38.

29. Quoted in Robert W. Rydell, *World of Fairs: The Century-of-Progress Expositions* (Chicago: University of Chicago Press, 1993), pp. 48–49.

30. See *Denver Daily News*, May 1, 1913.

31. See Letter, December 31, 1924, in the Davenport Papers, APS Library, "Mary T. Watts" file. Also quoted in Rydell, *World of Fairs*, p. 51.

32. *Fortune*, July 16, 1937, p. 106, quoted in Kevles, *In the Name of Eugenics*, p. 114.

33. Quoted in Kevles, *In the Name of Eugenics*, p. 114.

34. See Edwin Grant Conklin, "Some Biological Aspects of Immigration," *Scribner's Magazine* LXIX (March 1921), p. 258.

XII: The Making of a Master Race

1. G. Stanley Hall, "Flapper Americana Novissima," *Atlantic Monthly* 129 (June 1922), p. 771.

2. Letter, June 26, 1895, Davenport Papers, APS Library, "Mrs. Gertrude Crotty Davenport" file.

3. Letter, July 3, 1923, Davenport Papers, APS Library, "Millia Davenport" file.

4. Letter, October 1, 1916, Cold Spring Harbor Laboratory Archives, Davenport/Harris Papers, Box 3.

5. *The Quill*, June 30, 1917, p. 3. (The only extant copies, as far as I know, are housed at the New York Public Library.)

6. Davenport Papers, APS Library, "Family Records" file.

7. Letter, undated, Davenport/Harris Papers, Cold Spring Harbor Laboratory Archives, Box 3.

8. See Charles Davenport, *Heredity in Relation to Eugenics* (New York: Henry Holt & Co., 1911), p. 216.

9. Letter, January 3, 1923, to Elizabeth G. Britton, Honorary Curator at the New York Botanical Garden and Wild Flower Preservation Society of America, Davenport Papers, APS Library, "Mrs. Nathaniel Lord Britton" file.

10. Or so Millia told her aunt Fannie in an undated letter, Davenport/Harris Papers, Cold Spring Harbor Laboratory Archives, Box 3.

11. *The Quill*, March 1, 1918, p. 7.

12. *New York Times*, April 1, 1925, p. 21.

13. *New York Times*, October 16, 1921, p. 40.

14. Letter, no date, Davenport Papers, APS Library, "Millia Davenport" file.

15. *Heredity in Relation to Eugenics*, p. 219.

16. See James Truslow Adams, *The Epic of America* (Boston: Little, Brown, 1931). See also Jim Cullen, *The American Dream: A Short History of an Idea That Shaped a Nation* (New York: Oxford University Press, 2002), p. 4.

17. Letter, no date, Davenport/Harris Papers, Cold Spring Harbor Laboratory Archives, Box 3.

18. See announcement in the *New York Times*, January 19, 1924.

19. See *New York Times*, April 20, 1920. The *Times* included four articles this day on the San Remo Conference and the partition of the Ottoman Empire.

20. Ibid., April 20, 1920, p. 4.

21. See "A Common Government of the World," Harry H. Laughlin Papers, Truman State University (Kirksville, Mo.), Box E-2-2. See also Frances Hassencahl, "Harry H. Laughlin, 'Expert Eugenics Agent' for the House Committee on Immigration and Naturalization, 1921 to 1931," unpublished doctoral thesis, Case Western Reserve University, 1971, p. 47.

22. See "Biological Aspects of Immigration," in *Hearings Before the Committee on Immigration and Naturalization,* House of Representatives, 66th Congress, April 16–17, 1920 (Washington, D.C.: Government Printing Office, 1921), p. 394.

23. For excerpts from the 1917 Immigration Restriction Act, see http://www.spartacus .schoolnet.co.uk/USAE1917A.htm.

24. See Roberta Strauss Feuerlicht, *America's Reign of Terror: World War I, the Red Scare, and the Palmer Raids* (New York: Random House, 1971). See also John Higham, *Strangers in the Land: Patterns of American Nativism, 1860–1925* (New Brunswick, N.J.: Rutgers University Press, 2002).

25. "Biological Aspects of Immigration," p. 404.

26. Ibid., p. 409.

27. See correspondence between Laughlin and Davenport on April 13, 1920, November 26, 1920, and December 12, 1920, Davenport Papers, APS Library, "Harry Laughlin" file. See also Hassencahl, "Harry H. Laughlin, 'Expert Eugenics Agent,' " p. 178.

28. See Dr. Oliver Wendell Holmes, Sr., "The Brahmin Caste of New England," *Atlantic Monthly* (May 1860).

29. For a discussion of the influence of Teutonism on New England intellectuals, see Barbara Miller Solomon, *Ancestors and Immigrants: A Changing New England Tradition* (Cambridge, Mass.: Harvard University Press, 1956).

30. See ibid., p. 123.

31. Hassencahl, "Harry H. Laughlin, 'Expert Eugenics Agent,' " p. 165.

32. Quoted in the Immigration Restriction League pamphlet, *Publications,* No. 37 (1903). See also Solomon, *Ancestors and Immigrants,* p. 103.

33. Letter to Charles Davenport, May 14, 1911, Davenport Papers, APS Library, "P. F. Hall" file.

34. See Grant's obituary in the *New York Times,* May 31, 1937, p. 15.

35. See Madison Grant, *The Passing of the Great Race* (New York: Charles Scribner's Sons, 1916), chapter I, "Race and Democracy." The complete text of Grant's book is readily available online at many racist Web sites. See, for example, http://www .churchoftrueisrael.com/pgr/pgr-toc.html.

36. Ibid., chapter II, "The Competition of Races."

37. Ibid., chapter XIII, "The Origin of the Aryan Languages."

38. See also George McDaniel, "Madison Grant the Racialist Movement," *American Renaissance* (December 1997).

39. Frederick Adams Woods, book review in *Science,* October 25, 1918, pp. 419–20.

40. Calvin Coolidge, "Whose Country Is This?" *Good Housekeeping* (February 1921), p. 14, quoted in Daniel J. Kevles, *In the Name of Eugenics* (Cambridge, Mass.: Harvard University Press, 1995), p. 97.

41. *New York Times,* May 31, 1937, p. 15.

42. See Hassencahl, "Harry H. Laughlin, 'Expert Eugenics Agent,'" pp. 161-97.
43. See Carl C. Brigham, *A Study of American Intelligence* (Princeton: Princeton University Press, 1923), pp. 208-10.
44. For a discussion of the Army testing program and its effects on the immigration debate, see Allan Chase, *The Legacy of Malthus: The Social Costs of the New Scientific Racism* (Urbana: University of Illinois Press, 1980), chapters 10-11.
45. Quoted in Kevles, *In the Name of Eugenics*, p. 97.
46. Letter to Madison Grant, April 7, 1925, Davenport Papers, APS Library, "Madison Grant" file.

XIII: Neighborly Love and Beyond

1. Hitler's speech is quoted in the *New York Times,* January 31, 1934, p. 13.
2. See George Dock's obituary in *Bulletin of the Medical Library Association* 39 (October 1951), pp. 382-83.
3. See E. S. Gosney and Paul Popenoe, *Sterilization for Human Betterment: A Summary of Results of 6,000 Operations in California, 1909-1929* (New York: The Macmillan Company, 1931), p. 2.
4. Letter, January 31, 1934, in the E. S. Gosney Papers, California Institute of Technology Archives.
5. Dock was quoting Arthur Gütt, et al., *Gesetz zur Verhütung erbkranken Nachwuchses vom 14. Juli 1933, mit Auszug aus dem Gesetz gegen gefährliche Gewohnheitsverbrecher und über Massregeln der Sicherung und Besserung vom 24. Nov. 1933* . . . (Munich: J. F. Lehmann, 1934). The quote here, however, is taken from his letter to Gosney.
6. For a discussion of German social Darwinism and the influence of Alfred Ploetz, see Robert N. Proctor, *Racial Hygiene: Medicine Under the Nazis* (Cambridge, Mass.: Harvard University Press, 1988).
7. Letter, November 24, 1920, Davenport Papers, APS Library, "Erwin Baur" file.
8. See Letter, December 21, 1920, Davenport Papers, APS Library, "Harry Laughlin" file.
9. See Harry Laughlin's file of news clippings in the Harry H. Laughlin Papers, Truman State University (Kirksville, Mo.), Box E-1-4.
10. See "Eugenical Sterilization in Germany," *Eugenical News* XVIII (August 1933), p. 90. This article is not attributed to Laughlin, but given that this was his area of expertise, and that he directed the content of the newsletter, it is almost certain that he was the author.
11. See *New York Times,* January 4, 1934, p. 10.
12. See *New York Times,* December 31, 1933, p. 7; January 5, 1934, p. 10; and February 3, 1934, p. 12.
13. "Eugenical Sterilization in Germany," p. 90.
14. See *New York Times,* August 10, 1934, p. 8.
15. See *New York Times,* July 25, 1934, p. 7.
16. See *New York Times,* November 12, 1934, p. 12.
17. See Marie E. Kopp, "Legal and Medical Aspects of Eugenic Sterilization in Germany," *American Sociological Review* 1 (October 1936), p. 764.
18. See Proctor, *Racial Hygiene: Medicine Under the Nazis,* p. 103.
19. See *New York Times,* January 9, 1934, p. 9.
20. *New York Evening Post,* March 20, 1934, p. 1.
21. *New York Times,* August 5, 1934, sec. VII, p. 8.

22. Quoted in Edward J. Larson and Leonard J. Nelson, "Involuntary Sexual Sterilization of Incompetents in Alabama: Past, Present, and Future," *Alabama Law Review* 43 (1992), p. 417.

23. See *Time,* September 9, 1935, pp. 20–21.

24. See *Richmond Times-Dispatch,* March 2, 1980.

25. Quoted in *Richmond Times Dispatch,* November 26, 2000, p. A-1.

26. Kopp, "Legal and Medical Aspects of Eugenic Sterilization in Germany," p. 763.

27. Paul Popenoe and Roswell Hill Johnson, *Applied Eugenics* (New York: The Macmillan Company, 1918), p. 184.

28. See "Report to the Board of Directors of the Human Betterment Foundation," February 12, 1936, Gosney Papers, California Institute of Technology Archives.

XIV: Harry's Secret

1. See *New York Times,* May 4, 1934, p. 1.

2. Harry Laughlin, *Report of the Special Commission on Immigration and the Alien Insane* (New York: Chamber of Commerce of the State of New York, 1934).

3. See *Manchester Guardian,* May 3, 1936. See also A. J. Sherman, *Island Refuge: Britain and the Refugees from the Third Reich, 1933–1939* (London: Paul Elek Books, 1973).

4. See Rafael Medoff, "Kristallnacht and the World's Response," *The Jewish Week,* November 7, 2003. See also Roger Daniels, *Guarding the Golden Door: American Immigration Policy and Immigrants Since 1882* (New York: Hill & Wang, 2004), pp. 78–79.

5. Harry Laughlin, *Conquest by Immigration* (New York: New York State Chamber of Commerce, 1939), p. 4.

6. See Stefan Kühl, *The Nazi Connection: Eugenics, American Racism, and German National Socialism* (New York: Oxford University Press, 1994), pp. 48–49.

7. See the memo "Eugenics in Germany," Harry H. Laughlin Papers, Truman State University (Kirksville, Mo.), Box C-2-3.

8. Quoted in Kühl, *The Nazi Connection,* p. 50.

9. Letters, May 28, 1936, and August 11, 1936, Laughlin Papers, Box E-1-3.

10. Letter, December 3, 1934, Laughlin Papers, Box C-2-2.

11. See, for example, the critique by Herbert Spenser Jennings outlined in Frances Hassencahl, "Harry H. Laughlin, 'Expert Eugenics Agent' for the House Committee on Immigration and Naturalization, 1921 to 1931," unpublished doctoral thesis, Case Western Reserve University, 1971.

12. See Walter Lippmann, "The Mental Age of Americans," *New Republic,* October 25, 1922, pp. 213–15, quoted in Daniel J. Kevles, *In the Name of Eugenics* (Cambridge, Mass.: Harvard University Press, 1995), p. 129.

13. Lionel S. Penrose, *Mental Defect* (New York: Farrar & Rinehart, 1933), pp. 172–74, quoted in Kevles, *In the Name of Eugenics,* pp. 107–8.

14. See Garland Allan, *Thomas Hunt Morgan: The Man and His Science* (Princeton: Princeton University Press, 1978).

15. See Letter, July 3, 1935, Laughlin Papers, Box C-2-3. See also Garland E. Allen, "The Eugenics Record Office at Cold Spring Harbor, 1910–1940: An Essay in Institutional History," *Osiris,* 2nd series (1986), pp. 251–52.

16. See Thomas H. Roderick, et al., "The Records of the Eugenics Record Office, A Resource for Genealogists," copy held at APS Library.

17. See "Report of the Advisory Committee on the Eugenics Record Office," Laughlin Papers, Box C-2-3.

18. Letter, December 31, 1938, Charles B. Davenport Papers, Cold Spring Harbor Series, APS Library, "Harry Laughlin" file.

19. Letter, November 12, 1940, Davenport Papers, CSH Series, APS Library, "Harry Laughlin" file.

20. See *Proceedings of the American Philosophical Society,* Vol. 80, No. 2 (1939), pp. 175–355, and Vol. 83, No. 1 (1942), p. 215.

21. See Oscar Riddle, "Biographical Memoir of Charles Benedict Davenport, 1866–1944," *National Academy of Sciences of the United States of America Biographical Memoirs,* Vol. 25 (1947), p. 91.

22. See the Cold Spring Harbor Laboratory Web site at www.cshl.edu.History/symposium.html.

23. Letter, undated, Charles B. Davenport Papers, American Philosophical Library, "Millia Davenport" file.

24. Quoted in E. Carlton MacDowell, "Charles Benedict Davenport, 1866–1944, A Study in Conflicting Influences," *Bios,* Vol. XVII, No. 1 (March 1946), p. 34.

25. Or so she says in the preface to her book; see *The Book of Costume* (New York: Crown Publishers, 1948).

26. *New York Times,* December 21, 1948, p. 23; March 27, 1949, p. BR18.

XV: The Palace of Justice

1. For a discussion of the head as an icon of atavism, see Lawrence Douglas, "The Shrunken Head of Buchenwald: Icons of Atrocity at Nuremberg," *Representations* 63 (Summer 1998), pp. 39–64. Many of the ideas in this chapter (including many of the quotes from the trial) are taken from this brilliant study. See also *Trial of the Major War Criminals Before the International Military Tribunal* ("Blue Series"), 42 vols. (Nuremberg, 1949), Vol. 3, p. 516. This can also be found online at Yale University's "Avalon Project" at www.yale.edu/lawweb/Avalon/imt/proc/12 13-45.htm.

2. See *The Times* (London), November 30, 1945, p. 4, and December 14, 1945, p. 4.

3. Thucydides, *History of the Peloponnesian War,* Book V, ed. and trans. Richard Crawley. See online version at http://classics.mit.edu/Thucydides/pelopwar.html.

4. See Jackson's opening statement in *Trial of the Major War Criminals,* Vol. 2, pp. 98–155.

5. Ibid., p. 99.

6. For a discussion of this issue, see Samantha Power, *"A Problem from Hell," America and the Age of Genocide* (New York: Perennial, 2003), p. 49.

7. *Trials of War Criminals Before the Nuremberg Military Tribunals,* 15 vols. ("Green Series") (Nuremberg, 1949), Vol. 1, pp. 694–737. An online version can be found at www.mazal.org.

8. Ibid., Vol. 4, p. 610.

9. Ibid., p. 1123.

10. Ibid., pp. 1159–60.

11. See Allan Chase, *The Legacy of Malthus: The Social Costs of the New Scientific Racism* (Urbana: University of Illinois Press, 1980), chapters 16–18.

12. Frederick C. Thorne, "Brandon Training School," *Biennial Report of the Vermont Department of Public Welfare* (1946), p. 70, quoted in Nancy L. Gallagher, *Breeding Better Vermonters: The Eugenics Project in the Green Mountain State* (Hanover and London: University Press of New England, 1999), p. 174.

13. See Philip J. Reilly, *The Surgical Solution, A History of Involuntary Sterilization in the United States* (Baltimore and London: Johns Hopkins University Press, 1991), p. 138.

14. See J. David Smith and K. Ray Nelson, *The Sterilization of Carrie Buck* (Far Hills, N.J.: New Horizon Press, 1989), p. 216.

15. *Richmond Times-Dispatch,* April 6, 1980, p. 1.

16. See *Poe v. Lynchburg,* 518 F. Supp. 789, 1981.

17. For a discussion of *Poe v. Lynchburg,* see Reilly, *The Surgical Solution,* p. 156.

18. See ibid., p. 149.

19. See *Washington Post,* March 9, 2000, p. A3.

XVI: "What They Did to Me Was Sexual Murder"

1. All information about Lucille is based on personal observation and interviews, news clippings, court papers, and medical files. Her story was first reported by Mike Anton in the *Rocky Mountain News,* November 21, 1999, p. 1.

2. See James H. Baker, ed., *History of Colorado* (Denver: Linderman Co., 1927), Vol. 5, pp. 496–97. See also Abbott Fey, *Famous Coloradans: 124 People Who Have Gained Nationwide Fame* (Paonia, Colo.: Mountaintop Books, 1990), pp. 139–40.

3. See Hubert Work, "The Sociologic Aspect of Insanity and Allied Defects," *Proceedings of the American Medico-Psychological Association at the 68th Annual Meeting,* May 28–31, 1912, p. 141.

4. Ibid., pp. 127, 134–35, and 140.

5. See *The Daily News* (Denver), March 11, 1913, p. 5.

6. See *Denver Times,* July 13, 1898, p. 2.

7. See Gail M. Beaton, "The Widening Sphere of Women's Lives: The Literary Study and Philanthropic Work of Six Women's Clubs in Denver, 1881–1945," in *Women's Clubs of Denver, Essays in Colorado History,* No. 13 (Denver, 1992). See also Gloria Moldow, *Women Doctors in Gilded-Age Washington: Race, Gender, and Professionalization* (Urbana: University of Illinois Press, 1987), pp. 24, 78; and *Rocky Mountain News,* July 12, 1927, p. 10.

8. *Denver Times,* July 13, 1898, p. 2.

9. See Baker, ed., *History of Colorado,* pp. 546–50.

10. See *Greeley Daily Tribune,* September 30, 1910, p. 2, and October 2, 1910, p. 3.

11. See *Denver Medical Times,* vol. 32, no. 10 (April 1913), p. 448; see also *Denver Daily News,* May 1, 1913, p. 1, and May 2, 1913, p. 1.

12. See *Rocky Mountain News,* January 26, 1913, p. 3.

13. Letter, February 11, 1913, in the Davenport Papers, APS Library, "David Starr Jordan" file.

14. As she told the *Rocky Mountain News,* on July 12, 1927—see p. 10.

15. Beaton, "The Widening Sphere of Women's Lives," p. 38.

16. See Robert Allen Goldberg, *Hooded Empire: The Ku Klux Klan in Colorado* (Urbana: University of Illinois Press, 1981), p. 88.

17. See *Denver Post,* April 12, 1927, p. 1; see also *Denver Catholic Register,* March 24, 1927, p. 1, and April 14, 1927, p. 1.

18. Letters included in Pueblo District Court file, case #38407, Colorado State Archives.

19. See H.R. 136, 1929, "A Bill for an Act to Authorize the Sterilization of Certain Persons . . ." at the Colorado State Archives.

20. Letters included in Pueblo District Court file, case #38407, Colorado State Archives.

21. This is noted in Philip J. Reilly, *The Surgical Solution, A History of Involuntary Sterilization in the United States* (Baltimore and London: Johns Hopkins University Press, 1991), p. xiii.

22. See www.patterson-smith.com/mags.htm.

23. Letter copied into Lucille's medical records file.

24. The written transcript of the trial has been lost. The quotes that follow are taken from the depositions. Newspaper reports, however, indicate that similar questions and answers were given during Zimmerman's cross-examination.

25. As told to Mike Anton in the *Rocky Mountain News,* November 21, 1999, p. 1.

26. This quote is taken from Lucille's interview with Mike Anton in the *Rocky Mountain News,* November 21, 1999, p. 1.

27. Ibid.

XVII: Epilogue: The Apex of Civilization

1. Echoing Morpheus, the pseudo-priestly leader in the film *The Matrix.*

2. See Max Weber's influential 1920 essay, *The Protestant Ethic and the Spirit of Capitalism* (New York: Scribner's Press, 1958). See also Niall Ferguson, "Why America Outpaces Europe (Clue: The God Factor)," *New York Times,* June 8, 2003.

3. See *Trials of War Criminals Before the Nuremberg Military Tribunals,* 15 vols. ("Green Series") (Nuremberg, 1949), Vol. 2, p. 155. An online version can be found at www.mazal.org.

4. See Malcolm Gladwell, *The Tipping Point: How Little Things Can Make a Big Difference* (New York: Little, Brown, 2000).

INDEX

Page numbers in *italics* refer to illustrations.